The Darwinian Revolution

MICHAEL RUSE

The Darwinian Revolution

Science Red in Tooth and Claw

SECOND EDITION

The University of Chicago Press • *Chicago and London*

Michael Ruse is professor of philosophy and zoology
at the University of Guelph, in Ontario, Canada.
Winner of Guggenheim and Killam Fellowships, his
most recent books are *Monad to Man: The Concept of
Progress in Evolutionary Biology* and *Mystery of
Mysteries: Is Evolution of Social Construction?*

The University of Chicago Press, Chicago 60637
The University of Chicago Press, Ltd., London

© 1979, 1999 by The University of Chicago
All rights reserved. Published 1999

08 07 06 05 04 03 02 01 00 99   1 2 3 4 5
ISBN: 0-226-73168-5 (cloth)
ISBN: 0-226-73169-3 (paper)

Library of Congress Cataloging in Publication Data

Ruse, Michael.
    The Darwinian revolution : science red in
tooth and claw / Michael Ruse. — 2nd ed.
        p.   cm.
    Includes bibliographical references (p. 285).
    ISBN 0-226-73168-5 (cloth : alk. paper). — ISBN
0-226-73169-3
(pbk. : alk. paper)
    1. Evolution (Biology)—History.   2. Darwin,
Charles, 1809–1882.   3. Darwin, Charles 1809–
1882. On the origin of species.   I. Title.
QH361.R87   1999
576.8'2—dc21

99-23377
CIP

♾ The paper used in this publication meets the
minimum requirements of the American National Standard
for Information Sciences—Permanence of Paper for
Printed Library Materials, ANSI Z39.48–1992.

To the memories of my mother, Margaret Ruse (1919–53),
and my father-in-law, Hawthorne Steele (1903–77)

Contents

Prologue

In 1859 the eminent British naturalist Charles Robert Darwin published his best-known book: *On the Origin of Species by Means of Natural Selection; or, The Preservation of Favoured Races in the Struggle for Life.* In this work he put forward his solution to one of the burning scientific questions of the day: What precisely are the causal origins of the world's organisms, past and present—a category to which we humans ourselves belong? As is well known, Darwin argued that organisms are what they are because of a process bound by natural law, and that all types of organisms have descended through many generations, gradually being modified from one or a few humble initial forms. Moreover, Darwin suggested a mechanism for this process, something he called "natural selection." By this mechanism of survival and reproduction, certain useful "adaptive" characteristics are "selected" for future generations.

The arrival of the *Origin* changed man's world. Darwin was not the first to present a theory of "evolution"; but never before had such a theory had such convincing impact. At once, it was seen to have implications far beyond biology. It struck at beliefs and behaviors from the most trivial to the most profound. Consequently, as might be expected, Darwin, his work, and the whole series of ideas and events leading up to and away from the *Origin*—a series commonly known as the "Darwinian Revolution"—have been written about in depth and at length. About twenty years ago several good general books were published on various aspects of the Darwinian Revolution (Irvine 1955; Himmelfarb 1962; Greene 1959; Eiseley 1961; De Beer 1963); but since then there has appeared no really comprehensive work, in spite of a real explosion of serious interest in Darwin and the revolution, manifested in the publication of newly discovered primary sources and in scholarly articles and monographs. Hence, although much information is available, there is no up-to-date general treatment of the subject. Because of this lack I have been led to write this book as a synthesis of the Darwinian Revolution, using the most recent findings and interpretations, for readers like myself who have a serious interest in the history of science and want to dig beneath glib generalizations and stark dramatizations, but who do not have the specialized knowledge and aims of the professional scholar.

Let me explain some of the strategies I shall adopt. First, some matters are mainly but by no means exclusively terminological. Today we speak of a theory like Darwin's as postulating "evolution," meaning it hypothesizes a more-or-less gradual and regular change, through successive forms, from the origin to the present: that is, the forms themselves change in a regular way and there is more than just stepwise succession from one unconnected form to another. But the term "evolution" came into general modern use only in the time of Darwin (Bowler 1975). One often spoke then of "transmutation." Darwin himself tended to term his theory "descent with modification," and indeed in the *Origin* he never used the word "evolution," although the very last word of the book is "evolved." Nevertheless, after deliberation I have decided that we today have a perfect right to use our own language. So I shall speak of a theory as postulating "evolution," in the sense just given; and I shall say that such a theory aims to solve "the organic origins problem"—the problem of how organisms came to be as they are. I do not imply that all suggested forms of evolution are exactly the same; that all putative causes of evolution are the same; or that evolutionists are necessarily linked beyond sharing vaguely similar patterns. I should point out that despite the title of Darwin's book, the problem of the origin of species is not quite the same as the problem of the origin of organisms. Organisms fall into groups with similar characteristics, reproducing within themselves and not with other "species," and one might certainly expect someone concerned with the origins of organisms to pay attention to the reasons for speciation. But how this expectation is realized must be dealt with in the individual case.

Other terminological points will have to be solved as we go along. Toward a clearer understanding, though, let me mention two other problematic terms: "law" and "miracle." Under the most standard interpretation, by "law" we mean some regularity we feel *must* hold in this world of ours, and we speak of things covered by such law as being "natural." (I here confine myself to physical laws and exclude consideration of legal and moral laws.) A "miracle" is something not covered by law, though whether this involves a break with law is perhaps another matter. Generally, miracles are thought to involve immediate divine intervention and thus to be "supernatural." These same interpretations of law and miracle were in general use in the last century, though then as now some people held different interpretations (Hodge 1872), which I shall consider when necessary. I assume that under the standard interpretations an evolutionary hypothesis for organic origins in some way appeals to law and that a miraculous attribution is not evolutionary. Whether there are other possibilities, and how they would stand with respect to law and evolution, we shall have to see.

The glaring omission from this terminological discussion is the concept that is perhaps more crucial than any other in this book—the concept of cause. Because of its great importance, I am hesitant to attempt a premature analysis: it will get full treatment later. Here I shall rest with the preanalytic notion that a cause is something that somehow brings about something else, an effect. If a cause is potentially subject to law, it is natural. If it is a function of direct divine intervention, it is supernatural or miraculous.

We must now address the major problem any historian faces: Where precisely is one to set the time limits? In the Darwinian case answers are by no means obvious. If we take the broadest view and think our subject ought to be the whole of man's quest for the causes of organic origins, with Darwin no more than one link in the chain, we can easily go back to the Greeks and forward to the present. Indeed, we will have a story without end, for many aspects of the causes of origins are still highly controversial (Lewontin 1974). But if we want to draw our boundaries more closely and consider the main question to be the theory of evolution— When and how did people get converted to the idea of evolution?—we can narrow our study to about a twenty-five-year span. Consider: In 1851, when Cambridge University first offered examinations in science, one question was as follows. "Reviewing the whole fossil evidence, shew that it does not lead to a theory of natural development through a natural transmutation of species" (Cambridge University 1851, p. 416). By 1873, however, a question told students to assume "the truth of the hypothesis that the existing species of plants and animals have been derived by generation from others widely different" and to get on with discussing the causes (Cambridge University 1875, qu. 162). If one makes the reasonable assumption that by the time something gets into undergraduate examinations it is fairly noncontroversial, it follows that in no more than a quarter of a century the scientific community had made a complete about-face on the question of evolution. Therefore we could presumably justify confining our attention to this short period.

I myself shall aim somewhere between the broadest and the narrowest extremes, though since I hope to dig fairly deep I shall stay closer to the narrow dates. My concern will be less with the general history of the organic origins problem and more with the switch to evolutionism. In particular, I shall take seriously the notion of the *Darwinian* Revolution: my interest will intensify as Darwin comes into the story and fade as Darwin himself goes out. The reader might fear that I subscribe to a great man theory of the history of science, where all the significant events of the past are seen as due to a few solitary geniuses. But things will not be as biased as all that, though I am not sure that using Darwin as a guideline is so very heinous. After all, he wrote the *Origin*—by any criterion the key work in the whole organic origins controversy. But I shall be particularly wary of slipping into uncritical hagiography and shall be careful not to keep the spotlight on Darwin alone.

Moreover, there are independent reasons for letting Darwin structure our time period. He comes into the story about 1830, not as an established scientist but as a young man beginning his career, in the same year that the geologist Charles Lyell started to publish his great *Principles of Geology*. As we shall see, Lyell's treatment of the organic origins problem provided much of the background and inspiration for the work of Darwin and of others. At the other end, though Darwin died in 1882, one senses that by about 1875 his major contributions were over. Again, there is a certain naturalness about choosing a date near here, not because everything was settled—it was not—but because by then the major initial moves had been made and a period of calm or exhaustion had been reached. It was some years before the tempo of the organic origins debate picked up again.

Finally, although my main coverage will be roughly from 1830 to 1875, I shall temper this in two ways. I shall provide a background chapter leading up to the year in which the *Principles* appeared. And in later chapters I shall indicate how the story went on after our prime period, not so that we from our "superior" vantage point might make absolute judgments on earlier thinkers, but to throw light on our main period of concern by seeing how posterity has dealt with the questions asked and the answers given.

Restrictions of place are not independent from restrictions of time. By choosing the middle half of the century, our attention will fall fairly naturally on the British. This is not simply because Darwin was an Englishman or because the British were the best scientists of the time—I very much doubt they were—but because for that period the organic origins question did become a distinctively British problem. Unlike the Victorians, I feel a little uncomfortable at such seeming jingoism; but I shall try to give reasons why this happened. Also, as with time, I shall treat my restrictions casually. Before 1830 my main focus will be on France. In the course of the narrative it will be necessary to pinpoint some crucial Continental influences. And it will become clear that after the publication of the *Origin* the story starts to broaden out again and becomes less insular. In particular, the New World rises from the horizon.

With time and place fixed, the next questions involve the various themes, or "strands," around which I shall structure my narrative. Obviously, a major strand must be that of *science* in its narrowest sense. We must see what empirical facts were known, what theories were proposed, and why various suggestions were accepted or rejected by scientists. But there was a great deal more to the Darwinian Revolution than just this narrow science. We must consider at least two other major elements. First, there was a continuing debate about which theories actually fitted the criteria of good science, quite apart from truth or falsity. I shall call this the *philosophical* strand. Then there was the major controversy about God and man and how the various solutions to the problem of the origins of organisms affected them: the *religious* strand.

In the realm of ideas, therefore, these three elements run through our period, and (without being artificially rigid), once the main story gets under way I shall try to consider them in the order just given. I must emphasize that I by no means consider them entirely separate strands. We shall see much intertwining, and I adopt them mainly for ease of exposition. But besides these there is another dimension to the Darwinian Revolution, one less easy to categorize. It is clear that the organic origins problem as considered through our time span was not purely a matter of intellect; it also involved human relationships, attitudes, and influences, both between scientists and with society at large. Where appropriate, I shall discuss these matters—for convenience called *social and political* factors—though not so systematically as my other strands. I should add that both for ideas and for social and political factors I will focus primarily on influential thinkers rather than on the general population.

Let me add a note of caution. By exploring the Darwinian Revolution through these various strands, we shall cover much territory and should be able to piece together a coherent picture of one of the major episodes in the history of science. But I doubt it will be a simple picture, ultimately reducible to one essential change in man's beliefs or behavior. Many elements are involved in the Darwinian Revolution—some, of course more important than others. But only so far as our historical explanations pay full tribute to the many sides of the revolution shall we achieve full understanding.

Acknowledgments

In writing this book I have incurred many debts, both to institutions and to individual historians and philosophers of science. Clare Hall, Cambridge, and the Department of History and Philosophy of Science at Cambridge made possible my early research, and the Department of History and Philosophy of Science at Indiana University made possible the later research. Among the many people who shared with me their knowledge of nineteenth-century biology, particular mention must be made of Peter Bowler; Joe Burchfield; William Bynum; Frederick B. Churchill; Frank Egerton; Peter Gautrey; Michael Ghiselin; John C. Greene; George Grinnell; Sandra Herbert; Vincent Kavalowski; Malcolm Kottler; Camille Limoges; Ernst Mayr; Roy Porter; Martin Rudwick; Sydney Smith; Mary P. Winsor; and Robert M. Young. Especially warm thanks must be reserved for Jonathan Hodge and David Hull, who have criticized and encouraged me constantly. I cannot expect that two such independent thinkers will accept everything that I have written, but I would like to think that this book might find favor with our common archetype.

Thanks must also go to Judy Martin, who typed the several drafts of this work, and to Alice Swayne, xv who had the onerous task of copy-editing.

The Darwinian Revolution

William Buckland (1784–1856)
Robert Chambers (1802–1871)
John F. W. Herschel (1792–1871)
Joseph Dalton Hooker (1817–1911)

Thomas Henry Huxley (1825–1895)
Charles Lyell (1797–1875)
Hugh Miller (1802–1856)

Richard Owen (1804–1892)
Adam Sedgwick (1785–1873)
Alfred Russel Wallace (1823–1913)
William Whewell (1794–1866)

1 *Background to the Problem*

The Idea of Evolution Suggestions that the origins of organisms were due to some natural law-bound process go back to Greek and Roman thought.[1] If we restrict the term "evolution" to the idea that each set of organisms comes from another set, back to the first, then these ancient suggestions are perhaps better thought of as "protoevolutionary," for they usually involved the idea that organisms, even the most sophisticated, spring full-grown from inorganic matter. But all speculation pointing toward a genuine evolutionism was stopped abruptly by two things: first, by the metaphysical systems of Plato and Aristotle, often spoken of as "essentialist" systems, and, second, by the rise and spread of Christianity, bringing with it what Carlyle contemptuously referred to as "Hebrew old clothes" (Carlyle 1896–1901, pp. 29–30). Plato's theory of Forms postulates transcendent objects of ultimate reality, the Forms or Ideas. All objects of this world have their particular properties only because they reflect or "participate" in the Forms: it is the Forms that give objects their "essences." Because the Forms are eternal, unchanging, and unique, the whole theory has profound antievolutionary implications, since organisms on the border between Forms are *logically* precluded. Either something participates in a Form or it does not, and that is the end of the matter. Even if one

3

thought of evolution as occurring in jumps rather than by gradual changes, there would still be no way something participating in one Form could give rise instantaneously to something participating in another Form. Aristotelian metaphysics likewise bars evolutionism, though Aristotle did not postulate a Form external to the material object. These philosophies, combined with the Creation story of Genesis, which is the antithesis of evolutionism, were enough to ensure a static world picture until the end of the Middle Ages.

Two things loosened this picture's hold on the human imagination. First was the birth of the new physics, which put pressure on belief in the literal truth of the Bible—the account of the sun's stopping for Joshua, for example—and also gave rise to speculations about nonorganic evolutionism. Newton's astronomy in the *Principia* dealt only with the world as it is now; but speculations on origins were forthcoming. In particular, Kant, William Herschel, and Laplace formulated the so-called nebular hypothesis, which sees the universe as formed out of gaseous nebulae. The implications of such a hypothesis for the organic world are clear and, as we shall see, did not go unnoticed (Greene 1959). Second were the advances being made in the sciences of biology and geology. Increasing fossil discoveries, for example, threw doubt on the very limited age of the earth that the Bible seemed to allow, as did the new geological theories explaining the earth's strata. And adding to the difficulty of maintaining the biblical stories of Creation and the Deluge were the mounting numbers of facts about organic geographical distributions that world travelers were bringing home.

Relaxing the old way of seeing things was not an easy or straightforward matter. To us, for instance, fossils obviously imply that the earth is of great age. But this is obvious only because we see fossils as the remains of long-dead organisms. A neoplatonist in the Renaissance found it far more natural to interpret fossils as manifestations of the Forms in the inorganic world, just as living organisms are manifestations of the Forms in the organic world (Rudwick 1972, chap. 1). This interpretation makes no direct connection between fossils and once-living organisms, and only after years of bitter debate did another view prevail. The grip of the Bible also proved very tight. It may have loosened a little toward the end of the seventeenth century, but in Britain, particularly, it clamped down hard again in the eighteenth, a direct result of the evangelicalism sparked by John Wesley. The Bible remained a major factor in the nineteenth century.

Nevertheless, by the end of the eighteenth century and the beginning of the nineteenth, speculation on organic evolution, though not commonplace or in any way acceptable, was no longer particularly novel. One of the best-known theories (to be considered briefly later) was that of Charles Darwin's grandfather, Erasmus Darwin. But undoubtedly all previous speculations paled beside the systematic evolutionary attack on the organic origins question made by the French biologist Jean Baptiste de Lamarck.

Lamarck's Case for Organic Evolutionism

Thomas Kuhn (1970) has remarked perceptively that the scientist who makes a really innovative move, breaking with the past and opening up new and fertile fields of scientific exploration, tends to be fairly young. This is not fortuitous: the innovative scientist must grasp the essentials of past scientific achievements while sensing acutely the problems that flaw them; but he must not be emotionally and intellectually committed to the past—for example, through having himself made significant contributions to established theories. Obviously a younger man is more likely to fit this category, and when we come to consider Charles Darwin and his work we shall find that he is the exemplar both of this pattern and of its rationale.

Lamarck, however, is something of an exception. Although in becoming an evolutionist he was not greatly innovative, and although his evolutionism contained elements drawn from his predecessors, Lamarck's conversion to evolutionism was apparently neither a phenomenon of his youth nor a drawn-out process with early beginnings, but an event that came somewhat abruptly in his fifty-sixth year. Until virtually the end of the eighteenth century, Lamarck agreed with almost everyone else that organisms and their groupings remain essentially unchanged from their first appearance. Then between 1799 and 1800 he suddenly swung to a diametrically opposite position, arguing that organisms evolve and that this evolution is constantly refueled by new organisms as life is spontaneously generated out of inorganic matter.

We do not have a wealth of material helping us reconstruct Lamarck's route to discovery or conversion. (For details on what is known of his life and work, see Burkhardt 1970, 1972, 1977; Russell 1916; Hodge 1971; and Mayr 1972a.) But clever detective work by a recent scholar makes it fairly clear that invertebrate taxonomy was chiefly responsible for the path Lamarck took (see Burkhardt 1977, esp. chap. 5). In 1793 Lamarck was appointed to the Museum of Natural History in Paris as professor of "insects, worms, and microscopic animals." It was a good time to be a professor at this museum, for during that decade the collections were being drastically augmented both through scientific expeditions of discovery and through the rape of other museums in Europe, as French scientists spread their tentacles in the wake of their conquering armies. And thus, given his allotted subject and the necessary means, it fell to Lamarck to try to answer the question of growing interest and importance within the scientific community: Did species of organisms always survive indefinitely, or, as had been recently claimed on the basis of comparison between living and fossil forms, did some species finally become extinct? Lamarck was in a peculiarly favored position to answer this question, because the museum's large collection of shells gave him the perfect opportunity to explore whether fossil shells always had living counterparts.

It seems that from his studies Lamarck was indeed compelled to admit that they did not. Yet he was unwilling to agree with many of his contemporaries that this proved the reality of total annihilation. In earlier years, particularly in Britain, people had rejected extinction because they feared it was irreligious: an extinct organism, particularly one that died out before man, seemed a blot on God's good sense (Greene 1959). Lamarck, however, opposed extinction for almost the opposite reason. Other than by appealing to a supernatural cause, something he as a scientist was loathe to do, he could not see how any species could become extinct (except, he came to concede, in the special circumstance where man destroyed all its members). Shellfish particularly, Lamarck thought, protected in their watery homes, could not be depleted to the point of extinction. Hence, since some kinds of shellfish apparently no longer exist, evolution into other forms of life seemed the only solution. Curiously, for the very simplest forms of life Lamarck somewhat reversed himself. In their case he could not see how anything so fragile could endure all the harshness of nature: snow, frost, and so forth. Hence he felt that extinction would almost be expected; but since the simplest forms of life so obviously do not become extinct, he felt compelled to postulate the spontaneous generation of new life forms (Burkhardt 1977, pp. 138–39).

At the beginning of the nineteenth century, therefore, Lamarck became an evolutionist, and over the years he presented his ideas in various forms. But, partly because his ideas did not really change much and partly because this was the version best known to the British, I shall concentrate exclusively on Lamarck's ideas as given in 1809 in his *Philosophie zoologique.* Since this work has been translated, for convenience I shall refer both to the original and to the translation (in brackets).

The backbone of Lamarck's theory was the "chain of being" or "scale of nature" (Lovejoy 1936). Lamarck believed, with reservations, that all animals can be ranged on an ascending scale, with the lowest, infusorians, at one end, and the most complex and perfect, man, at the other end (1:102–29 [56–67]). (Lamarck believed in two separate chains of being, one for animals and one for plants [1:92–93 (51)]. Later he divided his animal chain in two.) This idea of a scale was not peculiar to Lamarck. Indeed, it has antecedents in the Platonic dialogues, perhaps proving, to extend Whitehead's aphorism, that the evolutionary debate, like philosophy, is just a set of footnotes to Plato. Lamarck differed from his predecessors in that for him the scale of nature was dynamic rather than static; and this, in the broad sense in which we are using the term, made him an evolutionist. He believed that organisms made a fairly constant progression up the scale as they changed in the course of many generations from the simplest to the most complex. New primitive organisms constantly appear at the bottom of the scale as they are formed from inorganic matter.

Lamarck tried hard to be a good materialist; he denied that life or mind involved special entities or modes of understanding radically different from the inorganic world. What drove organisms up the chain of being was, first, that they experience certain needs (*besoins*; 1:6 [11]) brought about by the constantly changing environment. Then in some way, perhaps involving new habits (1:68 [41]), these trigger the movement of various bodily fluids that create or enlarge organs. These fluids are not sensible, like water or blood, but are "subtle" fluids like electricity and caloric. In the higher animals, between the needs and the fluids, Lamarck sandwiched his famous—or perhaps notorious—inner consciousness (*sentiment intérieur*; 2:276–301 [332–42]), which acts as a causal link enabling the organism to respond physiologically to its needs. Although all his critics later accused Lamarck of postulating consciousness in animals that obviously cannot really think, it seems clear that the *sentiment intérieur* did not involve true thought but was a kind of "life force."

Despite his materialist intentions, however, matters were not quite as straightforward as they first appear. Sometimes Lamarck wrote as though everything happens in the normal causal manner, just as in physics. An environmental change occurs, a need is set up, and so on. When critics claimed that evolution was impossible because animals mummified by the ancient Egyptians are identical to today's animals, he retorted that there had been no environmental change in Egypt since ancient times (1:70 [42]). But at other times he implied that progression up the scale of being will occur no matter what happens or fails to happen. Thus he argued that the chain of being would be perfectly regular even if all organisms were in a uniform, demand-free environment (1:133 {69}). Moreover, ignoring for the moment a secondary factor that he added to his chain-of-being doctrine, Lamarck seemed to envision an inevitable passage as organisms moved up the chain. This leads one to suspect that, materialist or no, Lamarck saw things in the organic world as being end-directed, with the end in the animal world being man. In a sense, therefore, he was a *teleologist*, trying to explain in terms of ends rather than merely prior material causes.

On this main thesis about organic change, Lamarck superimposed another evolutionary mechanism, if we can so describe it without prejudice. This mechanism seems to differ from the first primarily, if not solely, in that it is supposed to lead to anomalies, branchings, and irregularities in the chain of being—for instance, causing the birds to be put off on one side. Sometimes this secondary mechanism was said to act directly through the environment, as when poor nutrition causes stunted growth, which Lamarck thought was heritable (1:133 {69}). Sometimes it involved habits, as in change through use and disuse. Lamarck drew attention to animal and plant breeding. We find, for example, that when ducks are not allowed to fly, they lose their power of flight permanently, and this appears to be inevitable (1:225–28 {109–10}). Lamarck suggested that the same applies naturally, thus disturbing the uniform climb up the chain of being (see fig. 1).

This, then, was Lamarck's evolutionary theory—at least as I understand it, for it must be confessed that he is the most confusing of writers. Indeed, we will learn that the conceptually fuzzy way Lamarck presented his ideas had interesting implications for our story. One suspects that Lamarck is confusing because he was confused. Certainly his secondary "mechanism" seems little more than an ad hoc device for getting around problems. But this much we can say. Lamarck accepted a chain of being with irregularities. He thought that the satisfaction of needs was a significant cause of heritable change. This inheritance of acquired variations or characteristics is what we today call "Lamarckism"—somewhat inaccurately, for it was not his whole theory, nor was it original with him. Yet I would not deny Lamarck his rightful place in history. It is one thing to have an idea for change. It is quite another to have the imagination to use it to support a full-blown evolutionary theory.

One way to make Lamarck's theory more plausible is to argue that only his second mechanism involves needs, habits, and inner consciousness.

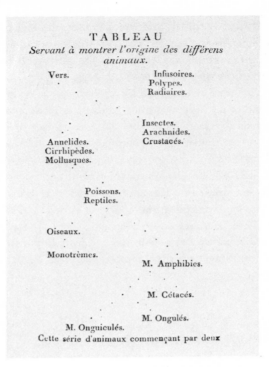

Figure 1. Lamarck's picture of evolution (from *Philosophie zoologique*). Compare figure 2, being careful not to confuse Lamarck's diagram with Darwin's superficially similar diagram (fig. 23 in chap. 7).

The first simply involves the body's fluids blindly carving out new paths, leading to new characteristics and thus driving organisms up the scale of nature. There is some justification for this interpretation in the *Philosophie zoologique,* and a recent commentor, basing his analysis on the whole of Lamarck's work, has read him this way (Burkhardt 1977, chap. 6). But, even ignoring the fact that there is no reason to think so undirected a mechanism as fluids making new channels can lead to so teleological a result as the drive up the scale of being, the *Philosophie zoologique* certainly claims that habits are involved in all permanent change. Consequently, if people see "Lamarckism" as the essence of Lamarck's theorizing, it is really nobody's fault but his own.

Lamarck's theory may have concerned the origins of organisms, but it was not a theory of the origins of species. He hoped to explain organic diversity, but species—distinct kinds of organisms, unable to breed with other kinds—were something of an embarrassment to him. Since he be-

lieved essentially in a gradual, continuous chain, he had to explain gaps
between organisms by various ad hoc hypotheses—that we have not yet
found the bridging organisms, that man has destroyed them, that the
secondary mechanism may have caused gaps.

Also, one must note that Lamarck's theory was in no way a theory of
common descent, supposing that all organisms descend from one or a few
common origins. We know that he thought simple forms of life are
constantly being spontaneously generated through the action of heat,
light, electricity, and moisture on the inorganic world (*Philosophie*
2:61–90 [236–48]). Then organic development continues on essentially
the same path it started on. Lamarck believed that lions and so on, if
destroyed, would be replaced in the course of time (1:368 [*187*]; see also
Hull 1967). There is therefore no reason to believe, for example, that
today's mammals and today's fish have common ancestors—they are
merely at different stages on the scale of being (see fig. 2).

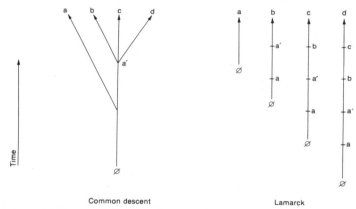

Figure 2. The difference between a theory like Lamarck's and one of common
descent. Life is supposed to begin at Ø, and *a, b, c, d,* are kinds of organisms
extant today.

We should note Lamarck's attitude toward the fossil record in the
Philosophie zoologique. If one believes in an evolution from simplest forms
up to today's most complex forms, one might expect the record to confirm
this sequence. Lamarck could perhaps not expect unambiguous progression, both because of irregularities and because he believed that through
the ages new organisms were constantly appearing and starting to climb.
But, assuming life did start at some first point, he might have expected
some kind of progressive record. (The notion of "progression" is as crucial

to our story as it is difficult to define. Although we shall see the idea evolve, for now let us understand it as a sequence from simple to complex, from the primitive to the most sophisticated, with man as the climax).

But Lamarck himself did not argue that life should have left a progressive record, nor did he bother to interpret the actual fossil record so as to support his position. Even though fossils may have triggered his evolutionism, in his theorizing, his reference to the fossil record was perfunctory at best. Noting that there are organisms in the record that apparently no longer exist, Lamarck argued briefly that since their potential rate of increase is far too great for them to have been eliminated, they must have evolved into today's forms (1:75–81 [44–46]). All Lamarck needed for this argument is that fossil forms differ from today's organisms; there is no inherent need for a progressive difference. Incidentally, given his escalator type of evolution, he found any kind of permanent complete extinction a problem—even the extinction where one organism evolves into another—and he went so far as to doubt that it occurs.

Finally, Lamarck presented himself as a kind of deist—as believing in God as an unmoved mover, creator of the world and its laws, who refuses to intervene miraculously in his creation (1:56 [36]). He made unselfconscious references to this God but made no stringent attempts to relate him to his creation or to read his existence and nature from it. Therefore Lamarck felt no compelling urge to prove that every useful characteristic of an organism—every "adaptation"—is evidence of God's beneficent design. Lamarck may have been an implicit teleologist in supposing some vague principle of progress, but he did not suppose that God keeps fiddling with his creation, constantly stepping in to mold organisms and their characteristics to some new ends.

Indeed, one of the curious features of the *Philosophie zoologique* is Lamarck's nonchalant attitude toward adaptation. If an organism needs something, apparently it will get it. As we have just seen from his reference to the fossil record, there is no question that a species will be wiped out because it lacks some characteristic or because it fails to change over time. Lamarck took it as more or less certain that organisms have or will attain what it takes to get along in their environments. Consequently, though it is obvious that in one sense adaptation was crucial to Lamarck, in that his whole theory was designed to show how organisms keep adapting as their environment sets up needs that trigger change, in another sense it was not of special note. Certainly adaptation ought not to be the focal point of any biological inquiry into why only some organisms have the adaptations they require. Moreover, although Lamarck was aware that organisms compete for resources, perhaps even killing and eating each other (though he doubted this happened within a species), except possibly when man interferes he considered this no real threat to the losers as a group—it merely keeps their numbers within bounds.

In history as in physics, we tend to find that for every action there is an equal and opposite reaction. Certainly, just as we looked to early nineteenth-century France to find the first major scientific defense of a version of organic evolutionism, so it is there we must look to find the first major scientific attack on organic evolutionism. Let us conclude this chapter with a brief consideration of this.[2]

Cuvier's Case against Organic Evolutionism

Paradoxically, it may have been Lamarck's contemporary, the great comparative anatomist Georges Cuvier, who first drove him toward evolutionism. Certainly it was Cuvier who in the 1790s made the strongest case for extinction, with his comparisons of living and fossil animals. If this was indeed so, and if Cuvier recognized it, he must have regretted it bitterly, for he was a lifelong opponent of evolutionism, ridiculing the evolutionists, particularly Lamarck, with all the authority, intelligence, knowledge, and venom he could muster: and Cuvier possessed them all in superabundance. It was, for instance, Cuvier who argued that since the mummified animals brought back from Egypt were identical to forms living today, there can be no truth in any evolutionary hypothesis—particularly one like Lamarck's (Cuvier 1822, p. 123). Much was made of this argument by British antievolutionists, but for Cuvier himself this was a surface argument—his opposition to evolutionary hypotheses stemmed from the most basic tenets of his view of biology.

Cuvier owed an immense debt to Aristotle's view of the organic world, in particular to the belief that organisms must be understood as *functional wholes*. Although one might also give a physicochemical explanation, what characterizes organisms is that all their parts, like the parts of a machine, serve some particular purpose. Unlike Lamarck, Cuvier began with an explicitly teleological view of the organic world rather than slipping it in later (though in any case their versions of teleology were quite different).

Cuvier expressed his concern with ends through his "conditions of existence" (Coleman 1964, chap. 2), a doctrine he believed made possible a science of anatomy, with certain laws applicable to the structure of organisms. Cuvier argued that everything must be related to ends, which imposes certain restrictions or "conditions" on the various parts of an organism—in particular, the various parts must harmonize, for any drastic alteration in one part will lead to violent deleterious effects in other parts. These conditions of existence, in anatomical terms, Cuvier called the "correlation of parts," since each part of an organism necessarily correlates with and depends on every other part. By this principle, Cuvier thought, he made the anatomical reconstructions for which he is still famous. Given just one part of an animal, say a bone or a tooth, Cuvier claimed to be able to "deduce" the other parts. Obviously, however, these "deductions" owed as much to Cuvier's profound knowledge of comparative anatomy, which enabled him to reason analogically from known to unknown animals, as to his metaphysical teleological principles (Coleman 1964, chap. 3).

But, given his particular teleological metaphysics, Cuvier could only be an opponent of evolutionary hypotheses, certainly of any hypothesis of gradual change. Although he was prepared to allow a certain amount of intraspecific variation, his correlation of parts doctrine implied that only certain basic forms are possible. Cuvier believed that if one altered the basic form of any species beyond a certain limit, its essential harmony would be so severely disrupted that the organism would no longer be viable. Thus, were the heart of an organism significantly reduced, the brain, the kidneys, and the liver would be disrupted. The only way to counter these effects would be to change all the other organs—in short, to change from one specific form to another. Hence transitional forms linking species (which Lamarck believed must exist) were not possible. To draw an analogy, since the internal angles of an n-sided polygon equal $(2n - 4)$ right angles, one can have many different polygons according to how one varies n. But one cannot have a polygon whose internal angles total, say, $(2n - 2)$ right angles.

Cuvier thus was violently opposed to the chain-of-being doctrine and argued that no organism (possibly excepting man) is more perfect than any other. By another derivative of his conditions-of-existence doctrine, the "subordination of characters," he claimed that some organic characters exclude others and that one can construct a hierarchy moving from those that allow other characters to those that do not. Instead of a linking chain, Cuvier divided animals into four distinct classes (*embranchements*): Vertebrates, Mollusks, Articulata, and Radiata (Coleman 1964, pp. 87–98). Against Etienne Geoffroy Saint-Hilaire (father of Isidore Geoffroy Saint-

Hilaire), a sympathizer with Lamarck's evolutionary ideas, Cuvier argued that no significant analogies could be drawn between members of different groups.

Though Cuvier's antievolutionism was essentially a function of his teleological world picture, he certainly had empirical arguments to support his position. He would have offered much evidence against a "saltatory" evolutionary theory—one postulating jumps between species. Besides referring to mummies and denying significant analogies between *embranchements,* Cuvier pointed out that, despite anything Lamarck might have said, domestic animals show no great possibility of change. He poured scorn on Lamarck's view that actions and habits might lead to heritable change. And he dismissed the possibility that there might be continuous spontaneous generation of new forms of life (Cuvier 1822, pp. 114–28).

Finally, there was the fossil record. Somewhat ironically, it was Cuvier's brilliant paleontological research that prepared the way for one of the major supports of evolutionism. He showed how the fossil record can be read as showing at least some progressive development of organisms. Reptiles and fish come before mammals in the record; mammals start with strange forms and continue to forms much like those we know today. Man comes at the very end of the record. Indeed, there are no genuine human fossils. Fish show a succession leading to today's forms and possibly precede reptiles (Cuvier 1822, pp. 114–28). But, although Cuvier brought these facts out, unlike Lamarck in the *Philosophie zoologique,* he himself did not see the record as essentially progressive—his metaphysical objections to any chain of being precluded this—and by concentrating on the gaps between different kinds of organisms he read the record as definitive evidence against any evolutionary hypothesis (Cuvier 1822, p. 117). If evolution occurred, he argued, there should be no gaps in the fossil record between organisms of different kinds. Yet such gaps exist. Hence, Cuvier felt able to argue against evolution as categorically as Lamarck had argued for it.

Cuvier was a sincere French Protestant, and his conclusion was obviously very satisfying from a religious viewpoint. But, like Lamarck, he tried to keep his religion and his science separate. Cuvier's world picture supposed frequent extensive floods—and there were strong hints that the most recent was that described in Genesis; but he felt no need to harp on the wisdom of God's design merely because he felt able to impose a teleological interpretation on characters in organisms. Nor was he concerned to promote the idea that God miraculously intervened in the history of the earth to create progressively new species up to and including man—a complement to the story of Genesis. Having refused to see genuine progression, Cuvier used the fossil record mainly to establish the actuality of extinction. He wanted to counter Lamarck's position, that fossil species with no living counterparts must therefore have evolved, with his own claim that such species had completely died out. He explained their successors in the fossil record not by new creations but by migrations of different organisms from other parts of the world. Cuvier was even prepared to push all creations back to original pairs of members of species in the unrecorded past (Coleman 1964, pp. 159–60; Bowler 1976*a*, pp. 16–22). Moreover, instead of hinting that God's creative powers were progressively revealed, Cuvier linked the changes we see in the fish fossil record directly to changes on the earth, and he probably had something like this in mind for other organisms as well. In short, although Cuvier wanted harmony between his science and his religion, he also wanted distance between them. The obsession with the relationship of science and religion that we shall see in future chapters was very much a British phenomenon. Indeed, I hope to show that this was one reason the Darwinian Revolution took place in Britain rather than elsewhere.

2 British Society and the Scientific Community

In 1830, the year William IV came to the throne, Britain was a land of paradox—in some respects the most progressive nation in Europe, in other respects the most retarded (see Beales 1969; J. F. C. Harrison 1971). The industrial revolution was far more advanced than in any other country. The British were applying technology to manufacture and to life in general with such sophistication and to such an extent that for the rest of the century Britain was the world's major power. Nor must we forget that there was also a revolution in agriculture as scientific methods were applied to farming, producing far greater yields. Thus food supplies were increasing to feed an exploding population that was rapidly becoming urbanized as the great new cities of the North and the Midlands drained the countryside in their need for cheap labor. In 1831 the total population of the British Isles was 24.1 million, of whom about a third were Irish. London was by far the largest city, with a population of 1,900,000 (13.5 percent of the population of England and Wales). By 1851 London had grown to 2,600,000, and in the same period Manchester jumped from 182,000 to 303,000, Leeds from 123,000 to 172,000, Birmingham from 144,000 to 233,000, and Glasgow from 202,000 to 345,000. Bradford grew from 13,000 in 1801 to 104,000 in 1851 (J. F. C. Harrison 1971).

But politically and socially Britain remained almost feudal. Power rested in the hands of a very small segment of the populace, most of whom were not leaders of industry but titled, aristocratic landowners (leaders of the Whigs) or lesser landowners, the gentry (leaders of the Tories). Most people did not have the vote, one house of Parliament was the exclusive province of hereditary aristocrats, and many of the seats in the House of Commons were totally under the control of particular individuals. Many seats, from the "rotten boroughs," represented only a few voters, who for various reasons, including fear of economic reprisal, would obediently return the representatives their patrons specified. New laws thus tended to favor the "establishment"—for instance, many concerned the preservation of game—and these laws, enforced by amateur "justices of the peace," were imposed upon the general population. The most notorious were the Corn Laws, enacted after the Napoleonic wars, which put high tariffs on foreign grain. The poor consequently had to pay higher prices for food; industrialists (who were not part of the establishment) had to pay higher wages; and landowners collected artificially inflated rents.

Part and parcel of the privileged group was the established church. Everyone, Anglican or not, had to support the state church, which had a monopoly on weddings and funerals (curiously, the Quakers were exempt); the bishops were considered lords and thus were entitled to seats in the upper house of Parliament; and though some of the lower-level clergy were abominably paid, for the upper echelons the remuneration was very good indeed and the work not particularly onerous. At the top of the scale, two prelates received 19,000 pounds a year at a time when a London policeman was paid only 50 pounds a year. There were close links between the lay and clerical elements of the establishment, for, as with parliamentary seats, many of the plums in the church were at the disposal of individual laymen, and appointments to higher offices (as now) were made by the secular government.

Not surprisingly, religion and the church were of more than incidental interest to certain members of the government. With good reason, they were considered essential theoretical and social underpinnings for the order and stability of society. Many warmly endorsed the poet laureate Robert Southey when he wrote in 1829 that "Nothing is more certain than that religion is the basis upon which civil government rests . . . . And it is necessary that this religion be established for the security of the state, and for the welfare of the people" (quoted by Beales 1969, p. 68). Consequently, attacks on religion and the church were seen not only as blasphemous and immoral, but as socially dangerous.

So far I have been describing a man's world. But a brief discussion of
the place and image of women in society will provide a valuable
touchstone. From today's viewpoint, one might argue with justice that in
the Victorian era middle- and upper-class women took a step backward.
Although never equal with men, they had previously been allowed to be
useful members of society—at least in the home. But then for a number of
reasons, not the least of which was a greatly increased pool of cheap
domestic labor, they were reduced in large part to constant childbearing,
personal helplessness, and an ornamental function—at least ostensibly,
the mindless needleworking dolls known to us through Victorian fiction.
As Tennyson, that great molder and mirror of contemporary thought, let
one of his characters put it (1847, 5, lines 437–41):

> Man for the field and woman for the hearth:
> Man for the sword, and for the needle she:
> Man with the head, and woman with the heart:
> Man to command, and woman to obey;
> All else confusion.

Few were quite this candid, even to themselves. Most preferred to soften
their assessment of woman's status by considering her not so much inferior
as different. Thus, at the conclusion of his poem, Tennyson gives what
perhaps he thought of as his own position. "For woman is not undevelopt
man, / But diverse." Hence: "in true marriage lies / Nor equal, nor
unequal. Each fulfills / Defect in each" (7, lines 259–60, 284–86). Man is
the creature of force, of power, of the intellect. Woman is loving, understanding the emotions and the heart. Rarely do these qualities cross, and
no more should they. Man is suited for the tough world of affairs; woman,
the "angel in the house," for the strength-restoring sanctuary of home and
family.

Up to this point I have presented a fairly static picture of British
society. But, partially in response to forces set in motion by the industrial
revolution, things were starting, very slowly, to change. Beginning in
1826 Dissenters (like Methodists) were allowed to hold public office without special acts of indemnity, and in 1829 Catholics were emancipated.
(This does not mean that every Catholic was allowed to vote, but Catholicism was in itself no longer an a priori bar to office.) Also, some of the
more barbaric laws were being removed from the books—a prudent move,
since juries, knowing the penalties, usually refused to convict. For instance, it was no longer a capital offense to impersonate a Chelsea pen-

sioner. And in 1832 the First Reform Bill was passed, abolishing some of the grossest inequities in the British way of electing a government. Unfortunately, *only* the grossest and most flagrant inequities went, and although somewhat curtailed, power still rested essentially with the old establishment. Most people still could not vote, including all women, and the industrial northern half of England, at all levels, continued to be substantially outside the pale.

Moreover, some of the change was of questionable value. At the end of the eighteenth century, in his *Essay on a Principle of Population,* the Reverend T. R. Malthus had argued that poor law relief is rather pointless. Indeed, he believed that it compounds the problem it aims to end, namely, the existence of an indigent section of the population. Malthus reached this rather gloomy conclusion by starting from the premise that whereas human populations tend to increase geometrically, food supplies can at best be increased only arithmetically. As a consequence, unless people practice restraint, there will inevitably be competition for existence, with some people unable to support themselves. And putting their collapse off through charity will only make matters worse in the end.

Those people who were not poor to the point of suffering took these arguments as almost a priori true (Inglis 1971). They dovetailed nicely with a popular theme in current political thought—that problems are only compounded if the state intervenes with massive welfare schemes. And the arguments had good religious backing. Did Christ himself not say "ye have the poor always with you"? Hence in the 1830s the poor laws were revamped and great new workhouses built—places so grim and unwelcoming that the poor would make maximum efforts to avoid applying for relief. Malthus would undoubtedly have been gratified to learn that as a consequence of these workhouses the numbers of the indigent did dramatically decrease. But, as Carlyle sardonically remarked (1872), one did not really need political economy for justification: "If paupers are made miserable, paupers will needs decline in multitude. It is a secret known to all rat-catchers."

The English Universities

The society I have been sketching was a strange fusion of old and new. A microcosm of the land as a whole can be found in the institutions of higher learning (see Hughes 1861; Adamson 1930; Campbell 1901; and the early chapters of Clark and Hughes 1890). Until almost 1830, there were only two universities in England. One could attend one of the several Scottish universities, and in medicine Edinburgh offered some of the finest training in the world. But apart from this, for an Englishman a university education meant either Oxford or Cambridge. And it presupposed that one was male and a member of the Church of England. In reaction to this latter restriction, a group of Benthamites in London had recently founded University College, and as a counterbalance a group of Anglicans had started King's College (London). But as yet these hardly counted against the ancient universities, and in fact King's College was little more than high-school preparation for Oxford and Cambridge.

Far from being the secular institutions they are today, Oxford and Cambridge were bastions of the Church of England. The universities were divided into colleges run by groups of fellows, nearly all of whom were required to be both bachelors and ordained ministers of the established church. The heads of the colleges were allowed to marry, but for most others marriage meant the loss of one's university post, replaced if one was lucky by a parson's living in a country parish as the gift of one's college. The colleges were the real power, not the universities as a whole. Indeed, New College Oxford and King's College Cambridge were so far autonomous that they granted their own degrees.

Most undergraduates were not at university for serious scholarship, nor was such scholarship demanded of them. They were expected to read a little mathematics, mainly Euclid, a little classics, and a little religion. Plenty of time was left for other pursuits, frequently involving dogs, horses, and foxes. University was looked upon as a gentlemanly way of passing from schoolboy to adult. For most graduates adulthood rarely involved anything so vulgar as trade (far more the province of Dissenters). If one were not to be an independently wealthy landowner, one would most likely be destined for a profession such as law or medicine or for the clergy. Obviously these demanded postgraduate training after university, for neither in practice nor in theory was English higher education expected to teach one anything that could be put to practical use.

However, not every undergraduate lived a life of ease and near idleness. There were at least limited routes of achievement for the serious student. Cambridge had two honors degrees, in mathematics and in classics. But before sitting for the classics honors degree, one had first to perform creditably in the mathematics honors program! At Cambridge those most successful in the mathematics exam were called "wranglers," and a high place on this list was usually a prerequisite to a fellowship at one of the large, prestigious colleges like Trinity or St. John's. Oxford had a similar system, which likewise led by a path of stiff examinations to coveted college fellowships. The major difference was that at Oxford one had first to perform well in classics, before turning to mathematics.

Like that of the country at large, this system seems hopelessly unsuited to the demands of a modern industrial state. There were no examinations in sciences like chemistry, botany, and geology; there was no real system for finding and sponsoring worthwhile needy students (there was some very limited sponsorship of the poor); and in any case the university was totally isolated from most men in industry—those who really needed science and technology—because of religious and social barriers. And yet, as in the country at large, we find seeds of change, inspiring hope for the future. For a start, in the second decade of the nineteenth century a number of the brightest young mathematicians, among them John Herschel, Charles Babbage, and George Peacock, had successfully effected a revolution in Cambridge mathematics, introducing the hitherto-excluded Continental analytical techniques—far more powerful than the traditional British methods stemming from Newton. British applied mathematics was being brought up to date. Second, and of great importance to our story, science professors were starting to take their duties seriously. Although the degree programs included no science requirements or options at all (except applied mathematics or very theoretical physics), Oxford and Cambridge had several chairs in science. At Cambridge these chairs included geology, mineralogy, and botany. In the eighteenth century it had been rare for a professor to bother to lecture in or even know about his subject—the stipends and privileges were the attractions—but by 1830 most made serious efforts to master and lecture on their subjects. Since these lectures were open to all members of the university, interested undergraduates had some opportunity to learn about science, though they could not be examined in the professors' subjects and also had to pay extra for the privilege of attending the lectures.

The essence of British society in the 1830s was reflected in the higher education it offered. Let us turn next to the contemporary men of science who figure in our story.

The Scientific Network

The professor of mineralogy at Cambridge was the Reverend William Whewell (1794–1866) (Todhunter 1876; Cannon 1964*a, b;* Ruse 1976).

Whewell came from Lancashire, of rather humble beginnings, and was a scholarship boy at Trinity. In 1816 he was second wrangler, probably the only time in his life he was second at anything. He stayed at Trinity until he died, becoming in turn fellow, tutor, and in 1842 master of the college. Although he was by no means a truly great scientist, he compensated by having an incredible range of interests—mineralogy, crystallography, political economy, astronomy (tidology), geology, chemistry—the list could go on. And it was Whewell who coined most of the new scientific terms his contemporaries needed. But his major works were his *History of the Inductive Sciences,* first published in 1837, and his *Philosophy of the Inductive Sciences,* published in 1840. In these he surveyed the sciences as had never been done before and produced a neo-Kantian analysis of the scientific method that spurred the empiricist John Stuart Mill to a lengthy reply in his *System of Logic.* (Philosophical discussions of Whewell can be found in Butts 1965; Laudan 1971; and Ruse 1977.)

Whewell was massive rather than handsome, knowledgeable about everything (Sydney Smith once said of him that "science was his forte, omniscience his foible"), and a dreadful bully, particularly toward his inferiors. It is perhaps plausible that his apparent self-confidence masked an insecurity because his father was a carpenter. He was a Tory, as were most of the churchmen, and as soon as he gained power he led the university opposition to the reforming royal commissions of midcentury. But for all his faults he was an impressive man, strong and respecting strength, and his abilities were much admired by his contemporaries.

The Reverend Adam Sedgwick (1785–1873), Woodwardian professor of geology at Cambridge, an ardent Whig and canon at Norwich, had a career that paralleled Whewell's in many respects (Clark and Hughes 1890). Also from the north of England (the Dales) and from a modest background, he came to Cambridge to spend his life at Trinity. In 1817 he campaigned for the Woodwardian professorship under the slogan that though hitherto he had turned no stone, if elected he would leave no stone unturned. He explained his success in the election by saying that while he himself knew no geology, his opponent knew a great deal but it was all wrong. One must add, however, that the major factor was that whereas Sedgwick came from one of the most powerful colleges, his opponent came from a small one with an unsavory reputation for evangelicism.

Despite his initial ignorance, Sedgwick rapidly became one of the best field geologists in Britain, concentrating on the strata of the Cambrian period (see chap. 6 for a geological chronology). Unlike Whewell, however, Sedgwick did not take easily to the pen, except when religion or university politics were at issue, and he left no major geological work. Also, unlike Whewell, Sedgwick was a pleasant man who, in the best Yorkshire tradition, was grand company. Though he was hot tempered, he did not harbor animosity. (The exception to this was a bitter feud with his sometime friend and collaborator Roderick Murchison over Murchison's absorbing Sedgwick's Cambrian strata into his own Silurian strata.) Upon the publication of the *Origin,* Sedgwick wrote his old student Darwin a most typically schizophrenic letter. In the first half he wildly berated Darwin; in the second half he sent warm greetings, referring to himself as a "son of a monkey" and discussing his health. The greetings were genuine, for Darwin and Sedgwick remained friendly (Clark and Hughes 1890). To the undergraduates Sedgwick was known as "Robin Goodfellow"—Whewell was "Rough Diamond" or "Billy Whistle"—and to his friends Sedgwick was "Old Sedge."

The Reverend John Stevens Henslow (1796–1861) was professor of botany at Cambridge and a fellow of St. John's (Jenyns 1863). When he assumed the chair, after a short period as professor of mineralogy, botany was almost a nonexistent subject at Cambridge. For the previous thirty years—the second half of his predecessor's sixty-year tenure—there had been no lectures, the herbarium had been allowed to rot, and the very small garden had degenerated into a wilderness. Henslow changed all that. He began to lecture with great success, and he was responsible for planning and financing the magnificent botanical gardens that are still a joy today. Henslow was not a great scientist, but he had a wide knowledge and deep love of the subject that he communicated to his friends, including the few undergraduates interested in science. One feels he would have been as kind and helpful to students who were immature and insecure as Whewell was intimidating. For the last twenty years of his life he was the incumbent of a living in Hitcham, Suffolk, and he appears to have been the exemplar of the cleric who believes he is obligated to serve his parishioners. Although this service took time from his science, he would have had no doubt which was more important.

Finally, let us mention Charles Babbage (1792–1871), Lucasian professor of mathematics at Cambridge (Gridgeman 1970). Babbage, a brilliant mathematician, was one of the group of friends responsible for introducing Continental mathematics at Cambridge. Although he held the Lucasian chair for eleven years (1828–39), he was not a professional academic in the sense of the men already discussed, and indeed he gave no lectures during his professorial tenure. Babbage is most famous for his invention of calculating machines, mechanical devices for solving mathematical problems. But though the government initially supported his work on these machines, such support dried up and he had to abandon his more grandiose plans. But this was not before Babbage's machines—surely a reflection of the industrial era—worked their way into the organic origins debate in, as we shall see, a most curious manner. Embittered by his failures and becoming morose, Babbage spent his declining years fighting the iniquities of organ-grinders, whose practices upset his high-strung nerves.

Two Oxford professors are of particular importance to us. First is the Reverend William Buckland (1784–1856), reader in mineralogy and in geology (although in the latter he seems to have been regarded as a de facto professor and so styled himself) (Gordon 1894). Buckland came from Devon and was educated at, then made a fellow of, Corpus Christi College. At Oxford science evoked far less support and interest than at Cambridge, and for a long while Buckland worked almost single-handedly to see that it had any existence at all. He was the first reader in geology and accumulated a large geological collection that he subsequently gave to the university. Buckland was supported in his lonely task first by his great knowledge of geology, particularly paleontology, and second because he was one of nature's great showmen and eccentrics. Academia's gain was a loss to the stage—or the circus. With his magnificent lecturing style, he could hold his audience spellbound while explaining the most difficult phenomena, and he kept everyone helpless with laughter as he acted out, for example, how a cock made impressions in the mud. Needless to say, such buffoonery did not always sit well with his fellow savants. Of our ordained scientists, Buckland was perhaps the most successful in ecclesiastical terms, for he was a canon of Christ Church and, later in life, dean of Westminster.

Also at Oxford was the Reverend Baden Powell (1796–1860), Savilian professor of geometry (Tuckwell 1909; Whately 1889). Powell was a brilliant mathematician; it was said that at his B.A. examination he worked out on the spot all the theorems he was expected to reproduce from memory. He did research on heat and optics, as well as writing books on the history and philosophy of the sciences (Powell 1834, 1838). But it was to religious controversy that he was really attracted, delighting in attacking the high-church excesses of his fellow brethren at Oriel College (such as John Henry Newman). He was no favorite of low church-

men either, and could be expected to hold forth at length on the
stupidity of rigorous Sabbath observance. Powell apparently had an incredibly languid temperament, but this did not stop him either from
teaching effectively or from fathering fourteen children, the youngest of
whom founded the scout movement. Although he was constantly involved
in religious disputes, reaching an appropriate climax just before his death
with a contribution to the notorious *Essays and Reviews* (see chap. 9), his
integrity was recognized and appreciated. After Powell's death, Cardinal
Manning wrote to his widow, saying that Baden Powell had treated
Catholicism with scrupulous fairness.

Leaving the universities, we must introduce four more scientists. John
F. W. Herschel (1792–1871), son of the famous astronomer William
Herschel, was an equally famous astronomer in his own right (Cannon
1961*b*). Herschel graduated first wrangler from St. John's College Cambridge in 1813 and was one of those responsible with Babbage for introducing European mathematics into that university. His scientific interests
were as broad as Whewell's—he had a significant part in spreading the
wave theory of light, wrote on crystallography, magnetism, and geology,
and was one of the innovators of photography. But his principal reputation came through astronomy. He completed his father's work of mapping
the nebulae of the northern heavens, then (in the 1830s from the Cape of
Good Hope) undertook the massive reciprocal mapping of the southern
skies.

Even more than Whewell, Herschel was in the public eye, to a great
extent because of his popular little philosophical work on the nature of
science, *A Preliminary Discourse on the Study of Natural Philosophy* (published in 1831), and because of his several general accounts of astronomy
(particularly Herschel 1833*a*). To be scientific, in the popular mind, was
to be as much as possible like Herschel, who was first knighted, then
given a baronetcy. Despite his fame, however, he seems always to have
been rather shy. Darwin (1969, p. 107) told a rather wicked story that
Herschel always came into a room looking as though his hands were
dirty—and that his wife knew they were!

Charles Lyell (1797–1875), though born in Scotland, was brought up
as an English country gentleman and sent to Exeter College, Oxford (Lyell
1881; Wilson 1972). He read for the bar, but as weak eyesight prevented
him from pursuing this profession successfully, his work switched more
and more to geology, in which Buckland had first whetted his interest.
His great *Principles of Geology* first began to appear in 1830, and all his life

Lyell revised and updated this work and offshoots from it. At the beginning of the 1830s, despite some episcopal misgivings, Lyell became professor of geology at the newly formed King's College, London. But although his lectures proved popular, he soon gave up the post because preparing and delivering them cut so deeply into his working time. Lyell, a liberal, was much interested in university education, and he incurred the wrath of his hitherto good friend Whewell when in the 1840s he presumed to suggest that education at Oxford and Cambridge stood in need of reform, compounding this slight by setting Germany and Scotland as models! Like many of the men discussed, Lyell was to become close to the prince consort, and he perhaps treasured the acquaintance of people of rank more than they deserved. Like Herschel, he saw his labors rewarded first by knighthood, then with a baronetcy.

The last of our main characters at this stage of the story is the only one with no connection at all with the ancient English universities (Owen 1894). Richard Owen (1804–92), a Lancastrian, was a schoolfellow of Whewell, with whom he remained close friends until Whewell's death. He was trained as a surgeon (through apprenticeship), briefly attended the University of Edinburgh, and in 1825 went to London. As a brilliant dissectionist already showing evidence of the ability that was to bring him fame in comparative anatomy, he became involved in cataloging the Hunterian Collection at the College of Surgeons—a task that necessitated many fresh dissections. In 1830 his knowledge of comparative anatomy was given a significant boost when he became acquainted with Cuvier. The ways Owen endorsed or rejected Cuvier's ideas will be discussed in a later chapter.

Owen first gained significant public notice with the brilliant anatomical descriptions in his *Memoir on the Pearly Nautilus* (1832). About this time he also began announcing the results of dissections he was making of animals that died at the London Zoo. His specialities came to be the primitive mammalian forms monotremes and marsupials, particularly their modes of reproduction and feeding their young. A fascinating paper on the kangaroo's method of suckling its offspring, which strongly supports divine design, was to receive repeated reference in the organic origins debate (Owen 1834). Owen's status as a leading comparative anatomist in Britain was consolidated in 1836 when he was appointed Hunterian professor at the Royal College of Surgeons, a post he held until 1856, when he became superintendent of the natural history departments of the British Museum and Fullerian physiology lecturer at the Royal

Institution. After 1835 Owen became far more interested in fossil organisms and in paleontological questions in general, probably in part a response to his relationship with Darwin.

It is not easy to assess what sort of man Owen really was, for he more than anyone else came into conflict with the Darwinian party, and with T. H. Huxley in particular (MacLeod 1965). Because Owen "lost" and the Darwinians "won," and because the Darwinians (particularly Huxley's son Leonard) lived to write the official histories, Owen has been painted as the bogeyman. No motive is too black to ascribe to him, no action too implausible. We shall learn far more of Owen as this story progresses, but it does seem fair to say that Owen was not a very comfortable man. He was touchy about his status and not generous about sharing possible glory. On the other hand, he was certainly not unfriendly to younger scientists— both Darwin and Huxley had reason to thank him for his solicitousness about their early careers. Perhaps he was one of those men, not uncommon, who can be helpful and friendly (in a rather distant manner) except in their own balliwicks.

With our dramatis personae onstage, let us see how they came together as scientists. (Good material is found in Cardwell 1972; see also Babbage 1830; and Becker 1874.)

The Scientific Societies

In Cambridge informal weekly meetings were held at Henslow's house, welcoming all university members who were interested in science (Darwin 1969, pp. 64–67). Undergraduates mixed freely with professors, united by a common love of science. More formal were the meetings of the Cambridge Philosophical Society, founded in 1819 by Henslow and Sedgwick (Clark and Hughes 1890, 1:205–8), which enabled Cambridge scientists to read and discuss papers and to publish volumes of proceedings. Whewell, Babbage, and Sedgwick figure prominently in the early numbers.

In London, the Royal Society was the oldest and most prestigious of the scientific societies, and its *Philosophical Transactions* were the most important of scientific publications. But in 1830 the society seems to have left much to be desired; it had become in large part a fashionable meeting place rather than a club where entrance was decided on scientific merit. Babbage (1830) in particular subjected it to scathing criticism. In 1831 an attempt was made to elect Herschel president, as part of an effort to bring the society back to the cause of true science. Unfortunately the attempt failed, and it tells us much about the state of the Royal Society (and of Britain in general) that instead of the best-known scientist in Britain, it elected the duke of Sussex, brother of the king. But all was not lost: the message behind the attempt to elect Herschel was noted, and the duke himself made strenuous and successful efforts to revitalize the society. When we reencounter the society later we shall find it transformed.

Partly because science was becoming so much more specialized and partly because the Royal Society was failing in its duties, a number of alternative specialist scientific societies had come into being—usually in the teeth of violent opposition from the powers in control of the Royal Society. There were the Linnean Society (devoted to botany), the Astronomical Society, and the Zoological Society. But undoubtedly the most vigorous (and hence to the Royal Society the most offensive) was the Geological Society of London, founded in 1807 (Rudwick 1963; Woodward 1907). Its members read papers at its frequent meetings, it had a strong geological collection, it published good series of *Transactions* and *Proceedings,* and once a year its president could be relied on to give a detailed and penetrating analysis of the state of the science.

Many of the men we have discussed were deeply involved in the Geological Society. One senses that it was here they gathered to exchange ideas, certainly those ideas bearing on the organic origins problem. Buckland was president from 1824 to 1826 (and again in 1839–41), Sedgwick from 1829 to 1831, Lyell from 1835 to 1837 (and again in 1849–51), Whewell from 1837 to 1839. Herschel was never president only because he refused Whewell's urging to succeed him. Apart from this, many of these scientists served long periods on the society's governing council. In 1830 Sedgwick was president, Lyell was foreign secretary, and Buckland, Whewell, and Herschel were all on the council. Before the end of the decade, Henslow and Owen were on the council, Babbage was a frequent and enthusiastic attender, and Baden Powell was certainly elected to the society. It is no exaggeration to say that it was in part because of the way these men and their ideas came together in the Geological Society that Charles Darwin was able to face and solve the problem of organic origins.

The Royal Institution, giving public scientific demonstrations and lectures and deservedly made famous by its support of (among others) the great scientists Davy and Faraday, lies essentially outside our sphere. But special mention must be made of the British Association for the Advancement of Science, founded in 1831 (Cardwell 1972, pp. 59–61). This annual scientific congress met each year in a different provincial town (not in London). In addition to general meetings, this association held sectional meetings devoted to specific sciences, and before long these became important on the Victorian scene. Professional scientists could meet there and exchange information. Moreover, some lectures were deliberately pitched at the popular level and the public was invited to attend—an invitation that was appreciated and accepted. The men of our network quickly became enthusiastic participants. In 1832, when the association met in Oxford, we find Buckland president and Baden Powell an active participant. In 1833, when it met in Cambridge, we find Sedgwick president and Whewell giving the keynote address.

Science as a Profession

We have gathered a number of facts about science and scientists in 1830. Let us try to put them together by asking a question that will be particularly pertinent when we come to Darwin: Can we in some sense at this time and place speak of science as a profession and of scientists as professionals? (See Ben-David 1971; Crane 1972.)

To answer, let us first ask a preliminary question: What makes us speak today of science as a profession and scientists as professionals? Certainly, an important factor seems to be that one can make one's living at science, in a university or in government service or, less often, in business. But there is more to it than this. One is usually expected to have certain qualifications—at least an honors degree and more often graduate training, and to be a member of professional societies. Also, one is expected to care about science—to practice it for self-satisfaction, not merely to put bread on the table. This caring will be shown in a desire to do research and to publish. Moreover, much of the publication is expected to be of a particular kind. Writing chiefly for popular magazines or journals, say *Scientific American*, does not count—indeed, too much popularization is a negative mark. In the recent sociobiology controversy, E. O. Wilson was criticized because too many of his references were to that particular magazine. The scientist is expected to publish in the "professional" journals, those at the forefront of the science, and his work, including books, is expected to be directed to his fellows—who are doing likewise. Being a professional scientist is a question not only of making a living but also one of attitudes: to whom one addresses oneself and whose respect one prizes.

The professional scientist directs himself to his counterparts who devote their lives and interests to science. A valued mark of respect is to be invited or elected to leadership in a professional scientific organization.

Turning back to 1830 and confining ourselves to the people and interests of our network, we cannot immediately say that we are dealing with a profession or with professional scientists. Indeed, it is easier to think of Lamarck and Cuvier as professionals in a modern sense, for both had government posts supporting their careers in science. But we have seen that in Britain, particularly in England, no formal system existed for training or employing scientists, though this began to develop as our story progresses and plays a pertinent role.

Take Sedgwick. He was certainly not paid much for his geology, receiving only a small professorial stipend (218 pounds a year) and lecture fees (Clark and Hughes 1890, 2:349). Most of his income came from his college and the church. However, he did teach geology, he helped administer geology as a discipline through societies and the like, he cared passionately about geology over a long period, he published in journals for fellow geologists, and, most important, he was recognized as a leading geologist. All these things contribute toward making Sedgwick a professional geologist. If the opposite of "professional" is "amateur," he was certainly not an amateur.

Of course at that time things were not tightly knit, and there are borderline cases. Consider Whewell's status. As a physicist—judged as we have just judged Sedgwick—he was a professional. He published extensively on tides in the *Philosophical Transactions* and was recognized as an expert. But as a geologist his status was more ambiguous. He did not do original research in geology and did not pretend to himself or others that he did. His writing on geology was at a level of general discussion and synthesis, and he often published where most readers would be laymen—for example, in the great Tory periodical *Quarterly Review*. On the other hand, as an official of the Geological Society Whewell was involved in science administration, and there is much evidence that he was respected as a thinker on geology by those with the best claim to be called geologists. Lyell went to some pains to ensure that Whewell would review him in the *Quarterly* (Lyell 1881, 1:351), and in fact Whewell's publishing outlets do not mark him irrevocably as an amateur. The popular reviews of that time carried substantial scientific discussions, written and read by scientists. They functioned somewhat as the nontechnical parts of *Science* and *Nature* do today. In short, it is a mistake to think of Whewell as outside the pale of professional geology. And on something like the organic origins question, involving so many aspects of science, Whewell's breadth of interest and knowledge made him eminently qualified to discuss it professionally as one whose opinion would be respected by all.

Though British science in the 1830s did not offer a full-blown profession, the seeds had been sown and were starting to send forth shoots. Certainly our network was more than a collection of amateurs.

Charles Darwin

We come now to the hero of the story—insofar as this story has a hero. Here I wish simply to introduce Darwin, in the context of his time and his contemporaries. His achievements, biological and otherwise, must wait until later. (The best biography of Darwin is de Beer 1963. Invaluable are F. Darwin 1887 and Darwin and Seward 1903.)

Charles Darwin (1809–82) was born into comfortable circumstances. His paternal grandfather, Erasmus Darwin, was a well-known physician, a leading scientist of the Midlands, a friend of industrialists, and the author of famous, perhaps notorious, prose and verse speculations about organic evolution. Darwin's maternal grandfather was the well-known potter Josiah Wedgwood, who introduced stunningly successful new techniques into the English china trade. Darwin's mother died while he was young. His father, Robert Darwin, though he did not match the glory of the previous generation, was a very successful doctor in the heart of agricultural England, at Shrewsbury in Shropshire.

We can begin to place young Charles Darwin in early nineteenth-century English society. First, his family was very rich. Medicine paid good dividends. On his mother's side, his Wedgwood uncles spent money freely (Meteyard 1871), for the pottery works of old Josiah Wedgwood ensured that none of his immediate descendents would ever want. And Charles tapped this source again when he married his cousin, Emma Wedgwood, in 1839. Consequently, young Darwin was at a great advantage in a society that made financial security a route to success. Second, though much of the Darwin money came from trade, it had become socially acceptable by the time it got to him. The Darwins were nominal Anglicans (unlike the Unitarian Wedgwoods), and his father was in a profession, not in business. The privileges of the British establishment were therefore open to him.

Following the course set by his father and his elder brother Erasmus,

Charles—an indifferent student at school—was packed off to Edinburgh to train in medicine. After two years, bored with the lectures and revolted by the operations, he had had enough. His father, fearing he might have an idle wastrel on his hands, took the obvious remedy for aimless young men and redirected Charles toward a career in the church. (This was as cynical as it sounds; privately, Robert Darwin was an atheist.) Since a church career required an English degree, at the beginning of 1828 young Charles enrolled at Christ's College Cambridge, an institution with a reputation for being fast, whose members were fond of visits to Newmarket (the mecca of English horse racing).

Academically, Darwin's progress at Cambridge was not distinguished. He did not take honors (nor did he attempt them), and although he did well among the nonhonors men, in later life he wrote: "During the three years which I spent at Cambridge my time was wasted, as far as the academical studies were concerned, as completely as at Edinburgh and at school" (Darwin 1969, p. 58). He brushed up a little classics, a little Euclid, and Paley's *Evidences of Christianity* and *Moral Philosophy,* and that was enough to get by.

The obvious conclusion is that with this kind of background Darwin would be a rank amateur at anything he tried: little more than a dilettante. And when one learns that he never earned his living but lived off the family fortune, his amateur status seems confirmed. But I think such a conclusion would be mistaken. Though he was independently rich and had no need of a paid profession, there are good reasons to believe that Darwin became as professional as any scientist at that time by the criteria we have set. Moreover, Darwin's work took place within a particular society and scientific community and grew out of the ideas, standards, and puzzles of that society and community. In short, the questions Darwin asked and the answers he supplied can be understood only in relation to his contemporary background, in large part because of his professional scientific status. He was not a solitary genius, indifferent to and unaffected by the currents around him. And the professional element, in various ways, was no less important in determining the reception of his ideas.

As a start toward establishing Darwin's professionalism, let us review the state of the English universities in the late 1820s. As I have mentioned, the route to formal academic success at Cambridge lay first in mathematics, then in classics. There was no other. Moreover, this was a route, by tradition and intention, designed for those who needed to get on (ignoring, for the moment, those like Herschel who were naturally bristling with mathematical talent). A young man with money and without a great deal of ambition would have neither motive nor pressure to take honors. Unlike Sedgwick, a person like Darwin would not have to push himself to avoid being banished to an underpaid church post in some remote and unpleasant part of England—perpetual curate of Hogglestock at 130 pounds a year.

But this explains only why we might not expect to see Darwin's scientific training reflected in his academic record. There were no degree courses in the natural sciences, and so he cannot be faulted for not taking one. Is there any positive evidence, starting with his time at Cambridge, to substantiate Darwin's "professional" involvement in science? There is indeed. It is clear that Darwin, who as a schoolboy had a love of science, entered the scientific community at Cambridge almost as soon as he arrived. Through his cousin W. Darwin Fox, who taught him to be an ardent beetle collector, Darwin was introduced to Henslow, professor of botany. He soon became a close friend of Henslow, regularly attending his scientific evenings, and consequently became close to some of the leading scientists of Cambridge, including Sedgwick and Whewell. Moreover, for the three years he was at Cambridge Darwin attended Henslow's lectures on botany, as did Sedgwick and Whewell (who appear on Henslow's lecture list in the Darwin Collection, University Library, Cambridge). This was of course a voluntary activity, for which Darwin had to pay extra.

By the time his Cambridge career ended in 1831, Darwin was hardly a professional scientist—or a professional anything. But for three years he had mixed with some of the leading scientists of his day, at a level far more intimate than would be possible for an undergraduate today. Much of what he missed in formal training he made up informally, particularly from the continued attention of Henslow. Moreover, by the time he left Cambridge, the hitherto undirected young man was determined to seek a place in science. In this determination he was fired by his reading of Herschel's newly published *Preliminary Discourse* and by the travel books of the great German explorer Alexander von Humboldt. Furthermore, it is clear that his elders had seen in Darwin a vital scientific spark. They wanted to help, guide, and encourage him. In the summer of 1831 Sedgwick took Darwin on an intensive geological field course in Wales; then, through Henslow's connection, an offer came to join the government's H.M.S. *Beagle* on a navigational trip around the world. After initial reluctance from his father Darwin was allowed to accept, and for the next five years he was de facto the ship's naturalist on the *Beagle* (Gruber 1968; Burstyn 1975).

Later we will explore the importance of this trip for Darwin. At any rate, he made great collections—geological, zoological, entomological, and botanical, among others—and worked continuously at his science, particularly geology. Moreover, he did not lose touch with the scientific community. He kept in close contact with Henslow (Barlow 1967), met Herschel (mapping stars at the Cape of Good Hope), and was remembered by Whewell, who sent him a copy of the address he had given in 1833 at the British Association (Barlow 1967, p. 87 n). Extracts from his letters to Henslow were read at the Cambridge Philosophical Society, then printed and distributed to the members. He was praised to his father by Sedgwick and obviously was well spoken of generally—Lyell, for one, was very eager to meet him (Lyell 1881, 1:460–61).

When he returned to England, Darwin considered himself a scientist by profession, in particular a geologist (not a biologist),[1] and he was certainly welcomed as such in the scientific community. To use a modern analogy, Darwin was regarded as one of the brightest new graduate students in the scientific network (perhaps by this stage postdoctoral) and was treated in this manner. Despite his own protestations, Henslow and Whewell urged him onto the council of the Geological Society, and then to the post of secretary (Darwin 1969, p. 83). He read papers to the society and received effusive praise from Whewell (1839) in his presidential address. He published a major paper on geology in the Royal Society's *Transactions* (Sedgwick was the referee arguing for publication), and he put together a highly regarded travel book based on his trip with the *Beagle* (Whewell gave him much advice and pressure about publishing this). Furthermore, Darwin became intimate with men of the scientific network whom he would not have met at Cambridge—Lyell, Babbage, and Herschel in particular. Finally, Darwin showed extreme care in arranging for his *Beagle* collections to be described and cataloged and was impatient when this was not done quickly. Buckland might have been the most diplomatic choice to work on the fossil collection; but Owen graciously offered his help and Darwin accepted, showing that he knew where to find up-to-date knowledge and efficiency (Herbert 1974).

In short, Darwin had as much scientific training as anyone in the network—more than most. From the time he left Cambridge his time was given entirely to science, more than can be said of the professors, who had university duties. He thought of himself as a geologist and was accepted as such. Most revealingly, in later life (Darwin 1969, p. 82) he admitted that the audience he hoped to impress was his fellow scientists, not the general public. In other words, insofar as we can speak of any man in the

1830s as a professional scientist, Darwin qualifies, and we can expect him to reflect the ideas and beliefs of his day. Darwin's background and training were a product of his society, and this pattern was repeated in more conceptual matters.

By the end of the 1830s Darwin was just beginning to show signs of the mysterious illness—headaches, heart palpitations, and the like—that was to make him an invalid for the rest of his life (Colp 1977). But at that point he had no real idea of what lay ahead, and so far fortune had been kind. He was tall, good-looking, personable, well-connected, rich, and newly and happily married to his cousin. Let us therefore leave him and turn from personalities to ideas. What exactly did the scientists of the day believe?

3 Beliefs: Geological, Philosophical, and Religious

To understand properly the positions that members of our scientific network took on the organic origins question, we must consider some of their more general beliefs. To this end we shall look in turn at matters of science, philosophy, and religion. Because of its key role, for the time being I shall confine my scientific exposition to geology. Throughout this book I shall introduce new ideas only as they enter the public domain. But, looking ahead to Darwin's producing a major evolutionary theory, I shall pay particular attention to his public work in the 1830s. By so doing, I hope to explore some of the important intellectual influences on Darwin and show his responses to them. Because these responses are crucial to understanding Darwin's evolutionary speculations, I shall go into greater detail than might seem warranted by Darwin's public importance in the 1830s—a time when he was a very gifted, but still rather junior, member of the scientific network.

The Geological Background

William Whewell had a word, or two, for everything. His terms for the geological factions in the 1830s were *catastrophists* and *uniformitarians* (Whewell 1832, p. 126). It will be useful for the moment to adopt these terms, although more subtle distinctions must be introduced shortly. The classic uniformitarian position was put forward by Lyell in his *Principles of Geology,* and it seems true that Lyell wrote against a background of catastrophism. Probably the best way to review these two geological positions is to consider them in terms of origins. Lyell, echoed by many of his followers and commentators, tended to portray himself as a complete revolutionary, whose ideas broke entirely with his past (for rather different evaluations, see Wilson 1972; Rudwick 1972). But without denigrating Lyell's very significant scientific achievements, it does seem fairer to recognize that both the uniformitarians and the catastrophists had roots in the past (Gillespie 1951).

The intellectual ancestors of the uniformitarians, Lyell in particular, were the Scottish geologists of the late eighteenth century who came to be nicknamed the Vulcanists. Inspired by the tortuous writings of Dr. James Hutton of Edinburgh (chiefly his *Theory of the Earth* of 1795), very necessarily popularized by his friend Professor John Playfair (1802), their argument was that geological formations come from a combination of weathering and heat. Wind and rain and such eventually lead to the deposition of silt on the ocean bed; heat from within the earth, combined with terrific pressure, causes this detritus to fuse into solid rock; then through heat expansion and volcanic action geological formations are thrown up, completing the cycle. We shall see strong reflections of all this in Lyell's work. The Vulcanists' vision of the earth as subject to a constantly repeating cycle implies its great age. With reason, Hutton's best-known geological claim was that "We find no vestige of a beginning,—no prospect of an end" (Hutton 1795, 1:200).

The Vulcanists' rivals were the ancestors of the catastrophists, the "Neptunists," who owed their inspiration to Abraham Gottlob Werner, professor of mineralogy at Freiberg in Saxony (Gillespie 1951). For them, water and precipitation determined virtually everything. At one point the whole earth had been under water; then, bit by bit, the various earth formations were precipitated out. Only recently did some coal deposits catch fire, thus bringing about volcanoes that formed certain localized igneous rocks. Essentially, the earth's rocks are sedimentary. Although by the beginning of the nineteenth century all serious geologists were starting to see the earth as fairly old, the Neptunists did not believe it was nearly so old as presupposed by Vulcanism. Moreover, unlike the Vulcanists, the Neptunists saw the world historically: for them it had an identifiable beginning and direction.

Neptunism, at least in some of its elements, was attractive to Cuvier, who brought it up to date (Coleman 1964) and who can therefore rightly be considered the founder of catastrophism, though Cuvier never used the term "catastrophe," from which Whewell coined the movement's name. Undoubtedly inspired by his biology, which saw organisms as functionally integrated units, Cuvier and his friend Alexandre Brongniart discovered that the Paris Basin shows distinct layers of freshwater and marine organisms, which he therefore interpreted as evidence of different floodings of the area (Rudwick 1972). Furthermore, argued Cuvier, these floodings (which he called "revolutions") must have been fairly rapid and both in nature and cause were unlike geological processes occurring within our experience. Apart from anything else, Cuvier pointed out, the perfectly preserved carcasses of frozen mammoths from Russia show that the past was interrupted by events that happened quickly and were of a kind and magnitude not experienced today (Cuvier 1822; the *Essay* was first published in Britain in 1813). Much of the past might have been like the present, but every now and then the uniform course of nature is shattered.

Cuvier probably did not have supernatural causes in mind for his upheavals, since he always downplayed possible religious implications of scientific speculations, but he had little doubt that the final deluge was the one described in Genesis. It is significant that Cuvier himself chose the term "revolution," with its hint of repetition according to law, rather than "catastrophe," with its faint odor of extraworldliness. We shall discover that the British did not always exhibit such sensibility about keeping science and religion separate, and indeed they would not always have appreciated it. But, speaking as geologically as possible, we find that William Buckland, Britain's foremost geologist of the 1820s, followed Cuvier in believing that there have been "successive periods of tranquility and great disturbance" (Buckland 1820, p. 29). Moreover, at the times of disturbance we have had "convulsions of which the most terrible catastrophes presented by the actual state of things (Earthquakes, Tempests, and Volcanoes) afford only a faint image" (p. 5). And of at least one of these convulsions, and presumably all, Buckland was prepared to speak of the impossibility of its having had causes "that are now or appear ever to have been in action" since it occurred (p. 38). Furthermore, Buckland agreed with Cuvier in thinking that at least this final catastrophe, a very recent event, involved flooding, although in his desire to confirm Genesis Buckland argued for a very quick universal flood, whereas Cuvier seems to have had in mind more limited and possibly somewhat slower phenomena. At the beginning of the 1820s a cave was discovered in Yorkshire, filled with the bones of extinct animals, which Buckland took as triumphant confirmation of this flood (Buckland 1823; and see fig. 3).

Figure 3. One of the illustrations Buckland gave (in his *Reliquiae diluvianae*) purporting to prove the existence of a universal flash flood. Skeleton *G* is of a rhinoceros, washed down into the cave and enclosed in diluvial rubble *(E)*. That there were bones to reconstruct only one rhinoceros proved, Buckland claimed, the extreme rapidity of the flood.

Buckland may have thought its cause was supernatural, but he was not at all clear on this. And Buckland's geology sympathetically endorsed Cuvier's interpretation of the fossil record, which was taken to prove a direction to the earth's history. In particular, Buckland was more than happy to report that man was not of great antiquity (Buckland 1820, p. 24). Buckland, unlike Cuvier, saw real progression (with evolution-barring gaps).

It would be a mistake to think that all of British geology immediately before Lyell came directly or indirectly from Cuvier. For instance, the overall directionalist view of the earth received strong support in the late 1820s from an argument purporting to show that the earth is cooling gradually, unidirectionally from an incandescent origin (Fourier 1827; Cordier 1827). This tied in nicely with fossil evidence showing that Europe in particular had progressed from a warmer to a cooler state, since many European fossils show characteristics now associated with tropical climates. Nor would it be true to say that Buckland was without critics. Certainly, some before Lyell thought that Buckland was rather too keen on catastrophes for explaining geological phenomena (Rudwick 1972). But it does seem true that, coming into 1830, most of the British geological community would have felt sympathy with a position much like Buckland's—periods of calm punctuated by catastrophic upheavals of a magnitude and (probably) cause no longer experienced, and some sort of overall direction to the world. It is against this background that Charles Lyell's theorizing must be judged.

Principles of Geology

The first volume of Lyell's *Principles of Geology* appeared in July 1830. We have seen already that Whewell's blanket term "catastrophist" is rather insensitive, for it is clear that the catastrophist position was more than mere advocacy of catastrophes. For example, the catastrophists argued for some direction to earth history. Similarly, Lyell's position was more than a mere denial of catastrophes. In the spirit of recent commentators and with an eye to the efforts of someone like Buckland, it will be useful to separate out three things Lyell tried to do (Rudwick 1969; Hooykaas 1959; Mayr 1972*b*).

First was what one might call Lyell's "actualism." He wanted to explain past geological phenomena in terms of causes of the kind now operating. This methodology is brought out clearly in the full title of Lyell's work, *Principles of Geology, Being an Attempt to Explain the Former Changes of the Earth's Surface, by Reference to Causes Now in Operation.* Second was Lyell's "uniformitarianism" (from now on we will restrict the use of this term). He wanted to explain past geological phenomena not only in terms of causes of the same *kind* now operating but also in terms of causes of the same *degree.* He wanted to avoid "catastrophes." Causes "never acted with degrees of energy from that which they now exert" (Lyell 1881, 1:234). Third, Lyell was committed to a *steady-state* view of the earth. He believed the earth was in a perpetual cycle of eruption and decay, where all periods were essentially similar. There is no sign of any direction or progression in either the inorganic or the organic world. Lyell did not deny absolutely that the world had a beginning (and might have an end), but as a geologist he, like Hutton, thought this irrelevant (Lyell 1881, 1:269–70).

Although Lyell was not always careful to keep his three aims separate, let us try to do so. As we look at the first volume of the *Principles*, we find it falls naturally into three parts. The first part attacked those Lyell took as his opponents and also prepared the way for his own system. To this end Lyell presented a highly selective "history" of geology, showing how his own geological position was a return to "true" geology, which has always had to combat a false tradition (Porter 1976). Of more direct interest to us is the second part of this first volume of the *Principles*, a two-pronged attack on directional views about earth history—in other words, primarily a defense of Lyell's steady-state thesis. Let us take the two aspects in turn.

First, Lyell felt he had to counter the claim that the earth shows a directional climatic change from hotter to colder. As we have seen, this claim is based essentially on the fact that European fossils have close analogies in living organisms from tropical regions, apparently implying that the world has become cooler over the years. As we have also seen, the claim ties in with theoretical physical considerations, supposing that the world had a very hot origin and has gradually been getting colder by radiation. Lyell's reply was his "grand new theory of climate," by which he explained temperature fluctuations "without help from a comet, or any astronomical change, or any cooling down of the original red-hot nucleus, or any change of inclination of axis or central heat, or volcanic hot vapours and water and other nostrums, but all easily and naturally" (Lyell 1881, 1:262). In brief, Lyell argued that temperature and climate are primarily determined by the distribution of land and sea, and that because this distribution is constantly changing (through erosion, earthquakes, and so on), we should expect fluctuations in climate of a kind that apparently have occurred in Europe. Hence, understanding "steady state" to allow fairly substantial fluctuations, the differences between past and present temperatures in Europe represent only oscillation about a mean. As the passage quoted makes clear, this climate theory is intended to serve at least two purposes. Obviously Lyell intended it to support not only steady-statism but also uniformitarianism. It eliminates the need for superintense forces. Indeed, it seems also to be actualistic—one explains past climate changes in terms of persisting causes, like erosion or the Gulf Stream, that cause peculiar climatic conditions (Ospovat 1977).

Second, Lyell wanted to counter the claim that the organic world shows progression from primitive forms up to the most sophisticated organisms. For, as he admitted to a geologist friend, Poulett Scrope, the probability of "proofs of a *progressive* state of existence in the globe . . . is *proved* by the analogy of changes in organic life" (Lyell 1881, 1:270; original italics)

Hence Lyell argued that any progression in the organic record is an illusion caused by the imperfection of the record and that there is no real reason to believe in anything but a steady state. Lyell excepted only man, whose origin he agreed was both recent and a "real departure from the antecedent course of physical events" (Lyell 1830–33, 1:167). But even then Lyell was at pains to argue that from a physical (as opposed to a "moral") viewpoint, man's appearance denotes no genuine progression or advance.

After these two antidirectionalist arguments, in the first volume of the *Principles* Lyell then attempted, as an actualist and a uniformitarian, to explain the phenomena of the geological past within a steady-state context, by the same causes operating today. Drawing on the work of the German geologist Karl von Hoff (1822–24), massive in its Teutonic thoroughness, Lyell divided inorganic geological processes into the aqueous and the igneous: an obvious reflection of the chief geological forces of the Neptunists and the Vulcanists. Thus, though Lyell's major debt was clearly to Hutton, he shared some roots with his catastrophist opponents. Given this division, Lyell set out to show how both processes could lead to the balancing phenomena of "decay as well as of reproduction" (Lyell 1830–33; 1:167).

First come the aqueous geological processes, which Lyell subdivided into running (fresh) water and sea currents and tides: fluvial and marine forces. In great detail he documented the power of water, both in cutting away existing rocks and ground (erosion) and in building up new land (deposition). We see a similar steady state in igneous forces, which Lyell divided into volcanic and seismic phenomena. (Lyell followed Hutton in believing that earthquakes and volcanoes are both manifestations of identical processes of liquid rock moving about deep in the earth.) Thus Lyell gave much information on the known elevating effects of volcanoes like Vesuvius and Etna; and he also gave detailed descriptions of known earthquakes and their depressing and elevating effects. Most significant, as the frontispiece for his volume Lyell chose a picture of the Temple of Serapis at Pozzuoli, near Naples (see fig. 4). Choosing this picture was not some neo-Wordsworthian romantic gesture; he felt it provided perfect evidence of land that had first sunk, then risen, for in no other reasonable manner can one explain why the lower parts of the pillars are not eroded. Lyell was so firmly committed to a steady-state thesis that he argued that aqueous and igneous forces compensate to achieve a balance. Hence, because he felt that volcanic action could only raise land, he "proved" by a somewhat convoluted argument that seismic subsidence exceeds seismic elevation. In the concluding words of the volume, the power of earthquakes is "a

Figure 4. The frontispiece of the first volume of Lyell's *Principles*. The columns perfectly illustrate a Lyellian steady-state earth. That erosion begins only about eight feet up shows that since they were built the land has first sunk eight feet (putting the columns under water), then risen again to its original level.

conservative principle in the highest degree, and, above all others, essential to the stability of the system" (Lyell 1830–33, 1:167; see also Rudwick 1969).

Actualism, uniformitarianism, and steady-state theorizing thus combine in the first volume of Lyell's geological synthesis. The extent to which his work represents a watershed in the history of geology is still controversial, but even the most churlish critic must allow that there is something rather magnificent in the scope and self-confidence of Lyell's attempt. Only a scientist of high order could have woven so many threads into so bold a tapestry. Nevertheless, for all the virtues of the first volume of the *Principles,* it did not immediately convert the catastrophists to Lyell's position. Let us consider the reactions of two fairly typical catastrophists, Sedgwick and Whewell. From now on I shall consider a "catastrophist" simply someone who, unlike Lyell, accepts major earth phenomena: "catastrophes." But in practice this always seemed linked with some sort of directionalism.

The Catastrophists' Reply to Lyell

Like just about everyone, Sedgwick found much to praise in Lyell's work. In particular he felt Lyell had shown that the methodology of actualism can be pushed much further than had hitherto been recognized (Sedgwick 1831). Nevertheless, when it came to details Sedgwick disagreed with all three of the points Lyell wanted to make. First, he felt that Lyell had no right to restrict past causes to the kinds we see in action today. We know so little of "those mysterious imponderable agents which co-exist perhaps, with gravitation, and unquestionably play their part in every change and every combination" (Sedgwick 1831, p. 301). Indeed, Sedgwick thought causes working in the past must have been of greater intensity, for he endorsed a theory of the French geologist Elie de Beaumont that supposed European mountain chains had been raised by "paroxysms of elevatory force" (Sedgwick 1831, p. 308).

Although endorsing past causes of a kind and intensity unknown today cuts off embarrassing questions about their exact nature, two things about Sedgwick's position seem clear. First, he had gone far beyond sole reliance on floods such as Cuvier posited. He thought there had been floods, but he also speculated widely about volcanoes and the raising of mountains. Second, although he believed in catastrophes of a kind and intensity no longer experienced, Sedgwick was rather inclined to believe that God did not intervene in the inorganic world. Understanding by "law" some kind of natural regularity, whereas a phenomenon not bound by law would be "miraculous" or "supernatural," Sedgwick rather supposed that catastrophes were nonmiraculous (although he did not entirely rule out the possibility of miracle; Sedgwick 1833, p. 28 n).

Finally, we find that Sedgwick opposed Lyell's steady-state hypothesis, in both the inorganic and the organic worlds. He referred to the cooling of the earth and also to Herschel's argument that since the earth's orbit is decreasing in eccentricity (the major axis remaining constant and the minor axis thus increasing), it follows that the amount of solar radiation the earth receives is also decreasing (Herschel 1832). This seems to imply that, all other things being equal, the earth's mean temperature is in general decreasing and therefore has a kind of direction. Sedgwick found this direction mirrored in the organic world, for he thought, contrary to Lyell, that "there has been a progressive development of organic structure subservient to the purposes of life" (Sedgwick 1831, p. 306).

In many respects, particularly in fieldwork, Sedgwick was perhaps the best geologist of his day. He and Roderick Murchison were doing sterling work on early parts of the geological record, providing strong evidence that fish appeared before reptiles—a matter on which Cuvier was somewhat ambiguous. In his response to Lyell, however, one senses a man defending ideas primarily because they are *his* ideas (a feeling one gets again, some thirty years later, when Sedgwick responds to Darwin). I do not mean to imply that Sedgwick was a professional reactionary. Certainly, even after the first volume of the *Principles* was published in 1830 it was not at all ridiculous for a geologist to be a catastrophist. The evidence for catastrophes was as strong as the evidence for many scientific theories. At the beginning of the 1820s, for instance, Brongniart (1821, 1823) showed unequivocally that the Alps themselves hold fossils unambiguously identifiable with very recent geological eras (see also Rudwick 1972). Hence, Sedgwick's accepting de Beaumont's suggestion that somehow the Alps had exploded upward because of stresses on the earth's crust, and that this had happened in a geological instant, was far from unreasonable.

Nevertheless, in spite of all this, in reading the *Principles* and Sedgwick's replies one feels that, whatever the truth of the matter, Lyell outclasses his opponent. Lyell has moved the discussion to a more theoretical level: those who disputed him primarily by citing empirical examples of would-be catastrophes were left slicing at the air. Lyell himself was simply not going to accept empirical refutation. In short, right or wrong, Lyell had an imagination that was large, subtle, and perhaps a little shifty. His ideal critic would need an imagination that was equally large, subtle, and shifty. Such a critic was Whewell.

Like Sedgwick, at one level Whewell had the highest praise for Lyell's work. He commended Lyell for almost single-handedly creating the whole new science of "geological dynamics," dealing with the causes of geological change. But, also like Sedgwick, at another level Whewell was totally unconvinced by Lyell's position. And in a series of critiques extended over the decade he countered all three aspects of Lyell's thought (Whewell 1831c, 1832, 1837, 1840).

Thus with respect to actualism—the attempt to explain using only causes of known kind—Whewell replied that such a limitation was altogether arbitrary. He wrote, "Limited as our knowledge is in time, in space, in kind, it would be very wonderful if it should have suggested to us all the laws and causes by which the natural history of the globe, viewed on the largest scale, is influenced—it would be strange, if it should not even have left us ignorant of some of the most important of the agents which, since the beginning of time, have been in action" (Whewell 1832, p. 126).

We are fortunate in having extant Lyell's criticism of this particular argument by Whewell and Whewell's reply.[1] These are worth looking at not so much for their own sakes as because they illustrate two related items: the extent to which the members of our network used the physical sciences as guides and ideals when working in the nonphysical sciences, and the way underlying philosophical commitments crucially influenced positions that on the surface appeared purely scientific. Recognition of both these points is essential to understanding the Darwinian Revolution in general and Darwin's own work in particular.

Lyell and Whewell based their arguments both for and against actualism on analogies drawn from astronomy. In particular, Lyell tried to pin down Whewell by an analogy with the tides, an area in which Whewell had done significant research. Lyell (1881, 2:5) argued that were one faced with some strange tidal phenomenon from the past—great inequalities, suspension of tidal oscillations, or such—one would not think much of explanations based on totally new and unknown causes like "a supposed periodical increase or diminution in the quantity of matter contained in the sun or moon, or in both of those heavenly bodies." It would make better scientific sense to suppose there must have been known causes at work, particularly since one can "recollect that several problems in the former state of the tides, once referred to violations in the ordinary conditions of our solar system, can now be explained without resorting to such expedients, and are, in fact, found to be the consequences of known and regular causes" (Lyell 1881, 2:6).

Whewell's reply to this argument is most interesting. Rather haughtily acknowledging that the "analogy of other sciences has been referred to, as sanctioning this attempt to refer the whole train of facts to known causes," Whewell (1837, 3:617) denied that there is any merit in the astronomer's (and by implication the geologist's) trying always to explain in terms of causes of a known kind. And rhetorically he asked whether "there is any praise due to those who assumed the celestial forces to be the same with gravity, rather than those who assimilated them with any other known force, as magnetism, till the calculation of the laws and amount of these forces, from the celestial phenomena, had clearly sanctioned such an identification?" (Whewell 1837, 3:618).

One might think Whewell surely missed a golden opportunity to argue for some kind of catastrophism, starting with Lyell's own actualistic analogy. In astronomy we find isolated phenomena caused by known causes—eclipses for example. Whewell might thus have argued analogically from astronomy for catastrophes, even given Lyellian actualism. But it is always dangerous to suggest what people should have said, and Whewell's actual reply makes one thing crystal clear. Whewell was not interested in defending catastrophism (causes of greater than present intensity) only by conceding actualism (that all causes were of the present kind). He wanted to go right to the heart of Lyellian geology and argue for unknown *kinds* of causes in geology, thus violating Lyell's actualism.

But even if we grant that Whewell did want to counter Lyellian actualism, it is by no means obvious *why* he wanted to. This suggests that there were underlying reasons, and I believe there were: reasons that came from his philosophy of science. I shall therefore for the moment leave Whewell's attack on Lyell's actualism. Let us move on to his criticisms of the rest of Lyell's position. In opposition to Lyellian uniformity, Whewell made it clear that were one restricted—which Whewell denied—to causes of the same kind, there was still no obligation to stay with the same intensity: "if by *known* causes we mean causes acting with the same intensity which they have had during historical times, the restriction is altogether arbitrary and groundless" (Whewell 1840, 2:126). The uniformitarian feels free to invoke unlimited *time* to explain phenomena; why should not the catastrophist invoke unlimited *force?*

Finally, we find Whewell in opposition to Lyell's steady-state thesis.

He made a friendly reference to Herschel's earth-directionist argument and similarly suggested that the earth's heat implies an original incandescence from which it is now cooling (Whewell 1840, 2:118). And, rounding out his attack, Whewell referred to the nebular hypothesis and indicated that if we are determined to argue analogically from astronomy, we might well start here. Since we have an astronomical hypothesis that intimates direction in the universe as a whole, perhaps we should argue analogically to direction on the earth by itself (Whewell 1837, 3:618–19). In sum, Whewell felt no commitment to a steady-state system, and he rejected it as he had rejected in significant respects Lyell's actualism and uniformitarianism.

But it is worth noting that, unlike Sedgwick, Whewell did *not* criticize Lyell's concept of organic direction. He agreed with Lyell that any such direction of the fossil record is apparent rather than real—an illusion caused by the imperfection of the record (Whewell 1832, p. 117). Precisely why Whewell took this position may perhaps be revealed in the next chapter.

In view of the attacks made on Lyell, one might wonder where his support lay. Somewhat paradoxically, since Herschel provided a major piece of ammunition in the catastrophists' attack on Lyell, he was Lyell's most enthusiastic advocate among the older members of the network—no small thing given Herschel's important position in the scientific community. But just as Whewell's opposition to Lyellianism was intimately connected with his philosophy of science, so was Herschel's advocacy. It will therefore be worth our while to postpone considering Herschel's geological views until we discuss philosophy. For now let us look briefly at the geological work of a man who was more Lyellian than Lyell himself: the new member of the scientific network, Charles Darwin.

Darwin as a Lyellian Geologist

Darwin came from the best of catastrophist backgrounds. In his second year as a medical student at Edinburgh he had attended the geological lectures of Robert Jameson, Cuvier's British editor. Nevertheless, though Darwin admitted that he "was prepared for a philosophical treatment of the subject" (Darwin 1969, p. 52), Jameson so bored him that he determined never again to study geology. He thus bypassed Sedgwick's lectures, but fortunately Henslow rekindled his interest. After finishing his degree, Darwin studied geology for some six months in 1831, climaxing this with his trip to Wales with Sedgwick. Since Sedgwick combined a warm and sympathetic character with his status as one of the best field geologists in Britain, it is likely that young Darwin could not have had a better crash course in practical geology. Judging from the results, it was fine training for his five years of solid geological work on the *Beagle* voyage.

When he set off for South America, Darwin took with him the first volume of Lyell's *Principles* (the second volume was sent out to him in 1832). He took the volume at the suggestion of Henslow, who—being himself a catastrophist, though broad-minded—warned Darwin "on no account to accept the views therein advocated!" (Darwin 1969, p. 101). The advice, regrettably or fortunately, was not heeded, for as soon as he started work, at Saint Jago in the Cape Verde Archipelago, Darwin started thinking in a Lyellian fashion.[2] In particular, reasoning from a general layer of sedimentary rock some sixty feet above the ground, Darwin concluded that the whole island had been elevated in a gradual Lyellian manner, and he was excited to find that there had also been later subsidence around volcanic craters, which he inferred from the way the sedimentary layer dipped below one of the inactive volcanoes (Darwin 1910, p. 172; see fig. 5). Thus, by using the actualist methodology he was able to avoid positing catastrophes and to push toward the steady-state view advocated by Lyell.

Figure 5. Diagram taken from Darwin's *Geological Observations on the Volcanic Islands* (1844), showing the geological structure of the coast of Saint Jago, Cape Verde Islands. *A*, ancient volcanic rock; *B*, limestone (deposited by the sea and elevated); *C*, recent basaltic lava.

In his *Autobiography*, written late in life, Darwin gave the impression that his conversion to an extreme Lyellianism occurred rather suddenly as soon as he got to Saint Jago. This is a little misleading. For a start, as we know, the gap between Lyell and other geologists like Sedgwick was not total. With respect to practical geology, both sides were generally committed to the actualist methodology and to positing few catastrophes, even though Lyell obviously wanted to push both actualism and uniformitarianism much further than did others. Certainly, Darwin's geological fieldwork on Saint Jago was not greatly alien to a Sedgwickian geologist, even though a Sedgwickian might not consider the elevations and subsidences quite so significant. But it is clear that as the *Beagle* went on to South America, where Darwin was to spend the next four years, he had by no means already given up all traces of his belief in catastrophism. From his notebooks, we know that he first thought South American geology gave evidence of a monstrous flood, and consequently he referred to such evidence as diluvial deposits (see Herbert 1968).

But it was not long before he began to use quotation marks with the word "diluvial," and it is clear that his belief in extreme catastrophes had begun to fade. Throughout Darwin's South American stay Lyell's hold on his mind kept growing, and by the time he left in 1835 Darwin was a complete convert. He was a Lyellian actualist; he no longer believed in catastrophes; and he was utterly committed to a steady-state earth. What finally tipped the balance is perhaps not that important. The full story must certainly wait until we get to philosophical matters. But of crucial importance for Darwin was the evidence that South America was being elevated, which he determined by examining shell-bearing deposits at various levels and comparing their composition with types still living or known to be extinct. He argued, in what we shall see was Lyell's best manner, that the higher the ratio of extant to extinct types, the more recent were the deposits. Thus, as figure 6 shows, Darwin felt safe in concluding that the Plain of Coquimbo in Chile must have been raised at least 252 feet in the not-too-distant past, because the shells on top of the plain are so similar to those on the beach. More generally, Darwin thought he could show widespread elevation down both the east and the west coasts of the continent. Indeed, he thought he detected elevation of up to 1,300 feet in the vicinity of Valparaiso (Darwin 1910, p. 307).

Of course, a catastrophist might explain all this elevation just as handily as a Lyellian, though it would not have quite the same significance for him. But Darwin believed it was all happening gradually, through causes of the same kind and intensity as those at work today. He was thus a complete actualist and uniformitarian (in the limited sense specified earlier). Darwin thought with reason that he had the best of all actualist evidence, for he himself was present during an earthquake (20 February 1835, in Chile) that raised the ground anywhere from two or three to ten feet (Darwin 1839c). Darwin thought that relatively small causes like this, undoubtedly similar to those still in action, were sufficient for South

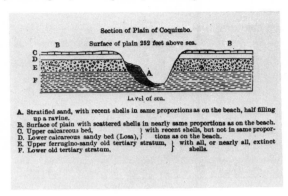

Figure 6. From Darwin's *Geological Observations on South America* (1846). This illustration shows clearly why Darwin thought he could argue for elevation in South America.

American elevation. And he thought that some phenomena, such as certain sloping coastlines, absolutely excluded catastrophic elevations (Darwin 1910, p. 294). Finally, Darwin saw South American elevation as an integral part of a world picture of elevation and subsidence: the very picture Lyell had so vigorously advocated.

Since this is not a book on Charles Darwin in himself, it would be inappropriate just to list his geological achievements, although they were not inconsiderable. He wrote a stream of papers in the years immediately after the *Beagle* voyage, and in the 1840s he completed no fewer than three books on geology. My primary aim here is to show how much Darwin was influenced by Lyell, and how strongly he advocated Lyell's position, bearing out a remark Darwin later made that "when seeing a thing never seen by Lyell, one yet saw it partially through his eyes." To underline this influence, let us look briefly at two pieces of geological work Darwin did after his conversion to Lyellianism.[3]

The first is generally regarded as Darwin's greatest geological triumph—his theory of coral reefs, formulated in the mid-1830s (Darwin 1838, 1842). Coral grows only at certain depths in the sea, close to the surface. In the second volume of the *Principles,* Lyell argued that coral reefs, those circular islands of coral found in warm seas, are built upon extinct volcanoes that were elevated close to the surface. Noting, among other points, that it was improbable that so many volcanoes would rise to almost the same height, Darwin revised Lyell's position and argued that coral reefs (except for one class) are the products of *subsidence.* In particular,

Darwin argued that the first stage of subsidence (as shown in fig. 7) will lead to islands surrounded by coral reefs, a class known as barrier reefs. Continued subsidence (as shown in fig. 8) will lead to rings of coral with no central islands, a class known as atolls.

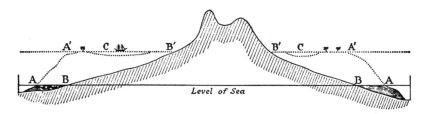

Figure 7. An atoll forming through subsidence. As the land sinks, the coral *(horizontal shading)* grows upward. From Darwin's *Structure and Distribution of Coral Reefs* (1842).

Although this argument attacks a specific part of Lyell's theory, it could not itself be more Lyellian—indeed, Lyell himself immediately dropped his own hypothesis and adopted Darwin's (Lyell 1881, 2:12). Actualistically, we see coral in action today. Uniformly, we have the gradual creation of reefs. And most important, here we have evidence of gradual subsidence, so crucial for the steady-state view. On this last point, Darwin thought he could do better than just provide evidence of subsidence. One class of reefs, shore or fringing reefs, he explained in terms of elevation; then, plotting all the reefs on the same map, he felt he had clear evidence of a steady state involving both elevation and subsidence. Darwin was particularly excited by the distribution he found, because by superimposing volcanic sites onto the map, he could show that active volcanoes occur only in areas of elevation (Darwin 1842, p. 104). This distribution tied in perfectly with Darwin's other beliefs, for he already had a theory to explain elevation and subsidence. In particular, he hypothesized that elevation involved shifts in molten rock not far beneath the earth's crust. Elevation is caused by pressures like centrifugal force as the globe tries to achieve a stable form when, owing to surface changes, more molten rock has been forced beneath the crust, floating it upward. In this situation one would expect volcanoes, giving evidence of increased pressure beneath (Darwin 1840).

Darwin did a second major piece of geological work in 1838, some two years after the *Beagle* voyage (Darwin 1839*a*)—his attempted explanation

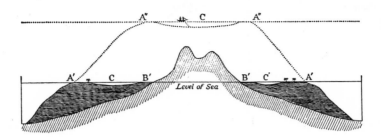

Figure 8. Here the atoll has fully formed, with the sea having gone from *A*, (as shown in fig. 7), to *A'* and finally to *A''*. From Darwin's *Structure and Distribution of Coral Reefs* (1842).

of the parallel "roads" of Glen Roy. Though often thought to be his greatest scientific failure, it also exhibits his geological fertility and his thoroughgoing Lyellianism. Given its similarity to the coral island work, I think both merit equal praise or criticism (see Rudwick 1974). Darwin set himself the problem of explaining the celebrated parallel ledges or terraces running around the sides of Glen Roy (and associated glens) in Scotland (see fig. 9). It was generally agreed that these "roads" are natural, being the beaches formed by water that once lapped the sides of the glen. Previous explanations had favored the idea that a lake, dammed at successively lower levels, was responsible. Darwin argued for a marine hypothesis, suggesting that the sea had caused the roads and that the successive road levels and the present location of the glen above sea level attested to the fairly gradual rise of that part of Scotland. But not long after Darwin's work the great Swiss expert on fish paleontology, Louis Agassiz, revived the lake hypothesis, suggesting that the lake had been dammed by glaciers. Agassiz's hypothesis eventually won out, converting even Darwin some twenty-five years later.

Although Darwin later wrote of his work as being "a great failure," of which he was "ashamed" (Darwin 1969), here we see his Lyellian viewpoint most clearly. Take the actualist methodology of arguing analogically from causes of a kind we know are now in action. In rejecting barrier-contained lakes, Darwin was being highly actualistic; for had there been such barriers, given what we know of them today, we might expect to find some traces—debris and so on. Since there were none, the way was open for supposing the action of the sea. But Darwin was also actualist in his positive arguments for marine action. Here his South American ex-

Figure 9. The parallel roads of Glen Roy. From Darwin's 1839 paper in the *Philosophical Transactions*.

periences were crucial, for in Glen Roy Darwin argued directly from causes he thought were in action in South America—in particular he cited some similar terraces in Chile, the "parallel roads of Coquimbo" (Rudwick 1974, pp. 114–15). These terraces he knew were of marine origin, caused by elevation. As he had found out personally, they were littered with marine shells, they adjoined the sea, and they were in an area where Darwin had had direct experience of earthquake elevation. Moreover, Darwin noted the similarity between the *Beagle* channel in Tierra del Fuego (where the sea comes right inland) and the Great Glen of Scotland, of which Glen Roy is an offshoot and from which the sea would have come into Glen Roy (Darwin 1839*a*, p. 56). Lyellian actualist methodology was therefore satisfied; and in the way it was satisfied, Darwin was being doubly Lyellian, for Lyell himself (in the *Principles*) had suggested that the roads of Coquimbo might be of marine origin (rather than of lake origin, as had previously been suggested). Furthermore, Lyell had drawn the analogy between Coquimbo and Glen Roy, without speculating on the causes of Glen Roy's roads (Lyell 1830–33, 3:131–32). Darwin was therefore merely completing Lyell's research.

That Darwin's Glen Roy work was anticatastrophist is obvious. He explicitly pointed out that his marine hypothesis required no catastrophes to break down barriers, as earlier proponents of the lake hypothesis had had to suppose. Moreover, Darwin drew attention to "erratic" boulders or blocks scattered over the Glen Roy area—rock not originating in the area that hence must have been brought there. Explicitly rejecting their having been brought by a debacle, he gave his own actualist, uniformitarian hypothesis that the rocks were carried in icebergs. He had seen such phenomena in the southern seas (Darwin 1839*b*).

Finally, Darwin clearly saw the elevation of Glen Roy as a critical piece of evidence for his steady-state theory of elevation and subsidence. Having demonstrated such phenomena in the southern hemisphere, he now had evidence that they occurred in the north. Moreover, he admitted to Lyell that he took the precise horizon orientation of the roads as a key feature supporting his speculation that elevation involves floating on rock (F. Darwin 1887, 1:297), and in his paper on Glen Roy he included an explicit discussion of causes. All in all, therefore, Darwin's work was little more than an extension of the *Principles of Geology,* and even in the most minute details Darwin remained Lyellian. Thus, when he failed to find the expected marine fossils on the roads of Glen Roy, Darwin went into a highly Lyellian discussion of why the fossil record is so imperfect and why we should expect the absence of such fossils.

One could go on illustrating Darwin's Lyellianism in geology. For instance, Darwin showed that when Europe was hot, South America was cold, which he took as definitive evidence in favor of Lyell's climate theory and as a blow against directionalism (Darwin 1910, pp. 408–9). But examining Darwin's two best-known pieces of geological theorizing surely makes the case. Darwin was Lyellian through and through. Moreover, like Lyell's, Darwin's tastes ran toward speculation. For him the joys of geology lay less in the nitty-gritty of working out the geological record than in making sweeping causal hypotheses. It is therefore no surprise to see him doing the same in biology.

As we turn to the next important facet of the intellectual milieu of the 1830s, we should forestall a possible misconception. Since Darwin became so thoroughly Lyellian while most members of our scientific network did not, would he not in some sense be cut off from the main group and come to be looked on as an outsider? Apparently this did not happen. The uniformitarian-catastrophist debate (using the terms now in the broad sense) seems not to have created any of the emotional rifts that the organic origins debate later brought on. Darwin was enthusiastically welcomed into the geological community and respected as a geologist: not just by Lyell but by catastrophists also. Sedgwick refereed the paper on Glen Roy for the Royal Society, writing that "it contains much original research, much ingenious speculation, and some new and very important conclusions" (Rudwick 1974, p. 181).

Philosophy of Science

In the literature of the period one finds a constant demand that scientific work be philosophically adequate, conforming to the canons of the "best" kind of science.[4] In this context, the names of two philosophical mentors appeared over and over again—Bacon and Newton. Everyone wanted to show that he alone was a true "Baconian." Everyone wanted to show that he alone was a true "Newtonian." Of course the precise meaning of these terms tended to vary, and the name-dropping reached ridiculous proportions. One Granville Penn (1822), who wrote a massive work showing that nothing in geology refutes a most literal reading of Genesis, invoked three authorities in his support—Moses, Bacon, and Newton.

Nevertheless, there were thinkers who wanted to go beyond mere labels and explicate what they believed was the true spirit of Bacon and Newton: to demonstrate the proper scientific methodology and show the criteria to which the best science ought to, and indeed does, conform. It is not too much to say that the 1830s was the best decade ever for philosophizing about science. In this section I shall look briefly at the work of Herschel and Whewell, who provide significant contrasts. In the next section I shall show how this philosophizing had significant repercussions in the kinds of geological work people carried out and endorsed.

Herschel's highly popular little book on the philosophy of science, *A Preliminary Discourse on the Study of Natural Philosophy,* appeared right at the beginning of 1831. Not surprisingly, Herschel took as his paradigm physics, particularly Newtonian astronomy, and we find that his philosophical reflections on science generally exhibit this bias. For Herschel, the model of a completed scientific theory—that toward which all science should aim—was what today we call a "hypothetico-deductive system." Herschel saw scientific theories as ideally being "axiom systems," where everything can be deduced from a few axioms, and scientific systems are distinguished from other such systems (like those of geometry) because their axioms (and the theorems derived from them) are *lawlike.* Scientific systems make universal claims about the world. Though these are not logically necessary, they are believed to specify connections that *must* hold in some strong sense. As Herschel wrote succinctly (1831, p. 36), "Every law is a provision for cases which *may* occur, and has relation to an infinite number of cases that never have occurred, and never will." (Note the similarity, if not identity, between Herschel's notion of law and what I am taking as the standard notion.)

In supporting the hypothetico-deductive model, although he threw in the blessing of Bacon, Herschel was showing the influence of his own scientific background, for the classic case of such a system is Newtonian astronomy, where Kepler's laws are deduced from the laws of motion and from Newton's law of gravitational attraction. But Herschel took more from physics than the ideal structure of theories. In several other respects he argued that mature sciences ought to have characteristics possessed by physics. For example, he argued that the best kinds of laws are *quantitative*—that they imply and demand exact measurement, just as do the laws of physics. "Indeed, it is a character of all the higher laws of nature to assume the form of precise *quantitative* statement" (Herschel 1831, p. 123). Throughout the *Discourse* there are similar comments making physics the touchstone for good science.

At the heart of Herschel's philosophy of science lay his doctrine of *verae causae,* and through this we can start to unravel that central but problematic notion of "cause." Herschel posited two basic kinds of laws. On the one hand we have mere *empirical* laws—laws that connect things and point to regularities without really showing why things occur as they do. Paradigms of this class are Kepler's laws, which prove the regularities of the planets without really explaining why such regularities exist. But the aim of the scientist must be to explain the reason for empirical regularities, and this involves considering higher laws—laws that refer to *causes.* Regretfully, Herschel was anything but precise about what he meant by "cause," but essentially he seems to have had in mind the idea of one phenomenon, the cause, in some way leading to or "making" occur another phenomenon, the effect. This may all seem somewhat anthropomorphic; but fundamentally Herschel's notion of cause *was* anthropomorphic. For Herschel the highest form of cause was force— indeed, he suspected that all cause was in some way reducible to force (Herschel 1831, p. 88). Moreover, Herschel suspected that all force was will-force (Herschel 1833*a*, p. 233): if not man's, then presumably God's.

Imprecise as Herschel may have been, he found reference to causes essential. Furthermore, he thought that one ought to refer if possible only to causes of a particular kind—*verae causae* (true causes). But how can we be sure that we have *verae causae?* The answer is simple. We can and should argue analogically from our own experience: "If the analogy of two phenomena be very close and striking, while, at the same time, the cause of one is very obvious, it becomes scarcely possible to refuse to admit the action of an analogous cause in the other, though not so obvious in itself" (Herschel 1831, p. 149).

We can bring out more clearly the significance of the *vera causa* doctrine for Herschel through a brief contrast with Whewell As is well known, Whewell was something of an anomaly—a British rationalist or, more precisely, a British Kantian (Butts 1965) arguing that it is possible in science to achieve something close to logical necessity, mind-imposed rather than experientially given. On this point Whewell differed from most of his fellows, including Herschel, but his views on many important aspects of the nature of science were very close to Herschel's. Like Herschel, Whewell saw Newtonian astronomy as the Platonic ideal of science; like Herschel he endorsed the Baconian/Newtonian hypothetico-deductive thesis; and like Herschel he upheld the distinction between empirical laws and causal laws. Whewell spoke of "formal" and "physical" science and of "phenomenal" and "causal" parts of theories.

But Whewell did break with Herschel over the doctrine of *verae causae,* showing himself a rationalist (wanting to argue to experience) rather than an empiricist (wanting to argue from experience). Whewell did not deny *verae causae* or their importance. But he did deny Herschel's empiricist interpretation of the *vera causa,* where one argues analogically from the experientially known to the unknown. Whewell (1840, 2:442) felt that this interpretation unduly limits any methodological rule based on it, because "it forbids us to look for a cause, except among the causes with which we are already familiar. But if we follow this rule, how shall we ever become acquainted with any new cause." So what was Whewell's understanding of *verae causae?* Whewell tied his interpretation to what he called the "consilience of inductions" (see Laudan 1971). In particular, he argued that the mark of the best kind of science—the definitive evidence that one is starting from true axioms—comes when different areas of science are brought together and shown to spring from the same principles. Newton did this in astronomy when he showed that the motions of the planets, the sun and moon, the tides, and so on, are all based on the same set of principles. Here, argued Whewell, one has a consilience—the guarantee of truth—particularly if some of the things explained were initially thought hostile or irrelevant to one's principles. Because the explanations were not "built in" to one's hypotheses but involve an element of surprise, one must be dealing with reality. Moreover, any causes referred to in one's starting principles (and only such causes) are entitled to the name *verae causae*, even though one might have no direct experience of them. The criterion is that *verae causae* be adequate to explain experience rather than that they be derived from experience.

Two points should be noted. First, Herschel did not deny the value of a consilience. In the *Discourse* he repeatedly stressed the importance of relating many areas to one central element, particularly in explaining the unexpected and hostile. But he does not seem to have tied it to his doctrine of *verae causae* in the way Whewell did. Second, although Whewell's approach was that of a rationalist—the important point is that one's principles can explain a phenomenon, rather than that one has had immediate experience of it— he did not simply stake his position on a metaphysical preference. Whewell held a high card in the wave theory of light. Herschel accepted the theory—indeed, he was one of its strongest advocates (Herschel 1827)—but we hardly can sense waves, even indirectly. Consequently, through the 1830s we find Herschel rather desperately trying to link light waves to experience by means of analogies (Herschel 1833*b*). Basically, however, he had to agree with Whewell that we have confidence in the theory because of its consilient nature rather because it contains a *vera causa* as Herschel understood it (Herschel 1841, p. 234). As a last resort, therefore, Herschel seemed to agree with Whewell that a theory can be first-class because it is consilient, even if it does not contain an empiricist *vera causa*.

In a nutshell, one can say that the central dictum of British philosophy of science in the 1830s was: "The best kind of science is that most closely modeled on physics, particularly astronomy" (see Wilson 1974). This dictum left room for both empiricist and rationalist interpretations, but even here there was much sympathetic cross-agreement. Thus a new entrant to the scientific network, though he may have inclined to one side or the other, would surely show influences from all. This being said, let us turn back briefly to geology and demonstrate what was intimated earlier—that people's philosophical commitments influenced in key ways their geological preferences and achievements.

Geological Systems *and* Verae Causae

Herschel responded enthusiastically to Lyell's *Principles*. But at first sight it seems somewhat implausible that this might have had anything to do with Herschel's philosophy. The very opposite seems so. For however highly one may regard Lyell's theorizing, and however much one may agree that in the *Principles* he pursues a deliberate, sophisticated, and successful strategy (Rudwick 1969), there is nothing very hypothetico-deductive about what Lyell produced. But Herschel did react favorably to Lyell's theorizing, and it is clear that his enthusiasm was indeed a function of his own philosophy. In particular, Herschel's liking for the *Principles* came immediately from his own notion of a *vera causa*. (See Kavalowski 1974. In a letter to Lyell in 1836 Herschel spoke most warmly of the *Principles;* see Cannon 1961*a*.)

It is perhaps easiest to document this claim by turning back to the threefold division of Lyell's aims. First we have the question of actualism, without a doubt attractive to Herschel, for the Lyellian drive to be an actualist is merely the Herschellian drive to find *verae causae* translated into geological terms. The Lyellian is trying to find and use causes of a kind already experienced—Herschel's empiricist program. Moreover, we find that Lyell's climate theory, obviously actualist since it relied on phenomena like the Gulf Stream, was seized upon by Herschel and used in the *Discourse* as an example of a *vera causa* (1831, pp. 146–47). And Lyell himself was not unhappy to jump on the bandwagon, for he said (perhaps significantly, to Whewell) that "the reiteration of minor convulsions and changes, is, I contend, a *vera causa,* a force and mode of operation which we know to be true" (Lyell 1881, 2:3).

Similarly, Herschel's *vera causa* doctrine made him leery of catastrophes. He spoke scornfully of those who felt the need to go beyond causes of normal intensity and invoke extraordinary conditions such as the approach of comets and other such "fanciful and arbitrarily assumed hypotheses" (Herschel 1831, p. 285). Remember, Lyell boasted that his climate theory did away with the need for such comets. Moreover, in endorsing Lyell's climate theory as a *vera causa,* Herschel (1831, p. 285) drew attention not only to its actualism but to its uniformitarianism (in the restricted sense), noting that we ought not to suppose times were once warmer because of catastrophic volcanoes.

Finally, although we know Herschel provided the directionalists with an important argument, and although he endorsed a progressive fossil record, methodologically he thought the geologist ought to assume a steady state (Herschel 1831, pp. 282–83). Moreover, Herschel put forward his own suggestion about the cause of the world's elevation and subsidence. This suggestion no doubt influenced Darwin's speculations, and Herschel explicitly thought it met the *vera causa* criterion. "Lay a weight on a surface of soft clay. You depress it below and raise it around the weight.—If the surface of the clay be dry and hard it will crack in the change of figure" (Cannon 1961*a*, p. 307). And this, Herschel added, is just what happens with the earth. Deposition in one place causes elevation in another.

Hence it seems clear that Herschel's philosophy significantly influenced his attitude toward Lyellian geology. Conversely, it seems clear that Whewell could and did attack Lyell's position and argue for causes of unknown kind and intensity (as well as for direction) precisely because of his philosophy—in particular, because of his version of the *vera causa* doctrine. Whewell would have considered it a philosophical mistake to rule out a priori the possibility of catastrophic causes. This is not to say that he had much idea about the nature of the causes he thought caused catastrophes—he did not—and he certainly advanced no consiliences to explain catastrophes. But Whewell's views about *verae causae* demanded an open-mindedness that Herschel's did not allow. Seeing such things as upturned strata in the geological record, Whewell was not philosophically committed to explaining them in terms of present causes. On the contrary, seeing what he took to be large upheavals, he was at least directed toward large causes, for he thought only these would supply adequate explanations.

Let us refer once again to Darwin. I have explained how, in view of his catastrophist training, it is a bit puzzling that Darwin became so ardent a Lyellian. We have seen the geological evidence that persuaded him, and there is no doubt that the months on H.M.S. *Beagle,* with Lyell's attractive book and away from his mentors, must have done something. But the intensity of Darwin's Lyellianism seems to call for a further cause, and philosophy offers us a key part to the puzzle. We know that Darwin read Herschel's *Discourse* at the beginning of 1831 and was wildly enthusiastic. He urged his cousin, W. Darwin Fox, to read it at once (unpublished letter, 5 February 1831, at Christ's College, Cambridge), and late in life he spoke of it as one of the two books that had set him on the path of science. (Darwin 1969; Humboldt's *Personal Narrative* was the other). In short, even as he began his geological training, Darwin was influenced by the man who was advocating Lyell's ideas as the truest way of looking at geology. It is therefore not so surprising that when faced with Lyell's work itself, Darwin should have succumbed to its charms. (In his Glen Roy paper, Darwin referred to the action of water as causing beaches, something known and from which he could argue analogically, as a *vera causa*.)

We know that, despite their differences, there was considerable overlap between the philosophies of Herschel and Whewell. Indeed, Whewell reviewed Herschel's *Discourse* very warmly in the spring 1831 issue of the *Quarterly Review* (Whewell 1831b), and he may have been responsible for recommending the book to Darwin. Since there was this overlap, and since we know Darwin was close to Whewell and respected him (one of the best conversers "on grave subjects to whom I have ever listened"; Darwin 1969, p. 66), we might wonder if other philosophical elements, less obvious in Lyell, can be found in Darwin's work—elements held jointly by Herschel and Whewell. For example, did Darwin pay due attention to hypothetico-deductive systems and to consiliences? There does seem to be evidence that, independent of his Lyellianism, as a geologist Darwin strove to fit the canons of the philosophers. When we consider Darwin as a biologist, I shall offer detailed evidence to substantiate my claim that these elements are essential.

Consider Darwin's analysis of coral reefs. (See Ghiselin 1969; I agree with most of Ghiselin's conclusions, although he implies that Darwin's philosophy was original, whereas I think he derived it from others.) Darwin certainly did not lay out his discussion in a tight deductive network of laws. But once he had presented his main idea that reefs were formed through subsidence, he showed that many consequences follow from his suggestion, much as a physicist shows how phenomena can be inferred from causal axioms. And Darwin claimed that as these consequences turn out to be true, we have confirmation of his hypothesis. It is interesting and significant that Darwin referred to these consequences as "deductions" (Darwin 1910, p. 88) and at least pretended to be surprised by some of them (1910, p. 105). Darwin pointed out, for example, that because he looked upon atolls and barrier reefs as merging into each other and as similarly caused by subsidence, he expected them always to occur together. And he was pleased to be able to show this on a map. Similarly, because he thought fringing reefs did not involve subsidence—if anything, they were formed by elevation—he did not expect to find them with the other kinds of reefs. Again, he was pleased to show this on a map. In short, Darwin presented his theory (and the evidence for it) as one might expect of a geologist who, though working in a science with a fairly informal style, was not insensitive to the philosophizing being done by Herschel and Whewell. One last point: since this matter will prove most important later, I must emphasize that it is plausible that Darwin, though an extreme Lyellian geologist, took seriously the notion of consiliences, despite its being the center of a *vera causa* interpretation that opposed Whewell to Lyell. Herschel proposed the empiricist *vera causa,* and it tipped him toward Lyellianism; yet he always supported consiliences and even came to agree with Whewell that they made for the best kinds of theory. Given Herschel's broad-mindedness, his influence on Darwin, and Whewell's independent influence on Darwin, such eclecticism is almost to be expected from Darwin. And time may have been a factor: Herschel and Darwin became Lyellians, then through the wave theory Whewell showed that a consilience is more necessary than previously thought. More cynically, whatever Darwin's private beliefs, it would have been good tactics to incorporate Whewellian methodology in his presentation where possible. But on this point I doubt Darwin was a hypocrite.

Religious Commitments

I have explained that the men of our network were, ostensibly, Protestant Christians, members of the established church. But it is important to recognize just what this means. As might be expected in the society we are dealing with, a frank avowal of disbelief was not likely to bring social or other success. As we know, one had to be a member of the Church of England to graduate from Oxford or Cambridge, and usually one had to be ordained to remain a university fellow. But though the members of the network were Anglicans, it does not necessarily follow that they were sincere Christians. We do rather expect, however, that—sincere or not— the British social and educational system would encourage them to spend a good deal of time worrying about the relationship between science and religion, defending science against charges of impiety and, where possible, showing that science was a true supporter of religion. Lamarck and Cuvier could ignore the science-religion relationship, for they were professors at a state-supported secular institution. Because of their training and indoctrination, if not their naked self-interest, such an option was not likely to attract the British. That a professor of geology in an English university had to be a clergyman or (in the case of Lyell) vetted for orthodoxy by the archbishop of Canterbury and the bishops of London and Llandaff was bound to have some effect (Lyell 1881, 1:316).

Actually the sincere Christian belief of the clergymen in the network seems beyond real doubt. It is customary to make a threefold division of the church at that time: high church, those like John Henry Newman who saw the Church of England as a quasi-Catholic organization (without a pope, of course); low church, or evangelical, those who were most affected by the Methodist revival; and broad church, the men in the middle. Not surprisingly, most of the men in our network belonged to this last category (Cannon 1964b). The nonclergymen are more difficult to categorize. But it does seem that some of the doctrines of Christianity lay more lightly on them than on their clerical brethren. Lyell, for instance, felt sympathy for the Unitarians and later in life worshiped with them. Nevertheless, one must not thereby conclude that they were not religious. Among the senior members of the network we shall see that Lyell, Herschel, and Babbage were, in their way, as religious as any.

Toward a fuller understanding of the scientists' religious beliefs, let me introduce two useful terms. By "revealed" religion or theology I shall mean that aspect of religion given through Christian revelation and faith: the Bible as the word of God, Jesus Christ as the Son of God, man's promised hopes for immortality. By "natural" religion or theology, I mean knowledge of God derived through reason and the senses: the evidences of divinity directly discernible in the world. Without pretending that these two aspects of religion can or should be kept absolutely separate, let us analyze in more detail the religious beliefs of our network.

Revealed Religion

To this point we have been dealing with ideas, geological and philosophical, for which the members of our network were personally responsible. They were not, however, theologians; at least they did not pretend to revealed theology. We must therefore cast our net more widely at this point, to see what kinds of revealed religious beliefs men of their background and station would be likely to have. We should keep in mind the proviso that some of our secular scientists were not particularly keen on revealed religion in any case, but that even they would have had orthodox training and would therefore tend to think of revealed religion as consisting of what other Christians accepted.

The segment of society we are dealing with (with these limitations) took the truth of the Bible, both the Old Testament and the New, very seriously indeed. The Bible is God's word to man, telling him of his past, of his special relationship to God, of how man fell into sin, and of how man can yet save himself through his own moral effort and his acceptance of Christ's sacrifice. It was a segment of society that used a standard textbook written to explain why it was reasonable to be a Christian. *Evidences of Christianity,* from the pen of Archdeacon Paley at the beginning of the century, was certainly highly enough regarded to have been a major and obligatory part of undergraduate education at the time we are considering. Darwin, for example, may not have studied much formal science at Cambridge, but by the time he had finished he "could have written out the whole of the *Evidences* with perfect correctness" (Darwin 1969 p. 59).

Paley's argument was rather simple—crudely so, to the modern reader. The mark of true revelation, he claimed, is miracle—some phenomenon not covered by ordinary laws of nature. If indeed Jesus Christ was the Son of God, he must have performed miracles. Since it is maintained that Jesus did indeed perform miracles, the problem now is authenticating them. Suppose twelve men, whom we know existed, not only claimed to have seen such miracles, but suffered and died rather than deny their belief—"if I myself saw them, one after another, consenting to be racked, burnt, or strangled, rather than give up the truth of their account" (Paley 1819*a.* 3:9). Then the authenticity of the miracles is beyond doubt. Having thus set up his problem, Paley set to work with gusto to solve it. And on the basis of such evidence as non-Jewish reports of the existence of Christ's disciples and their sufferings, Paley concluded that the apostles really did exist, that the miracles they reported must have been genuine, and that Jesus Christ must therefore be the Son of God. In short, Christianity is just good common sense.

Of course, nothing is ever quite so straightforward as it appears in textbooks. Even by 1830 the pernicious influence of Germany was starting to trouble the British acceptance of revealed religion. The major advance in German religious thought in the early nineteenth century was "higher criticism"—the attempt to understand the Bible in natural terms rather than as an entirely supernatural record. It was not until later in the century that the full force of this movement hit the British coast; but even by 1830 there was some warning, particularly in the form of the Reverend Henry Hart Milman's *History of the Jews,* published in that year. In this work, put out as a series for the younger reader, Milman calmly described Abraham as a sheikh, showed by detailed anthropological discussions that the marital relations between Abraham and Sarah were the norm for tribes of that place and time (thus defusing the supernatural aspect of their union), and speculated on natural ways that Lot's wife might have been turned into salt.

It does not take much imagination to see how an approach like this would threaten a Christianity that raised Paley's *Evidences* to the status of a textbook. Abraham today, Christ tomorrow. But though Milman's work shows that some British religious thinkers were moving well beyond a crude biblical literalism (quite irrespective of moves made for other reasons, like science), and though higher criticism had its own role to play in the Darwinian Revolution, in 1830 it was little more than a beachhead. Indeed, Milman's work proved so controversial that the publisher prudently canceled the rest of the series. Among reviewers, Newman (1830) was outstandingly strident, taking particular umbrage at the description of Abraham as a sheikh.

Striking a somewhat different note, but also showing the influence of Germany, was that remarkable seer, Thomas Carlyle. His religious tour de force, the difficult but magnificent *Sartor Resartus,* was completed in 1831, appeared in serial form in 1833–34, and was published in England as a book in 1838. Purporting to be an editor reconstructing and interpreting fragments of a work on clothing by a fictitious German professor, one Diogenes Teufelsdröckh, Carlyle put forward an account of his own religious journeys and his own unorthodox answer to the mysteries of life. Plunged toward atheism and faced with the "everlasting no," salvation came through what was undoubtedly a mystical experience. Not surprisingly, therefore, Carlyle dismissed most of the trappings of orthodox Christianity and found the true spirit of religious belief in an almost pantheistic (and Goethean) recognition of a God "felt in my own heart," who is virtually to be identified with nature—at least "the universe is but one vast Symbol of God" (p. 220). This, by Carlyle's own admission, comes close to regarding God as Plato regarded the Good. Carlyle translated this concept into his celebrated doctrine of "natural supernaturalism," where the whole of nature is to be understood as sacred— where an obsession with miracles is misconceived, if miracles are taken as the basis of religious belief. In short, where the very constancy of the laws of nature is evidence that the world transcends the ordinary and is in some sense supernatural.

It is not difficult to see what a contrast and threat these ideas offered to a religion that made evidence of the miraculous nature of Christ's deeds the cornerstone of faith. Again, we do not see an immediate impact on our story, though we shall see effects later. At any rate, revealed religious belief, at the intellectual level with which we are concerned, does not present a monolithic front—cracks were appearing that were destined to widen.

But of course the greatest threat of all to revealed religion—and certainly a major factor in 1830—was science.

Science and Revealed Religion

Undoubtedly, about 1830 many people, particularly those at one end of the religious spectrum, cared not at all for science and ignored or perverted it, assuming that when science conflicts with the Bible it must be wrong (Millhauser 1954). But others could not accept this, including, of course, the members of our network. They wanted to give a place to science and were fond of quoting Herschel's *Discourse* (1831, p. 9) to the effect that "truth can never be opposed to truth." Later we will make somewhat finer distinctions, but for now the two groups can be labeled "conservatives" and "liberals" (Ruse 1975*b*). These terms broadly correspond to catastrophists and uniformitarians (opponents of and sympathizers with Lyell). I use the different terms to emphasize that religious beliefs did not necessarily follow from geological beliefs—if anything the opposite was true—and because the older terms blur points I have been at pains to make clear. Loosely, the conservatives were "theists" and the liberals "deists," although here deism means little more than belief in God as unmoved mover, without the trappings of Christianity. The term is not really sensitive enough for our purposes, for Lamarck was a deist, as were some (to be discussed) who opposed evolutionism on religious grounds. Prominent among our conservatives we find most of our clergymen scientists—particularly Buckland, Sedgwick, and Whewell. Prominent among the liberals are Herschel, Lyell, Babbage, and (above all) Baden Powell. Let us take these in turn.[5]

The conservatives were treading a very fine line—and they knew it. They wanted to accept the findings of modern science; yet even consenting to catastrophism (1830 version) was going to involve pushing the age of the earth back thousands, even millions, of years, as well as believing in the existence of hordes of animals that were now extinct. But they were believing Christians, with a faith centered very much on the Bible, and the Bible spoke of six days of Creation, which was widely thought to mean that the world began some few (traditionally, six) thousand years ago. Even had they been hypocritical enough to make a living in the church while treating its central doctrines lightly, and I do not for a minute think they were, there were watchdogs of the faith ready to set up a howl of disapproval at apparent heresy. The compromise they formulated was, basically, to allow science fairly full rein up to the appearance of man, then let the Bible take over. Thus, for example, Whewell argued (1831c, p. 206) that "the time is now come when . . . the condition and history of the earth, so far as they are independent of the condition and history of man, are left where they ought to be, in the hands of the natural philosophers." The Bible has to do with God, man, and the relationship between the two. It is not intended to be a book of science. Thus, when one is faced with embarrassments after man has arrived, such as the sun standing still for Joshua, one is entitled to explain them away. The ancient Jews did not understand physics, hence it would have been inappropriate for God to speak to them in those terms and unlikely that the writers of the Bible would record events in terms always consonant with physics or geology (Whewell 1840, 2:137–57).

But even though one might not be committed to a literal six thousand years as the age of the earth, though some time span of this order seems to have been thought proper for man (Sedgwick certainly believed this; see Lyell 1881, 2:37), there was still the problem of explaining the beginning of Genesis. It could not just be dismissed as false. Some favored supposing that the six "days" of Genesis refer to six long periods of time. Others, including Buckland (1820; see also Millhauser 1954), followed the Scottish divine Thomas Chalmers and supposed there was an unrecorded gap between the "beginning" referred to in Genesis and the succeeding six days. Either way, one got masses of time for geology.

The attitude of the conservatives to the relationship between science and revealed religion was far from entirely defensive. As I intimated at the beginning of the chapter, some early catastophists were much attracted to that geological doctrine precisely because they thought it dovetailed with the Bible and gave it independent support. At one point in the 1820s some, particularly Buckland (1823), certainly thought there was definitive geological evidence of a recent flood—thus confirming the story of Noah and also vindicating the study of geology. By about 1830 it became clear that there was no such evidence. But it was still considered legitimate to take the Bible not as a work of science but as man's oldest records (Whewell 1837, 3:602), and on this basis the possibility of finding evidence of some limited flood was allowed (Sedgwick 1831, p. 314). Yet it does not appear that the conservatives rushed in (those that accepted it) and used the progressive fossil record—fish, reptiles, mammals, man—as confirmation for the Creation story of Genesis. That man was created last, and recently, was what really counted (Bowler 1976a). Certainly the conservatives saw the world as created primarily or exclusively for man.

Certainly they were happy to see cosmological direction, particularly earth-cooling, as preparation for man; and (Whewell excepted) certainly they were happy to see organic progression as a mark of this direction. But, publicly at least, this was as far as things went. Yet it must be confessed that Sedgwick later spoke warmly of one thinker who did indulge freely in such religion/science synthesizing; so perhaps on Sundays one or two of our conservatives let their imaginations run wild (Clark and Hughes 1890, 2:161).

One gets the feeling from Whewell's treatment of the science/religion problem, as always when he was dealing with a tricky subject, that one is reading a document drawn up by a clever lawyer—one, that for all it seems, will allow him to do precisely what he pleases. Having apparently given science virtually all it could ask, he then suggests that science has nothing to do with origins. At such points "the thread of induction . . . snaps in our fingers" (Whewell 1830–33, 2:145; this phrase came from Cuvier). Whewell was far too wily to spell out precisely what this entailed for the early books of Genesis, but critics would find it difficult to hurl charges of heresy. Also, Whewell turned this position to good advantage with the question of organic origins.

I have suggested that, privately, some of the liberals were probably not overly committed to the dogmas of Christianity. It seems clear that this was in large part a function of their scientific and philosophical beliefs. An empiricist *vera causa* looks askance at miracles, understanding these as divine interventions countering or overruling natural law. Miracles certainly do not involve causes of the kind or intensity we experience today. But, since Christianity was being presented with a heavy emphasis on miracle, it was difficult for liberals to accept it while approaching science as they did. Of course the type of geology one subscribes to has little to do with whether one believes in the miracle of the loaves and fishes. But one's geological beliefs do make a difference in respect to the Old Testament, and here the liberals seem much more inclined toward the autonomy of science, particularly where geology conflicts with Genesis. Lyell's historical introduction to the *Principles* was a polemic against anyone who would let revealed religion influence his science.

One must take care here. Conservatives like Sedgwick inclined toward the view that the geological past, catastrophic or not, did not involve miracles—at least not in the inorganic world. And we know that by the 1830s conservatives were not catastrophists (in the broad sense) merely because the Bible told them so. Moreover, Lyell, in arguing against the influence of revealed religion on science, was undoubtedly motivated less by a desire to keep science pure than by the desire to tar scientifically respectable catastrophists like Buckland with the same brush as extreme scriptural apologists. Incidentally, through prudence or conviction, in the third volume of the *Principles* (1830–33, 3:270–74) Lyell denied that he had said anything contrary to a flood, properly interpreted. (As a professor at King's College, Lyell was striving to assure the bishops on his governing board that he was orthodox.)

Nevertheless, the liberals did go further than the conservatives. For instance, although Lyell considered man's arrival fairly recent, he was more willing than the conservatives to push back this date. To give the newly developing science of linguistics enough time, Herschel, writing to Lyell, was quite prepared to grant that the patriarchs lived 50,000 years. And for the days of Creation he was prepared to talk in terms of 50,000 million years (Cannon 1961*a*, p. 308). And at least one person, paradoxically a sincere Christian, was prepared to give science anything it liked. Baden Powell, with an audacity that is breathtaking from one who was a member of the same college as Newman, spoke openly of "absolute contradictions" between science and parts of the Old Testament, and he dismissed claims about the universal flood and the like as "rhetorical illustrations and poetical imagery" (Powell 1833, p. 35). After Whewell's intellectual jesuitry, this comes as a breath of fresh air. And Powell proved to have the courage of his convictions, though he had the prudence not to be converted to extreme liberalism until after he was elected to his professorship. Without praising or condemning, it is clear that in 1830 at least one man was as advanced in this thinking about science as any modern Christian thinker. His faith did not rest on violations of natural law.

Natural Religion

I come finally to natural religion or theology—the theology that involves man's knowledge of God through reason and the senses. Those who know of David Hume's devastating critique of the argument from design for God's existence (1779) may find it hard to realize that the argument was still taken seriously in the nineteenth century. But it was indeed. One can truly say that in the first half of the century British support and enthusiasm for the argument was at its height. (Obviously, Hume's criticism had fallen on unresponsive ears.)

Again Paley provided the standard text, a work that drew the classic

see !

Ayer says Hume never rejects
the argument f. design !
See Ayer 2000.

analogy between a watch and the eye, arguing that just as a watch evidently must have a maker, so we must suppose that the eye also had a maker (or, rather, a Maker). But by about 1830 Paley's *Natural Theology* (first published in 1802) began to seem dated (see Ruse 1977). Fortunately, the argument from design gained fresh life with the publication in the 1830s of the "Bridgewater Treatises"—eight works, commissioned in the will of the eighth earl of Bridgewater, aimed at demonstrating "the Power, Wisdom, and Goodness of God, as manifested in the Creation" (for details see Gillespie 1951).

Probably the most popular of these works was written by Whewell, officially commissioned to cover the subject of God's magnificence as evinced by astronomy (Buckland wrote on geology). He began by arguing that the world runs according to laws and that the effects of these laws are instances of apparent design. Thus by law our earth has a year lasting exactly twelve months, and by law plants have a year lasting exactly twelve months. This, Whewell pointed out, is a coincidence essential for the well-being of plants. Were plant cycles eleven months and the earth year twelve months, we would soon have flowering in January, which would spell doom. Having set the stage, Whewell drew his conclusion. "Why should the solar year be so long and no longer? or, this being of such length, why should the vegetable cycle be exactly of the same length? Can this be chance? . . . No chance could produce such a result. And if not by chance, how otherwise could such a coincidence occur, than by an intentional adjustment of these two things to one another?" (Whewell 1833, pp. 28–29). God (understood as an all-wise designer) must have matched the lengths of the solar year and the vegetable year. More specifically, since a disjunction would not make much difference to the sun but would be fatal to a plant, there must be a God who looks out for the interests of plants.

Since I shall refer repeatedly to the argument from design, some comments are in order. Although Whewell's commission ostensibly was to show God's design in astronomy, and though he did pay some note to this restriction, he at once brought the problem of design around to apparent intentions as they relate to *organisms*. If the solar and plant years did not coincide, plants would not survive. In short, design is shown by plants, having adaptations that enable them to survive and reproduce—in this case adaptation to the fixed length of the earth year. To use other popular terminology, design is shown as plants evince teleology or "final causes" (Whewell 1840, 2:78).

This maneuver, raising organic adaptations, was not accidental. Nearly everyone would have agreed with Whewell that design was shown most keenly in organisms, specifically in the adaptations that enable them to survive and reproduce. Whewell certainly saw design in the inorganic world, but he readily conceded that the impression of design was not nearly as strong as that shown by the organic world (Whewell 1833, pp. 148–49). Whewell himself pushed his neo-Kantianism to such a pitch that he argued (in his *Philosophy;* 1840, 2:78) that we cannot even observe organisms without reading in evidence of design. Most would not have gone this far with him, but they would certainly have concurred with his sentiment that the organic world most clearly exhibits design.

But this concentration on organisms raises two questions. Does every facet of organisms therefore show design? And what kind of design can one find in the inorganic world, if (as seems usual) design is primarily directed toward survival and reproduction of organisms? Whewell had to allow that some facets of organisms do not seem very purposeful. For instance, male nipples seem fairly useless, as do the skeletal similarities between different species. But, at least in the 1830s, Whewell and other Britons were inclined to play down the importance of these phenomena. Loath to allow outright that God might have created without intention, Whewell did suggest that such anomalies prove God had other aims besides mere survival. Clearly he sometimes created for the sake of symmetry, or to bring about similarity or order, or to produce beauty. These ends, Whewell thought, tie in nicely with the inorganic world and help answer the second question posed above. The inorganic world exhibits intention in such things as the shape of snowflakes, which again point to God's designing for the sake of beauty and order (Whewell 1833; 1840, 2:86–89).

For all his apparent confidence in dealing with difficult cases, Whewell thought it prudent to invoke a variant of the design argument that side-steps all anomalies in the organic and inorganic worlds. That things obey law—natural regularity—is in itself a sign of intention. "To most persons [including Whewell himself, it transpired] it appears that the mere existence of a law connecting and governing any class of phenomena, implies a presiding intelligence which has preconceived and established the law" (Whewell 1833, pp. 295–96). It is not necessary to show direct purpose to prove Mind. As we shall see, a variant of this variant was to play a key role in the organic origins debate. (Because of these variants, the standard argument from design, centering on organic adaptation, is sometimes labeled the "utilitarian" argument. Unless the usage is qualified, I refer to the standard version.)

A third point about the argument from design relates to man. Subscribers to revealed theology were automatically assured that man is special. But from the viewpoint of reason and the senses alone (from a natural theological viewpoint), it was thought that a case could still be made for man's distinctiveness. As we know, although Lyell thought man was physically on a par with other animals, he also thought that morally man was special. Something along these lines seems to have been generally accepted: as an intellectual and moral being, man shows he is more than a brute. Thus it seemed obvious that man must be an object of God's special attention and favor. But this raises an awkward question. Do lower organisms survive and reproduce in their own right or only because they benefit man? There was no unified answer to this question, though no one doubted that man was God's primary concern. Indeed, from most of our English writers one gets the distinct impression that God, being himself an Englishman, designed the world primarily for the benefit of the English. And some rationalizing was required to excuse God's not creating all races equally favored. Fairly typical was Buckland, who showed considerable ingenuity at proving that God had humans, particularly Britons, in mind when he placed coal strata and iron deposits in such convenient places (Buckland 1820, p. 11; 1836, 1:63–67). (Obviously, neither Buckland nor his God ever worked in a pit, nor did their children have to.) At the same time, Buckland (1836, 1:99) was prepared to concede that animals sometimes exist just for their own enjoyment.

Finally, though the clergymen of the network were most public in parading the design argument—not surprising, since by emphasizing how science reveals design they could justify their own existence—everyone shared the urge to see the world as a product of design. It is clear, for instance, that a major reason Lyell was drawn to a steady-state world picture was that he thought this harmonized best with an all-wise God who set his creation going like a perpetual motion machine and then had no need to interfere with it (Lyell 1881, 1:382). We shall dig more deeply into this in the next chapter. Certainly, not everyone saw design in the same way. Sedgwick, for instance, saw design in the progressive way God created organisms, leading up to man (1831, pp. 305, 315–16). For Lyell, it was the lack of progression that counted. But, one and all, the members of our scientific network were teleologists.

At this point it might seem appropriate to examine Darwin's religious beliefs. But in the 1830s Darwin seems not to have differed publicly from other liberals like Lyell and Herschel. Privately, his religious beliefs became significantly different; but we will delay a discussion of their origins and nature until we reach the time when their implications become a matter of public concern.

Enough has been said by way of background. Let us turn now to the burning scientific issue of the day—the problem of organic origins.

4 The Mystery of Mysteries

In 1832 Charles Lyell published the second volume of his *Principles of Geology*. In this volume he turned his full attention to the organic world, thus establishing the British background to the organic origins question that persisted right up to the publication of Darwin's *Origin*. Even in the first volume of the *Principles*, Lyell had broached the question of the organic world. There he had argued three things: that there is no genuine historical progression in the appearance of new species but that all such apparent progressions are illusions caused by the imperfections of the fossil record; that man is of very recent appearance; and that, though man is very special at the intellectual and moral level, at the physical level he represents no special progression because he is much the same as animals. But in the first volume the organic world was discussed less for its own sake than as part of Lyell's strategy to convince the reader of steady statism. It was only in the second volume of the *Principles* that Lyell's discussion of the organic world came into its own.

In this second volume, Lyell turned first to the question of organic evolution, taking Lamarck's theory as a paradigm. After summarizing Lamarck's claims that habits give rise to forms, that new life is constantly being spontaneously generated, that man evolved from lower forms (specifically the orangutan), and all the rest (Lyell 1830–33, 2:1–21), Lyell attacked Lamarck's theory, explicitly using and acknowledging many of Cuvier's arguments. He denied the claims about spontaneous generation, and he pointed out that although we get some variation within species, domestic forms prove that there are limits to such variation and show that when organisms return to the wild they revert to their original forms (2:32–33). He noted that Egyptian mummies are not different from contemporary man and interpreted this as evidence against Lamarck's speculations (2:28–31). Nor, Lyell warned, should we be misled because some varieties show more variation than species—the former simply encounter a wider range of conditions, and this implies nothing about evolution (2:25). And so he concluded that species have a stable existence: "it appears that species have a real existence in nature, and that each was endowed, at the time of its creation, with the attributes and organization by which it is now distinguished" (2:65).

One aspect of Lyell's treatment of evolution is particularly worthy of note: he thoroughly misrepresented the relationship between Lamarckian evolutionism and the fossil record. Lyell presented Lamarck's doctrine as an answer to the progressive fossil record (2:11); then denying, as in his first volume, that the record is indeed progressive, he went on to use the non-progressive record as evidence against Lamarckism (2:60). In fact, Lamarck had little interest in the fossil record and certainly did not take its supposed progressiveness as a major support for evolution. But, thanks to Lyell's misconstrual, British thinkers would have trouble not thinking of evolutionism as an answer to a progressive fossil record.

Having denied organic evolution, it was up to Lyell to explain what he thought was the true state of affairs. Toward this end, Lyell's treatment of the organic world in the second volume of the *Principles* involved a detailed analysis of geographical distribution. In particular he pointed out that though climates and ecological backgrounds in different parts of the world are often very similar (different "habitations" have similar "stations"), the inhabitants frequently are very dissimilar (2:66–69). Since God did not create exactly the same forms for the same stations, we must seek an alternative hypothesis to account for geographical distributions. As a preliminary, Lyell undertook an extended discussion of the ways plants and animals are distributed by natural means around the globe. For instance, animals unintentionally pick up seeds in their coats, carry them some distance, then drop them. On the basis of this discussion Lyell advanced the hypothesis that species of organisms are monogenetic: there is no need to suppose them created piecemeal, already distributed around the world. "Each species may have had its origin in a single pair, or individual, where an individual was sufficient" (2:129). For Lyell, therefore, the origins of new species were a direct function of the origins of new organisms. Different species would be a response to different external demands. But why should the same species never be repeated? I find no answer in Lyell.

Following this conclusion, Lyell presented evidence of the relationship between species and explained how one would expect species to become extinct—in particular, how one would expect species to become extinct on a regular, steady-state basis, thus obviating the need to invoke wholesale catastrophes that wiped out vast numbers of species simultaneously (2:128–75). He discussed the struggle for existence between species (a term Lyell used earlier in discussing the inferiority of hybrids [2:56]) and quoted the French botanist Augustin Pyrame de Candolle to the effect that "all the plants of a given country are at war one with another" (2:131). He pointed out that terrific population pressure is constantly forcing such a war—organisms always have a propensity to explode in number if unchecked (2:133–35). Hence he concluded that should one organism get the upper hand, perhaps by invading a new territory, other species may become extinct. In support he cited the dodo, the classic and well-documented case of extinction (2:150), and used this whole struggle as an argument *against* Lamarckian transformationism (a remarkable irony given the use British evolutionists later made of it). Of two competing groups, the less successful would be wiped out before it could respond to the situation and change for the better (2:173–75).

Lyell then brought up the introduction of new species, which was needed to keep up his steady-state balance and to counter extinction. Here Lyell suddenly became very hazy and concentrated on why we should not expect to see the creation of new species—in a sense, why actualism would not work at this point. In an argument based on the total number of extant species and the number of replacing species that would be required, and incorporating his earlier conclusion that new species are created only once, he claimed that new speciation would occur so infrequently that we would hardly ever see it in action (2:182–83).

But hazy though Lyell was at this point, the reader was free to assume that he favored some sort of natural (albeit nonevolutionary) species-making mechanism. Having argued at length that species become extinct by regular and natural means (subject to law), Lyell supposed that the reader "will naturally inquire whether there are any means provided for the repair of these losses" (2:179). Although the question went unanswered, a reader would have been obtuse had he not inferred that Lyell himself favored a mechanism involving the causes and laws of this world rather than direct supernatural interventions. This is how Lyell's friends and critics read him, and he agreed with this interpretation. But I feel that Lyell's position may have been clarified and sharpened in his own mind by the various responses. And his creative laws were very odd.

In the *Principles* Lyell denied organic progression, though he allowed that man is a moral advance and is recent. His position was antievolutionary and specifically anti-Lamarckian. He believed that species (though one individual or a pair) are created only once and are transported by natural means to their present places on the globe. Consequently, for Lyell the problems of organic origins and of species origins collapsed into the same question. He believed that extinction is a natural and regular process, and he implied that species are also created (using creation in a nonbiblical sense) by a natural, regular process. Given the theoretical dichotomy of the time, one assumes that at the phenomenal level Lyell would have believed that new species would exhibit regularities of appearance that are discernible at least in principle, and that beneath the surface one could find causes subject to law. After all, this is what is meant by "natural" rather than "miraculous." But Lyell did not in any way hint at what nonevolutionary natural speciation would be like: Would new species be formed from organic material? As an exception, man as a moral and intellectual being probably did require divine intervention.

Nevertheless, one feels that in the *Principles* Lyell was very much like a large iceberg, with only the tip visible. One suspects that his opinions on organisms and their origins went beyond what some of his rather cryptic remarks revealed. Hence, toward better understanding Lyell's position, I shall offer a number of explanatory points. Though some of these have been suggested as supporting rival explanations of Lyell's position on organisms and origins, and though some are far more important than others, I myself regard them interlocking, mutually supporting parts of the story (see particularly Bartholomew 1973).

Reasons for Lyell's Position on Organisms

I have suggested that the organic origins debate was not a matter of "pure science" (whatever that might mean). Quite apart from social and external factors, we find that along with science it incorporated important elements of philosophy and religion. Moreover, although for convenience of exposition one may identify various elements in someone's thought, such separation tends to be artificial, for science, philosophy, and religion are closely meshed. Certainly it is hard to deny that Lyell's stand on organisms involves all three. It will therefore prove convenient to structure the discussion with these separate aspects firmly in mind, though it would be grossly misleading to pretend they do not overlap.

Beginning at the more scientific end of the spectrum, let us examine Lyell's denial of organic progression. One must grant that in 1830 there was some good scientific evidence against progressionism (see Wilson 1970, pp. xxiv–vi). In particular, some fossil beds at Stonesfield in Oxfordshire provided evidence of mammal life far earlier than the time the progressionists normally dated the first appearance of mammals. Although this evidence had first been uncovered in 1814, crucial pieces were at once mislaid and were not rediscovered until 1828. But once recovered, this evidence was indisputable; it could not be explained away as an error of cataloging or the like.[1]

On the basis of these Stonesfield mammals, Lyell concluded that "the occurrence of one individual of the higher classes of mammalia . . . in these ancient strata, is as fatal to the theory of successive development, as if several hundreds had been discovered" (1830–33, 1:150). Obviously progressionists did not concur; they thought it was simply necessary to revise the time when mammals entered the (still-progressive) fossil record. But it does show that Lyell, who argued that these mammals demonstrate the fragmentary nature of the record, had at least some good scientific basis for his antiprogressionist case. We shall see how this fared over time.

Of course, the actual fossil record was not all there was to Lyell's denial of organic progression. Still dealing with science, though starting to veer toward the philosophical, we cannot doubt that Lyell saw organic progression as a threat to his general stand on the steady-state nature of the world. Lyell admitted this to his friend Scrope: once allow organic progression and one is on the slippery slope to general progression (Lyell 1881, 1:270). But why would Lyell have had such a fear of a progressive world picture? In part it was clearly because he thought—rightly or wrongly—that his steady-state thesis was intimately connected with his overall scientific/philosophical attitude toward geology. Jettison steady-statism, and you jeopardize actualism and uniformitarianism. Even if one might feel, as some of his critics did, that Lyell was mistaken in assuming so close a tie between the various parts of his program, he was certainly right in seeing some connection. It is difficult to sort out the different elements in many of his arguments. The climate theory, for example, is actualistic and a *vera causa*, uniformitarian and a major plank in the steady-state cause.

But there was more than science (and philosophy) behind Lyell's enthusiasm for the steady-state thesis. And here we see with a vengeance how artificial it would be to attempt to segregate science, religion, and philosophy. Undoubtedly Lyell found a steady-state world religiously attractive. He believed the world was in a kind of self-sustaining perpetual motion, neither building up nor running down, and thought it testified to the greatness of God by being perfectly formed and requiring no additive or corrective actions by God (man perhaps excepted). Lyell openly admitted that he found this picture of God the most satisfying (1881, 1:270), and he was proud that his own geological system was most in tune with such a religious viewpoint. Hence in this respect Lyell had religious, as well as scientific and philosophical, reasons for opposing organic directionalism.

Just as many strands of Lyell's thought came together in support of his overall steady-state belief, so this belief influenced many of his conclusions. In particular, the stand Lyell took on organic origins was a function of his steady-statism in at least two ways. First, having tied evolutionism so tightly to progressionism, Lyell clearly saw in evolution a threat to his nonprogressive world picture. Hence, in arguing against evolution he thought he was striking a blow for a nondirectional world (not to say that his antievolutionism was solely or even primarily a function of his desire to defend steady statism). Second, Lyell's stand on extinction was explicitly a plank of the steady-state thesis, and it is clear that his hints toward a converse, regular, natural mechanism of species creation were an integral part of the same thing. Lyell saw species appearing and becoming extinct as part of his overall uniform picture of the world. Although he thought man a recent addition to the world scene, he believed that extinction had certainly occurred since man's arrival, and that creation likewise was an ongoing phenomenon. His argument that we are unlikely to see such creation did not screen a belief that creation is finished (Lyell 1881, 2:36).

Among those factors that were as much scientific as anything else, at least one other was of great importance in Lyell's stand on organisms. Lyell needed stable species for a purpose revealed in the third volume of the *Principles* (which he had originally intended as the second half of the second volume). Lyell had argued that the creation and extinction of species are fairly regular processes. He therefore inferred that as we go down through the strata, by considering the ratio of extant to extinct species we can determine the relative ages of the strata. There is no need to look for characteristic fossils—the ratio will tell all. Given two groups, "There might be no species common to the two groups; yet we might infer their synchronous origin from the common relation which they bear to the existing state of the animate creation" (Lyell 1830–33, 3:58). But this use of a faunal chronometer relies on stable species, not species that change from one form to another almost at will. And we certainly do not want Lamarckian evolutionism, where the same species keep repeating themselves and throw the crucial ratios right out of line. Here too we find that Lyell had a scientific reason for his stand on organisms, specifically for his attack on Lamarck. (And here too we have a major reason why he introduced his whole extended discussion of organisms into the *Principles*—something that might strike the modern reader as a trifle extraneous in a book on geology.)

Turning to more philosophical issues, it is overwhelmingly clear that one of Lyell's strongest objections to Lamarckism was that such a theory fails the test of actualism. To Lyell, Lamarck's evolutionism was scientifically and philosophically unacceptable because, far from basing his speculations on *verae causae,* Lamarck supported a position that present evidence shows to be false. There is no good evidence of spontaneous generation, and breeders know only too well that they cannot change one species into another. Hence, by analogy one ought to argue that species are stable. This is the only move consistent with good philosophy.

As a kind of corollary, although this methodology of science may have led Lyell into opposition to Lamarck, it nevertheless inclined him toward the view that species origins and their causes are natural, in the sense that they are subject to normal laws of nature (albeit nonevolutionary). Lyell saw about him things subject to law: as an actualist, he had to be inclined toward law-bound species origins. We cannot entertain the possibility of *verae causae* that are not law-bound, even though we might not have much idea what the real *verae causae* are like.

Coming now to religious factors influencing Lyell's stand on organisms, we find that he was very much a child of his time. Take first the question of man. We know that Lyell was no orthodox Christian in the sense of subscribing to a doctrine bristling with miracles. But, though it is convenient to speak of Lyell as a liberal, implying something akin to a deist, it is clear that even if his ultimate motivation may have stemmed more from natural than from revealed religion, Lyell shared his contemporaries' obsession with the importance of man. And this was a major influence on his position on organisms—in particular, his attitude toward evolutionism.

Admittedly, in the *Principles* Lyell never came right out and said that his religious fears for man inclined him against evolutionism—after all, the *Principles* was a scientific work condemning those who mix science and religion! But, anticipating Lyell's reaction to Darwin, we know that a major reason Lyell found Darwin's theory unpalatable was his justifiable fear that such a theory brings man down to the level of other organisms (see particularly Wilson 1970). Man becomes no more than another item in the natural world. Since, in the first volume of the *Principles,* Lyell went out of his way to highlight man's special nature and his recent arrival, it seems certain that one of his major objections to Lamarckism was what he read as its tendency to downplay or eliminate man's special status. Lyell certainly gave prominence to the purported connection between the orang-utan and man in his exposition of Lamarckism. Moreover, it seems equally probable that his denial of any progression in the fossil record was closely linked with his concern for man. In the early 1820s Lyell, like everyone else, was a progressionist. It was not until he read Lamarck in 1826 that he ceased to be one and suddenly swung round to violent opposition (for details see Bartholomew 1973). With prescient insight Lyell undoubtedly saw geological progressionism as a natural first step to evolutionism. Although he overlooked the fact that Lamarck made little of it, he recognized that the evolutionist almost demands such a progression. Determined to preserve man's status, therefore, Lyell had a major reason to deny progression—and he did. Paradoxically, however, in his eagerness to defend man Lyell drew a far closer bond between a progressive fossil record and evolutionism than Lamarck had ever drawn.

But, man aside, natural religion in the broader sense also probably influenced Lyell's stand against evolutionism. Lyell was fully committed to the traditional argument from design, as we see from repeated comments throughout the *Principles*. Indeed, he went so far as to make the remarkably British suggestion that the dog's instincts were provided directly for man's benefit and indirectly benefit the dog, who thus gains a protecting master! (Lyell 1830–33, 2:44). Since Lyell was hazy on just how new species originate, it is difficult to be certain, but, given his commitment to design, his species-making mechanisms would have to reflect this design. But one of the major objections Lyell's contemporaries made to evolutionary theories like Lamarck's was that, dependent on "blind" law, they could not adequately account for design. They felt that unguided laws lead to randomness: hence any position postulating such law must be biologically inadequate because it will fail to explain intricate adaptation. And any position relying on such law must be theologically suspect because in jeopardizing adaptation one undercuts the major evidence of design. Certainly, if blind law is adequate for adaptation (which they doubted) the pressing need to infer a designer is much reduced. One feels no need to call on a designer to explain the patterns the wind makes in the sand. (Actually, if my earlier claims are well taken, Lamarck's laws are not quite so blind as he and others assumed.)

It therefore seems highly probable that, sharing his contemporaries' views on design, Lyell found Lamarckian evolutionism unacceptable for the same reason. Some thirty years later, Lyell found this very question of design a major difficulty with Darwinian evolutionism (Wilson 1970). Even on the internal evidence of the 1830s, it seems likely that man and design turned Lyell against evolution. Given the open reactions of the 1860s, the importance of man and design seem indubitable.

If one does allow that design influenced Lyell's stand on species origins, one must concede that any species-making mechanism he would allow would have to involve rather peculiar laws, specially guided to effect the design and in some sense unlike normal natural laws. If one understands by "miracle" something not covered by normal (natural) law, either at the phenomenal or the causal level, one might feel that Lyell got dangerously close to requiring the miraculous for organic origins, as he certainly seems to have done for man as moral being. More light may be thrown on Lyell's position when we compare him with his critics, but it is hard to deny that here we have an unresolved tension in his thought (not unconnected with his haziness). In part he wanted normal natural laws for organic origins, in part he was repelled by them. As we shall see, for Lyell (and others) this tension was to grow as the years went by.

The points we have discussed are interlocking rather than exclusive factors in Lyell's stand on organisms, though one suspects that the most important item is his concern with the dignity and importance of man. But, whatever the exact ranking, it is clear that his attitude toward the organic world was a compound of science, philosophy, and religion. Let us turn now to contemporary reactions to Lyell, keeping these points in mind.

Herschel and Babbage on Lyell on Organisms

Herschel's response to Lyell on organisms (in a letter written in 1836 from the Cape of Good Hope) was noteworthy less for the new ground it broke than for the vigor with which it embraced crucial aspects of Lyell's position. It is true that Herschel was a progressionist; but, this said, the only difference between Herschel and Lyell was that Herschel was less ambiguous, perhaps partly because he was not writing for publication—although in fact a couple of years later Herschel's views became public knowledge when they were published as an appendix to a book by Babbage (1838). Referring to the problem of organic origins as the "mystery of mysteries," Herschel read Lyell as advocating some kind of law-bound, nonevolutionary causal mechanism for new species, and he endorsed Lyell's position enthusiastically, writing that "all analogy" leads us to suppose that God "operates through a series of intermediate causes and that in consequence, the origination of fresh species, could it ever come under our cognizance would be found to be a natural in contradistinction to a miraculous process" (Cannon 1961a, p. 305).

Two comments should be made about Herschel's response. First, as he made very explicit, Herschel was guided to his position that organic origins must have natural causes by his general scientific/philosophical methodology. We must argue as best we can from our experience, and we experience law-bound mechanisms rather than miracles. Hence, understanding Herschel's overall methodological strategy of actualism makes his position here immediately explicable, as it did for Lyell's position. The deistic beliefs I have ascribed to both Herschel and Lyell would also have led Herschel to find Lyell's position on organic origins highly acceptable. (Lyell's position, that is, minus the antiprogressionism. Herschel, like most others, saw the essence of Lyell's position as a natural origin for organisms and found no reason to tie this to antiprogressionism, though, if reactions to Darwin are anything to go on, Herschel was no less worried than Lyell about man's special status.)

Second, Herschel's reaction may well have helped Lyell to clarify his own position and to feel confident that the right solution to the species problem lies in appeal to natural processes. In reply to Herschel, Lyell wrote (1881, 1:467): "In regard to the origination of new species, I am very glad to find that you think it probable that it may be carried on through the intervention of intermediate causes. I left this rather to be inferred, not thinking it worth while to offend a certain class of persons by embodying in words what would only be a speculation." But one suspects that Lyell's fuzziness in the *Principles* was as much a function of personal confusion as of prudence.

Baden Powell also endorsed Lyell's position, minus antiprogressionism (1838), as did Babbage (1838), who added a most interesting fillip to the argument. Taking umbrage at what he took to be a slight by Whewell on the possibility that mathematics and natural science could bring one closer to God, Babbage provided his own unofficial addition to the Bridgewater Treatises. In this, he drew an ingenious analogy from his own work on calculating machines, showing that he could set such a machine so that it would exhibit the natural numbers in sequence from 1 right up to 100,000,001, at which point it would start exhibiting a different sequence, the next number of which would be 100,010,002 (Babbage 1838, p. 36). Now, argued Babbage, God's laws could well be like this— absolutely regular and discernible for times almost without number, but having the built-in ability and necessity to do unexpected, anomalous things. Moreover, argued Babbage, we have a far more exalted conception of the Creator if we think of his laws in this way, where the anomalous is just as much a planned function of law as the regular, than if we think of God as setting up regular laws and then sometimes having to intervene directly. In other words, Babbage set the argument from law right on its head, arguing that the more anomalous something seems, the more it shows the magnificence of God's laws!

What kinds of things would call for a cause that is anomalous yet covered by law? Babbage (1838, pp. 44–46) made no secret of the fact that he had organic origins in mind. He wanted to suggest that organic origins were a function of law, implying that they had causal mechanisms covered by law; but he also wanted to suggest that they may be anomalous events like those illustrated by the machine. It was therefore no wonder that he was eager to print as an appendix Herschel's letter to Lyell, for, as Babbage himself said (1838, p. 225), "the almost perfect coincidence of his views with my own, gives additional support to the explanations I have offered"—particularly given Herschel's prestige.

Babbage's argument, far-fetched though it may strike us, is most important for two reasons: first, for its direct, crucial influence; second, because it illustrates what I think was an important element in the whole Darwinian Revolution, broadly construed. Let me pick up on this second point, leaving the first until later.

As we move toward organic evolutionism, in some loose sense we are moving toward law and away from miracle. A major influence in this move, as Babbage's argument hints, was the success of the industrial revolution, in particular the success of the machine. For fifty years or more before our story opened the British had with incredible success been harnessing the forces of nature: to man's end, they used the laws that the elements obey, getting things done by machines without direct human intervention, more rapidly and efficiently than preindustrial man had dreamed. Moreover, this industrial progress continued right through the period we are concerned with. The 1830s and 1840s, for example, were the time of the railway, immeasurably speeding travel through Britain. All this was bound to have its effect on the Victorian frame of mind. We find the Victorian prophet Carlyle obsessed with power and machines (Houghton 1957). And we see its effects in our story too. Britons conquered nature: they used its laws to effect things mechanically without need of human intervention. Therefore God, since he has shown his love for the British in letting them do this, must himself be able to do no less. In short, God is the supreme industrialist. If Thomas Arkwright can show his strength by making thread automatically, God can certainly make species automatically, thereby showing his strength. For many Victorians, the more they could see the world as like James Hargreaves's spinning jenny, the greater God would seem—and this influenced their attitude toward organic origins.

I do not want to exaggerate. The metaphor of God as engineer certainly did not convince every pre- or early Victorian to throw out miracles. But in the conservatives' "argument from law" we see traces of how the machine metaphor subverted the traditional argument from design. And, even as he introduced the argument from law, Whewell pointed out that one consequence of laws is to make every atom of one kind act alike. He was happy to quote Herschel's *Discourse* to the effect that this in itself proves a God, since mass-produced articles imply a machine, which in turn implies a designer and manufacturer. The action of laws on matter proves God "by giving to each of its atoms the essential characters, at once, of a *manufactured article* and a *subordinate agent*" (Whewell 1833, p. 302, quoting Herschel 1831, p. 38; his italics). But for Babbage and those who picked up his ideas, we see how one version of the metaphor influenced positions on organic origins. I suggest that the whole success of the industrial revolution worked its way into the organic origins debate, turning people toward evolution. Furthermore, it worked itself in by way of religion, making highly suspect any general claim that religion was absolutely opposed to the spread of evolutionary ideas.

Enough of this point now. In passing, let us note that Lyell (1881, 2:10) complimented Babbage, thinking "favorably" of his work. He stated explicitly that he found Babbage's conception of the Creator more acceptable religiously than any position postulating intervening miracles and noted that "the argument of changes of laws comes home to some of my geological speculations." This it certainly did, for Babbage was arguing that organic origins, although law-bound, call for something rather special, and that they testify to the glory of God into the bargain!

Sedgwick and Whewell on Lyell on Organisms

In his presidential address to the Geological Society in 1831, and again in a discourse to undergraduates in 1833, Sedgwick took a far more orthodox position than did Lyell. Agreeing with Lyell that the transmutation of species is "a theory no better than a phrensied dream," Sedgwick (1833, p. 26) opted flatly for genuine miraculous interventions by God to create new species, and so that there might be no mistake that by "miracle" he meant something outside the normal laws of nature, he spoke explicitly of "an adjusting power altogether different from what we commonly understand by the laws of nature" (Sedgwick 1831, p. 305). Every now and then God created new sets of organisms, and although these creations probably (certainly logically) occurred after naturally caused catastrophes, the creations themselves required supernatural intervention. Furthermore, although Sedgwick saw man as of very recent origin, unlike Lyell he also saw him as the culmination of a long progressive series. Moreover, Sedgwick criticized Lyell severely for even suggesting that species are still being created (Lyell 1881, 2:36). Man was God's last creation, and that is that.

Sedgwick's opposition to evolutionism, though backed by what he considered scientific facts, was based rather explicitly on religion. Apart from his reluctance to see man's origin as part of the natural order, Sedgwick stated flatly that "contrivance proves design" (1833, p. 21) and that any creation by law jeopardizes this design. But, trying to have his cake and eat it too, he also argued that even were there law-bound origins there would still be design. Nevertheless, one might be forgiven for feeling that Sedgwick's desire to find a religiously satisfactory answer to the organic origins question contradicted his overall scientific/philosophical strategy, for he was no less adamant than Lyell that the world runs according to immutable laws (Sedgwick 1833, p. 18; 1831, p. 300). God's laws for this world do not chop and change or break down.

There seems therefore to be some inconsistency, like Lyell's but far more extreme, in the way Sedgwick invoked miracles to account for new species.

Prima facie this inconsistency is evident in Whewell's thought too. Whewell argued that though God might work according to his own laws (special, nonnatural laws), the coming of new species belongs "not to what we are accustomed to speak of as the laws of nature" (Whewell 1832, p. 125). But he repeatedly stated that the laws of nature allow no exceptions and that the world runs by law. As Darwin rather cheekily quoted Whewell at the beginning of the *Origin:* "But with regard to the material world, we can at least go so far as this—we can perceive that events are brought about not by insulated interpositions of Divine power, exerted in each particular case, but by the establishment of general laws" (Whewell 1833, p. 356).

We have already seen that Whewell was adept at defending his right to believe precisely what he pleased, and this was no less true for the organic origins problem. To justify his holding apparently contradictory positions—miraculous creation of species and universal rule of law— Whewell relied on a number of beliefs that he thought we must always apply to geological phenomena. One was the notion that everything in the geological world had a cause: "Every occurrence which has taken place in the history of the solar system . . . has been at the same time effect and cause;—the effect of what preceded, the cause of what succeeded" (Whewell 1840, 2:112). Another was the requirement that the cause be sufficient to produce the effect: "Our knowledge respecting the causes which actually *have* produced any order of phenomena must be arrived at by ascertaining what the causes of change in such matters *can* do" (Whewell 1840, 2:101). And a third belief was that we can postulate causes and effects only from what we experience ourselves: "If we cannot reason from the analogies of the existing to the events of the past world, we have no foundation for our science" (Whewell 1839, p. 89).

This last claim might raise eyebrows, for it seems to go directly against the spirit of Whewell's *vera causa* doctrine. But in Whewell's opinion it did not. His point is that, given some phenomenon we know occurred in the past, present experience might well convince us that there is no cause familiar to us now that will explain its occurrence. To stay faithful to the other beliefs about geological causation—namely, that a phenomenon must have had a cause capable of bringing about the appropriate effect—as well as to have any hope of a *vera causa*, we might need to invoke kinds of causes other than those we have experienced or that are acting today. These other causes might be supernatural or miraculous rather than subject to law as we know it.

> [With respect to] the formation of the earth, the introduction of animal and vegetable life, and the revolutions by which one collection of species has succeeded another . . . it may be found, that such occurrences as these are quite inexplicable by the aid of any natural causes with which we are acquainted; and thus the result of our investigations, conducted with strict regard to scientific principles, may be, that we must either contemplate supernatural influences as part of the past series of events, or declare ourselves altogether unable to form this series into a connected chain. [Whewell 1840, 2:116]

Whewell's position was rather like that of a professor who suspects a student of cheating on an exam. The professor knows that the student's brilliant answer must have had a cause; he knows also that the cause must have been sufficient to account for such apparent brilliance; and he knows from experience that nothing in the student's normal performance could be such a cause. Hence the professor searches for other causes, though he usually stops short of miraculous intervention by the Creator.

His general theoretical position staked out, Whewell now had to show that organisms and their origins call for causes outside law as we know it. First we find that Whewell quoted many of Lyell's negative arguments practically verbatim (Whewell 1837, 3:573–76). In particular, he followed Lyell (and acknowledged this) in arguing that the evidence of breeders and Egyptian mummies is against transformationism. Moreover, he noted that Lamarck's theory, with its assumptions about the creation of new life, seemed suspiciously ad hoc and complex, a grave fault in the eyes of Whewell, for whom simplicity was a major virtue (Whewell 1837, 3:579). Hence, Whewell concluded that no causes of a kind we now know are sufficient to create new species. Yet species must have had a cause, and so Whewell concluded that their originating causes must be of unknown kind.

So far, one might think that a Lyellian/Herschelian position would have been acceptable to Whewell—natural causes (causes subject to laws) of unknown kind. One might even feel that Whewell would have welcomed such an answer, for it seems to cause less tension for his *vera causa* interpretation than for the interpretation of Herschel and Lyell, who usually wanted not merely to know that *verae causae* are natural but to understand what they are like. Whewell never laid such a condition on his *verae causae*. But Whewell rounded on Lyell somewhat roughly, stating that the "bare conviction that a creation of species has taken place, whether once or many times, so long as it is unconnected with our organical sciences, is a tenet of natural theology rather than of physical philosophy" (Whewell 1837, 3:589). Whewell's reason for this position lay, as we might by now expect, in the areas of organic adaptive organization and design (Whewell 1837, 3:574). Whewell just could not see any way to reconcile blind, unguided law, and causes obeying this law, with the creation of this organization.

This might seem a bit hard on Lyell, for we know that he too wanted to pay his respects to design. But he therefore had to suppose rather strange laws, and Whewell would have nothing to do with such contrived laws, unlike other laws in that they can have design effects. For him it was a contradiction in terms to speak of these laws as like other laws. But since organisms do show organization, we must suppose a cause adequate to it, and the only thing that seems adequate is a directly intervening God: "We must believe in many successive acts of creation and extinction of species, out of the common course of nature; acts which, therefore, we may properly call miraculous" (Whewell 1837, 3:574). Whewell does not deny that God himself might obey his own private laws; but these are not the laws of this world, natural laws, that the Lyellians wanted imposed on species origins.

In short, Whewell's position was that to explain organic adaptation *as good scientists,* we must appeal to miracles, something outside natural law. In the material world there is organic adaptation. As scientists we must explain it. But we realize we must turn to miracles because ordinary laws (and causes obeying such laws) are inadequate. "Nothing has been pointed out in the existing order of things which has any analogy or resemblance, of any valid kind, to that creative energy which must be exerted in the production of a new species" (Whewell 1840, 2:133–34). This position fitted in nicely with Whewell's philosophy and religion: his beliefs about geological causation were not being violated; his rationalist *vera causa* doctrine did not bar miracles (as the empiricist doctrine seems to do); and he was acknowledging God as designer.

But did not Whewell's solution, appealing as it did to miracles, conflict with his belief in the uniformity of law? In Whewell's mind it did not, for he believed that in appealing to miracles one has gone beyond science:

geology "says nothing, but she points upwards" (Whewell 1837, 3:588). On the other hand, law (normal, natural law) holds in the material world, at the level proper to science. Hence Whewell's position was not that creative miracles violate law, but that in some way they stand outside law. Miracles have to do with creation, and only after this has occurred can laws take hold. Consequently, miracles of this kind no more violate the rule of law than does the fact that electrical phenomena cannot be subsumed beneath the law of gravity.

Some concluding points should be made. First, we know how important for Whewell, as for everyone at that time, was the distinction between empirical or phenomenal laws and reference to causes and the laws governing them. It seems that Whewell, and Sedgwick also, wanted to put organic creations right outside everything—phenomenal laws or causes of any natural kind. How this stood the test of time we shall have to see. Second, Whewell, unlike almost everyone else and most surprisingly unlike Sedgwick, endorsed Lyell's antiprogressive reading of the fossil record (Whewell 1832, p. 117), despite the fact that more generally he was a directionalist. One can only suppose that Whewell, who was brilliant at anticipating possible pitfalls, appreciated Lyell's fears about close links between progressionism and transmutationism, particularly concerning man, and prudently followed in Lyell's path.

Finally, a general point about teleology. In appealing to design to refute evolutionism or any kind of natural origin, one might feel that Whewell was showing a great debt to Cuvier, who also used teleology to argue against evolution (or, one suspects, against any natural origin). However, though Whewell repeatedly invoked Cuvier's authority and ideas (Whewell 1837, 3:472–76), there was a subtle difference between Cuvier's Aristotelian teleology and British natural theological teleology. For Cuvier, evolving intermediates were conceptually impossible. To the British, such an impossibility would have unduly limited God's powers (Ruse 1977). They located the impossibility in the *process* of creation: blind law cannot lead to teleological objects. This British teleology goes back to Plato and is often called "external," as opposed to the Aristotelian "immanent" variety favored by Cuvier. The way a Briton like Whewell used teleology to approach the organic origins question, though related, was not quite the way taken by Cuvier and the Continental tradition (Hull 1973*b*). (Obviously the teleology of Whewell and everyone else was also different and stronger than the pallid teleology Lamarck smuggled in to complete his evolutionism.)

Conclusion: Was There a Genuine Difference?

In retrospect, one might feel there was not really much difference between Lyell and his sympathizers and scientists like Sedgwick and Whewell. Both sides thought adaptation and its implications of design caused trouble for unguided, species-creating law, quite apart from their determination not to drag man down to a purely natural level. But though there is undoubted truth in this claim, as an overall judgment it is rather hasty. For all his equivocation, Lyell wanted organic origins governed by law of some kind. His opponents made a virtue of pushing the whole question of origins outside law, and indeed outside science. Despite the overlap, they definitely differed in their attitudes toward the origins of organisms. (The Lyellians, moreover, thought they could have both law and design.)

Indeed, this difference about organic origins is reflected in general attitudes toward organisms. Despite his religious motives, or perhaps because of them, Lyell wanted a world left alone by God, in which unaided organisms battle painfully for supremacy, always under the threat of extinction if they fail. His critics were not about to deny either extinction or the struggle—the dodo and Malthus's predictions were common knowledge—but they tended to see an immanent God hovering protectively over his creation (which is precisely what a good Christian God ought to do). As a passage from Whewell quoted above makes clear (Whewell 1840, 2:116), extinction is seen less as a matter of failure than as clearing the decks for God's next round of creation. We find Buckland arguing that animals eat other animals not in an unguided fight for survival but because God has decided that herbivores should not starve painfully through population growth outstripping resources—better a quick, unexpected death than a protracted and miserable end (Buckland 1836, 1:129–34). Any struggle is part of a more general, divinely ordained "balance of nature" (Gale 1972). In short, though Lyell's sympathizers and critics alike were religiously motivated and had many similar motives, and though there was much agreement on facts and interpretations, they drew different organic world pictures. The liberals wanted God to leave the world alone; the conservatives did not.

This is not to say that a bright young entrant into the scientific community like Darwin would find either side satisfactory. Lyell's critics took the whole question of organic origins right out of science, and Lyellians, with whom Darwin felt a far keener sympathy, could not bring themselves to make a clean break and allow that God plays absolutely no role in organic origins. As Whewell was ever ready to point out, they wanted to have it both ways, arguing that organic origins are bound by laws, but then imputing all kinds of divine direction to these laws. It is no wonder that a new star in the network would feel uneasy about the organic origins problem and try to do something about it.

But all this lies in the future—for us, at least. Although Darwin's great creative work was done in the late 1830s, it was two more decades before it was made public. We must therefore turn toward the 1840s: the decade in which the Irish starved, Newman went over to Rome, and the evolutionary debate blew open.

5 *Ancestors and Archetypes*

Despite their differences on the organic origins question, the men of our scientific network were not so far apart that they considered the bonds of friendship threatened or felt there was anything impious or inherently ridiculous about the opposing position. At most there was a cozy disagreement among friends. They were, after all, united both in opposition to some of the more outrageous speculations from France and in the feeling that any reconciliation between science and religion must not be purchased purely at the expense of science.

The gentlemanliness of the debate about organic origins ended abruptly in 1844 with the publication of the anonymous *Vestiges of the Natural History of Creation.* This full-length work, crammed with evolutionary speculations, proved immensely popular with the general public, was discussed at length in leading reviews, and provoked invective far more bitter than anything that followed Darwin's *Origin.* But those like Sedgwick and Whewell, who wanted to balance science and Christianity and for whom anything tainted with evolutionism was anathema, were not left entirely without consolation, for during the 1840s Richard Owen began to expound his theory of archetypes. This theory attempted to provide far more orthodox answers to those many questions that the

author of *Vestiges* claimed could be resolved only by his evolutionary hypothesis. Certainly, Owen's answers were far more acceptable to our Cambridge Christians, though it would be a mistake to think that his direction was completely different from that taken in *Vestiges*.

Before considering these two lines of development, however, it will be worth our while to return briefly to the Continent and look at certain speculations about embryology, paleontology, and the links between the two. As we shall see, ideas developed abroad on these subjects were to have a direct influence on the course of British thought. (Essential background is provided by Lurie 1960; Ospovat 1976; Bowler 1976*a*; Gould 1977; and Russell 1916.)

Embryos and Fossils

Two important embryological positions were conceived and articulated in the early part of the nineteenth century. The first sprang from the French and German "transcendentalists," philosophers and scientists who tried to see all organisms as at least conceptually linked: in some sense one. The second position was that of the great German embryologist Karl Ernst von Baer.

The transcendentalists developed the law of parallelism between the individual's stages of the development and the scale of being. They argued that, just as organisms can be ranged along a scale, so we find this scale reflected in the development of the individual. In a higher organism, the embryo passes in turn through the *adult* forms of the lower organisms: "An animal high in the organic scale only reaches this rank by passing through all the intermediate states which separate it from the animals placed below it. Man only becomes man after traversing transitional organisatory states which assimilate him first to the fish, then to reptiles, then to birds and mammals" (Russell 1916, p. 82, quoting E. Serres). This is known as the law of parallelism, also called the "biogenetic" law or, perhaps best, after two of its chief developers, the Meckel-Serres law. To accept this law, one does not necessarily have to be an evolutionist, though one can be. Evolutionism was the position of Ernst Haeckel (1883, 1:309), who later in the century was responsible for popularizing a version of this law with his slogan "ontogeny recapitulates phylogeny." But one can continue to hold to a static chain of being. Alternatively, one can postulate certain variations on the law. For instance, though one might put all organisms in the same chain, one might argue less ambitiously that the higher organisms pass through only a limited section of the lower stages: for instance, one might restrict the developmental history of man to some or all of the vertebrates. This latter position seems to have been held by Darwin's Glen Roy opponent, Agassiz, who thus reflected both his transcendentalist leanings and the influence of Cuvier, particularly Cuvier's division of animals into four exclusive *embranchements* (Agassiz 1849).

Although I shall contrast von Baer with the transcendentalists, it is important to recognize that his position was far from being completely different from the one just sketched. His own transcendentalist background deeply influenced his thinking. Like Cuvier, but by his own account independently, von Baer divided animals into four basic kinds—radiate, articulate, molluscous, and vertebrate—each with its own ideal form or type. This type or pattern was at the center of von Baer's embryological thought, for in individual development he saw divergence from a particular type. The type is a kind of simple, elemental ground plan, manifested in the early embryo; then the development of individuals is a progressive specialization away from the general, following various paths (von Baer 1828; see Ospovat 1976, p. 6, for a full statement of von Baer's multipart law of development).

Note two points. First, there is no need to be an evolutionist to accept von Baer's embryology. Von Baer himself never became an evolutionist, though many of those influenced by him did—most notably Darwin. Second, note the similarities and differences between von Baer's law and the Meckel-Serres law. Both sides see general resemblances in early embryos. Dealing with the vertebrates, for example, both see the early embryo in all forms as being like the adult form of the most primitive vertebrates (lowest down the scale or closest to the type), the primitive fish. But the transcendentalists see the embryo as being exactly like an adult fish; the von Baerian sees it as exactly like an embryonic fish and only approximately like an adult fish. Thereafter, differences between the two positions start to increase. The von Baerian finds no reason to expect that after their first stage all higher vertebrates will come to resemble adult or even embryonic forms of the next level of vertebrate organization—say, the reptiles. Indeed, there is no unique set of levels through which embryos could progress, though in the development of the highest vertebrates one might sometimes get a certain succession through embryonic forms of vertebrates lower on the same conceptual path of divergence. To reach advanced forms, it may be that all will follow obvious paths at least part of the way. There is therefore room for superficial overlap (and consequent confusion) between the views of von Baer and the transcendentalists.

Turning to paleontology, we see that the ideas of Agassiz were important in two ways. First, Agassiz was primarily responsible for a new emphasis on the progressiveness of the fossil record. Sedgwick, for example, certainly argued for the progressive nature of the record and for man as its culmination. But for Sedgwick the emphasis was on the world's being prepared for man (reaching the right temperature) and on associated organic changes rather than on organisms that foreshadowed the arrival of man. Everyone could recognize similarities (e.g., skeletal isomorphisms) between man and the other vertebrates; but the emphasis on design as adaptation devalued their importance. There was no question of these lesser forms building up in a deliberate crescendo climaxing with man. Agassiz, on the other hand, perceived such a progressive preparation for man. Influenced in his youth by transcendental thought, he looked for conceptual connections between vertebrates and, finding them, he argued that we can see a progression in today's vertebrate world—fish, reptiles, and so on, up to mammals and man. So he promptly read this progression into the fossil record too. "The history of the earth proclaims its Creator. It tells us that the object and the term of creation is man. He is announced in nature from the first appearance of organized beings; and each important modification in the whole series of these beings is a step towards the definitive term of the development of organic life" (Agassiz 1842, p. 399; quoted by Bowler 1976*a*, p. 49).

As we have seen, Agassiz was also influenced by Cuvier. He therefore saw no progression between *embranchements,* arguing that representatives of all four first appeared together, even though Sedgwick and Murchison were beginning to uncover prevertebrate fossils. And Agassiz was no evolutionist. Between the progressive classes he saw unbridgeable gaps, and, except for man, he saw no temporal progression within classes (e.g., there is no temporal fish progression). This antievolutionism lasted all Agassiz's life. He crossed the Atlantic in the mid-1840s, settled at Harvard, and became the leader of the American opposition to Darwinism. (His classic statement is the *Essay on Classification* [1859], first published in 1857 as part of the *Contributions to the Natural History of the United States of America.*)

Agassiz's second important move was to link paleontology with embryology. He was not the first to do this; but he does seem to have been the one who impressed the scientific consciousness with the importance of this link. Agassiz accepted a modified version of the Meckel-Serres law and thus proclaimed a threefold parallelism. Within vertebrates, for example, we find three parallel lines linking fish with man: the development of the individual, the progression of adult forms of organisms as they first appeared on the globe, and the scale into which extant forms can be put. It was quite open, however, for a follower of von Baer to accept with thanks and praise Agassiz's linking of embryology and paleontology yet quietly reject his transcendental interpretations, substituting something more in keeping with von Baer's embryology.

Although the developments just described occurred on the Continent, by the early 1840s they were well known to British scientists. In the first edition of the *Principles* (1830–33, 2:62–64) Lyell thought it politic to mention the Meckel-Serres law, even linking it with paleontology in denying that it had any important evolutionary implications. In 1836 and 1837 the Scottish embryologist Martin Barry published an extensive discussion of von Baer's ideas in a widely read Scottish journal (Barry 1836–37; cited in Ospovat 1976). These ideas were at once picked up and used publicly by Owen (lecture notes; MS. 42.d4, Owen MSS, Royal College of Surgeons). William B. Carpenter incorporated them into a popular textbook, *Principles of General and Comparative Physiology* (1839), an English rival to a textbook by the German biologist Johannes Müller, which also incorporated von Baer's ideas and was translated into English (1838–42). Agassiz and his ideas were also well known. He presented his theory in person at the British Association's annual meeting in Glasgow in 1840 (Agassiz 1840), a meeting Darwin attended. And in 1841 a Scottish amateur geologist, Hugh Miller, whose fossil finds Agassiz had admired and classified, published a nontechnical book popularizing Agassiz's views on the fossil record and embryology.

Vestiges of the Natural History of Creation

That the author of *Vestiges* was genuinely anonymous added greatly to the book's fascination. Speculation on his identity ranged from Prince Albert on down. But many fastened on one Robert Chambers of Edinburgh—Darwin, for one, had little doubt that it was he (Darwin and Seward 1903, 1:48–49)—and in 1884, long after interest had subsided, Chambers was indeed revealed as the author (this was announced in the twelfth and final posthumous edition).

On the surface, at least, Chambers (1802–71) had the kind of career Victorians loved to hold up as a model (Chambers 1872; Millhauser 1959). Though he was born to reasonable comfort, by his early teens his family had fallen on very hard times. Hence, although originally intended for the church, he followed an elder brother in becoming a bookseller, his whole stock in trade consisting of a handful of schoolbooks and a few cheap Bibles. By diligence, frugality, and enterprise, he dragged himself up from the most extreme poverty, until he and his brother formed one of the most successful publishing firms of the nineteenth century, chiefly famous for *Chambers's Journal,* a weekly magazine packed with edifying and useful information. Like his brother William, Rober Chambers was more than just a publisher. He wrote copiously, chiefly treating topics related to Scotland, in a form likely to appeal to the widest reading public. One such production, *Scottish Jests and Anecdotes* (1831), said to be aimed at showing the Scots to be a "witty and jocular" race (Chambers 1872, p. 209), was undoubtedly a formidable enough task to prepare him for the assault on the organic origins bastion.

Primarily to protect his family and business, Chambers decided to publish *Vestiges* anonymously, and things were so well concealed, at least initially, that we know little about his motivation: Why should a respectable Edinburgh businessman have wanted to involve himself in such wild speculations? But this mystery is perhaps less worrying than it might be for some works, for much of the argument of *Vestiges* was conducted less at the strictly scientific level than at the methodological/philosophical level. We can therefore get a fairly true idea of what motivated Chambers, though by his own later admission (in a preface to the tenth edition of *Vestiges*), he was sparked initially by a recognition of the possible analogy between embryology and the history of life (see Hodge 1972, pp. 136–38). Since this recognition apparently occurred in the mid-1830s, it may well have been Lyell's offhand remarks that did the trick. At least this explanation fits a general pattern among Victorian evolutionists—to accept Lyell's arguments but stand his conclusions on their heads. Be this as it may, let us now examine *Vestiges.* In what follows, unless stated otherwise, I shall always refer to the first edition of the work and to a short supplementary work, *Explanations,* published in 1845.

Chambers opened *Vestiges* by discussing and accepting the nebular hypothesis, and his reason for this was plain (Ogilvie 1975). At the very least he hoped to establish that since the inorganic, cosmological world is subject to unbroken law, by analogy it is reasonable to suppose that the organic world, including organic creations, is also subject to law. But he also hoped to persuade the reader of the stronger claim that because the cosmological world has evolved through law, by analogy it is reasonable to suppose that the organic world also evolved through law. But, although in the first edition of *Vestiges* Chambers stood firmly behind the stronger claim, in *Explanations* (1845, p. 5) he covered himself against criticisms of the nebular hypothesis by arguing that only the weaker claim was really important for his position: "It is a mistake to suppose this [nebular] hypothesis essential, as the basis of the entire system of nature developed in my book. That basis lies in the material laws found to prevail throughout the universe." Chambers therefore considered it really impressive that in the organic world, as physics demonstrates, laws hold, and that these "natural laws work on the minutest and the grandest scale indifferently" (1845, p. 6). If proved true, the nebular hypothesis would thus give a nice boost to his organic evolutionism; but if it were shown false nothing crucial would be lost.

In *Explanations,* having weakened his analogy, Chambers attempted to repair the damage by adding a refinement. He cited evidence by a Belgian investigator (M. A. Quetelet) that man too is governed by laws. At the individual level the laws may be impossible to discern, but taking man as a group we find regularities in rates of birth and death, height, weight, strength, and so on. Even man as moral being does not escape law; for, considering man en masse, all the moral qualities, "even the tendency to yield to those temptations which give birth to crime,—are proved to be of no less determinate character, however impossible it may be to predict the conduct of any single person" (1845, p. 25; drawn from Quetelet 1842; note his publishers!). Then, having set the outer limits on the range of law (planets and men), Chambers did not hesitate to argue that everything is subject to law, including the origins of species.

Chambers next discussed in some detail the nature of the fossil record. Without implying that this was the least contentious part of his work—indeed, in sheer volume it probably evoked as much reply as all the rest combined—it was perhaps the area Chambers tackled with most confidence, if not competence. In particular, he argued in the way that, fifteen years before, Lyell had feared an evolutionist would argue. He pointed out that we see in the record a progression from the humblest organisms right up to mammals, with man certainly one of the most recent additions. More specifically, Chambers started with invertebrates—zoophytes, polyps, mollusks, crustaceans—then moved up to primitive fish, with a quick reference to certain putative links between crustaceans and fish (Chambers 1844, pp. 54–75). He then moved on to reptiles, with a side mention of progression in plants. Next came a link between reptiles and birds, with Owen cited as the authority. Following this came the earliest mammals, significantly, the primitive marsupial forms, and so on up to the present, as forms become ever more recognizable, culminating in man. Thus Chambers felt he had demonstrated a relatively gradual progressive climb—just the sort of thing one might expect were organic origins evolutionary rather than miraculous.

There are three points to be made before we go on. First, in *Explanations* Chambers modified his version of a smooth upward climb, allowing that one gets some branching, these branches—called "stirpes"—then going upward separately but parallel with the main paths (1845, p. 69). Second, Chambers attempted to tie in the progressive record with changing earth conditions, and we shall see shortly that he had views on how environment could trigger changes. But there is some flavor of Agassiz's transcendental progression toward man. And this certainly ties in both with his notion of law, reeking of predeterminism, and with certain idealistic reasons for putting man at the head of things. Third, Chambers mentioned resemblances between embryonic fish and fossils, remarking that "these facts seem to hint at a parity of law affecting the progress of general creation, and the progress of an individual foetus of one of the more perfect animals" (1844, p. 71). Although we know Chambers started his evolutionary quest with this analogy, he seems to have gotten this particular example (though not the evolutionary interpretation) from Agassiz, either directly at the Glasgow meeting or indirectly from a follower. Since Chambers was also drawn to Agassiz's particular interpretation of the embryo/fossil-record analogy (albeit to von Baer's ideas also), we seem to have further reason for suggesting an odor of man-directed transcendental progressionism about *Vestiges*.

After this lengthy geological discussion, designed to show the reasonableness of transmutation, one might think Chambers would examine in more detail the exact nature of the transmutationary change. But he postponed this for a while, pausing to consider the questions of spontaneous generation and the creation of the organic from the inorganic. Chambers hypothesized that life is formed by a "chemico-electric" operation (1844, p. 204), and he gave great prominence to the alleged creation of an insect when electricity was passed through solutions of silicate of potash and of nitrate of copper (1844, p. 185). He also pointed out how (inorganic) frost-forms on windows resemble (organic) plants (1844, p. 165), how the organic substance urea can be manufactured from inorganic substances (1844, p. 188), and how the domestic pig, unlike the wild pig, can contract a form of measles, which proves the disease must have originated after the pig was domesticated (1844, p. 183). All these "facts" apparently point to the possibility of creating the organic from the inorganic. One might perhaps wonder how, with all this creation going on, one could ever see any progression in the fossil record, since at any period one might expect to find organisms at all stages of evolution. But apparently one needs just the right conditions to create life from nonlife: far back in time these conditions might have prevailed, and life and the evolutionary progression would have begun. But that beginning, by itself changing conditions, would have brought to an end most creation from the inorganic except when something special like a domestic pig comes along (1844, p. 184).

Chambers now turned back to the question of just how transmutation could occur. He showed clearly that Babbage's argument had been heard, for he demonstrated that a law (such as a law of generation) could hold in a certain way for 100,000,001 times, then without violation react differently the 100,000,002d time. Chambers therefore argued that although like normally produces like, it need not always do so. On rare occasions like might produce unlike, and thus by law new species could be formed (1844, p. 210). But note that Chambers apparently expected the move from one species to another to be a one-step affair rather than a very gradual process.

But, one might feel, he has not handled the mechanism of transmutation or evolution. Here, so far as he ever got to such a point—which was hotly disputed by some of his critics—Chambers turned to embryology. We have seen two different embryological positions being worked out, and Chambers showed a true ecumenical spirit by adopting elements of both. First he suggested that man's organs recapitulate adult forms of insects, fish, reptiles, birds, and lower mammals before they reach maturity (1844, p. 200). Then a few pages later, without acknowledgment, he swung round and endorsed a position containing elements taken from von Baer, suggesting that in the development of the embryo of higher forms, "the resemblance is not to the adult fish or the adult reptile, but the the fish and reptile at a certain point in their foetal progress" (1844, p. 212). And in trying to explain exactly how he thought evolution occurred,

Chambers again showed von Baer's influence (as much as he showed the influence of anyone), for his central suggestion was that sometimes, because of external conditions, the embryonic stage is prolonged so that the organism develops (how was not specified) into the embryo of a higher organism, and thus a new species is created.

To illustrate his position Chambers included a diagram (adapted from Carpenter's discussion of von Baer; see fig. 10). Restricting ourselves to vertebrates, everything advances up to A. Fish then split off and go on to the mature forms at F. Reptiles, birds, and mammals develop together up to C, at which point reptiles split off and mature to R. Birds leave the mammals at D, and mammals proceed up to M. To get evolution, all we need suppose, argued Chambers, is that in some cases embryos stay embryos and go on developing longer than is usual. An overdeveloped fish is hence no longer a fish but a reptile, and so on. "To protract the *straightforward part of the gestation over a small space*—and from species to species the space would be small indeed—is all that is necessary" (1844, p. 212). Evolution aside, this does not strictly reflect von Baer's position, for von Baer did not advocate a necessary recapitulation of embryos, though he argued for a similarity of initial embryos of the same type and a consequent divergence and specialization. Any further succession is incidental, and organisms certainly do not go through all stages as Chambers's diagram implies. Even here Chambers probably owed more to transcendentalism than to von Baer, for he seems to side with the transcendentalists in accepting change from one kind to another rather than from general to specific.

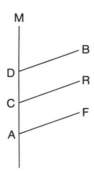

Figure 10. Chambers's diagram of the mechanism of evolution (see text for details). From Chambers, *Vestiges* (1844).

Chambers then backed up his suggestions with two important illustrations. The queen bee has a much shorter gestation period than the worker, and we all know how superior the worker is to the queen, since the worker is so industrious and the queen is distracted at every turn by sexual passion and jealousy (1844, pp. 214–16). Chambers may have been an atypical Victorian, but he was not *that* atypical. Then he noted that oats, if cropped repeatedly and thus not allowed to mature (not given so long a developmental period), have an irritating tendency to turn into rye (1844, pp. 220–22). Finally, Chambers suggested that light and oxygen might have something to do with the length of gestation and hence with transmutation—in particular, the supplies of light and oxygen may have increased, bringing in their wake the evolutionary development to present forms (1844, pp. 228–29).

This final suggestion, one might think, somewhat detracts from any man-centered progressionism in Chambers's work, but, having made this suggestion implying unguided change, Chambers at once started to speak of things having been "devised and arranged for beforehand" and of "preconception and forethought" (1844, p. 232). Of course, his whole use of Babbage's notion of law, let alone his embryological speculations, implies that the progression to man is contained in the first moves.

Before turning to his critics, let me make some final points about Chambers's position. First, note his "solution" to the origin of *species* as opposed to the origin of *organisms*. Species were not quite the embarrassment for Chambers that they were for Lamarck, but they were very much a side issue. Species were a by-product of Chambers's saltatory approach to evolution—an organism switches from one species to another in the act of evolving—rather than significant entities in their own right. He saw no inherent reason why species should be as they are, or for the particular differences between species; for example, why some species are very similar to each other and others very different. Species just happen because of the way evolution progresses.

Second, Chambers tended to speak harshly of his evolutionary predecessor Lamarck (1844, p. 231). Everyone in Britain was speaking this way,[1] so perhaps there was a tactical advantage in trying to dissociate oneself from him; but one suspects that Chambers's antipathy went deeper. In particular, Chambers allied himself with the "quinary" system of one William Macleay (1819–21), which classified all organisms into sets of five distinct types, each divided into five subsets, and so on (see fig. 11).

Figure 11. The basic outline of Macleay's quinary system (taken from Macleay's *Horae entomologicae* and reproduced in the *Treatise on the Geography and Classification of Animals* of William Swainson, Macleay's chief popularizer). Classification is in terms of nested sets, each containing five members. There are, however, "osculant groups" like the *Cirripedia* (barnacles) that fall midway between main groups.

Such a system sees a fairly rigid order within the organic world, and Chambers praised it precisely because of this characteristic. He argued that such order could not have occurred without law, and that hence the system is *"a powerful additional proof of the hypothesis of organic progress by virtue of law"* (Chambers 1844, p. 250; his italics). But, wrote Chambers, the Lamarckian hypothesis of needs and their fulfillment is inconsistent with the quinary system and would lead to great irregularity. Hence Chambers attacked Lamarck because he saw his speculations as threatening a major plank in his own program.

Third, though Chambers did not himself remark upon it, in one significant respect his theory was not unlike Lamarck's. We saw that Lamarck did not propose a common ancestry for all organisms—rather, we get continuous, parallel evolution, destined always to go through more or less the same stages. Chambers appears to have believed in something of this kind, though not necessarily that evolution constantly and continually starts over from scratch. He believed in a certain number of separate "foci of organic production" (1844, p. 259) and, like Lamarck, envisioned separate but parallel evolution from these points. Thus Australia housed one of these foci and was probably the most recent, since it had no indigenous true mammals, but only marsupials, which were primitive mammalian precursors (1844, p. 258).

Chambers's system is thoroughly progressive—again like Lamarck's, not to mention the nonevolutionary Agassiz's. We start with the lowest organisms and progressively evolve right up to the highest, man. One might wonder where everything ends. Is man the highest possible organism, or might we yet evolve into something higher? Always ready to take the more speculative alternative, Chambers assured his readers that evolution to something higher was indeed a live option—perhaps we are but the forerunners of a crowning race of superhumans (1844, p. 276). As we shall see in the next chapter, this speculation fell on very fertile ground, as it was incorporated into something far more orthodoxly Victorian than *Vestiges* might have hoped to be.

Intrinsically orthodox or not, Chambers certainly had no false modesty about his system, for with a confidence close to insouciance he calmly likened his own work to that of Newton.

> The inorganic has one final comprehensive law, GRAVITATION.
> The organic, the other great department of mundane things,
> rests in like manner on one law, and that is —DEVELOPMENT.
> Nor may even these be after all twain, but only branches of one
> still more comprehensive law, the expression of that unity which
> man's wit can scarcely separate from Deity itself. [1844, p. 360]

Scientific Criticism of Vestiges

It is difficult to choose from among the many offerings the most memorable phrases hurled against Chambers. A young man named Thomas Henry Huxley (1854) wrote what was probably the cruelest and most cutting review. Sedgwick certainly wrote the longest and the one that tried in most scientific detail to douse the evolutionary conflagration (Sedgwick, 1845, 1850). Typically, Whewell's contribution was the most pompous, a book that refused even to name its target (Whewell 1845; he named *Vestiges* in the second edition). My own favorite comes from the pen of Sir David Brewster, Scottish optical physicist, biographer of Newton, and general man of science: "Prophetic of infidel times, and indicating the unsoundness of our general education, 'The Vestiges . . .' has started into public favour with a fair chance of poisoning the fountains of science, and sapping the foundations of religion" (Brewster 1844, p. 471). And so on.

But entertaining as it is to pick out these more colorful pieces of Victoriana, we should be aware that almost every one of Chambers's scientific claims was singled out, examined, and found wanting for good scientific reasons. Yet in some ways the scientific reactions to *Vestiges* are the least interesting, because they are the least original. Many, if not most, of the rebuttals had appeared decades earlier, leveled against Lamarck. In the book just mentioned, Whewell did not bother to write much of anything. He collected antievolutionary extracts from his *History* and from his *Philosophy,* added a preface, and published them. Thus once again we get the breeders who have failed to cross species boundaries, Cuvier and his mummies, and all the other well-worn antievolutionary arguments.

This is not to say that no new arguments were brought against Chambers. Great scorn was thrown on his speculations about spontaneous generation (Sedgwick 1845, pp. 7–9). Almost every critic had a field day with the frost ferns. And the mite supposedly created by passing a current through chemicals fared little better—it hardly deserved to, since it came from the experimenters' dirty fingers. Equally castigated were Chambers's embryological speculations—hardly difficult, since with his catholic approach everyone could find something to disagree with (Sedgwick 1845, pp. 74–82). The subject that received most attention was his treatment of the fossil record, the scientific topic he himself examined in most detail. This was undoubtedly the field in which Chambers showed most competence (or least incompetence), and was of course the area that bordered most closely on cherished beliefs of the antievolutionists. Most people were happy to see progression in the fossil record, at least as evidence of God's preparing the world for man; and for anyone at all like Agassiz it was evidence of God's overall plan as he worked from the most humble organisms right up to man. The antievolutionists therefore had to show that though the fossil record is progressive, it contains features that absolutely preclude any evolutionary interpretation.

Just about everybody in favor of progression took a crack at Chambers's interpretation, and the standard counterarguments appeared with all the monotonous regularity of the mummies. But the assault on the geology of *Vestiges* that was generally considered the most popularly successful came from Hugh Miller (1802–56), a man who in important respects had a career that paralleled that of Chambers (Miller 1854; Bayne 1871). Like Chambers, he was a poor Scot who by prodigious effort raised himself from poverty and obscurity (he began as a stonemason) to a well-deserved position of esteem among his countrymen and indeed all Britons: in his final years he was editor of the *Witness,* the organ of the evangelical secessionist branch of the Church of Scotland. Miller had a lifelong interest in geology and, ironically, his first pieces on geology appeared in 1838 in the weekly journal edited and published by his good friend Robert Chambers of Edinburgh (Chambers 1872, p. 263). The irony probably runs much deeper than this, for the articles that I believe are Miller's give detailed discussions of geological theories (including Lyell's) and a full and favorable exposition (mainly taken from Buckland) of the progressive nature of the fossil record.[2] Moreover, in 1841 Miller published a work, referred to in *Vestiges,* endorsing the progressiveness of the fossil record, quoting Lyell on the Meckel-Serres law, and referring to Agassiz on the embryological/fossil parallels, with special reference to Agassiz's statements on fish at the Glasgow meeting of the British Association. That the trinity of Lyell, Miller, and Agassiz may have sparked—and certainly fanned—the flames of Chambers's evolutionism is delicious indeed. (Perhaps one can also add Whewell's name, for he suggested the nebular hypothesis as a good analogy for geology!)

But straight geology was only part of Miller's output—not surprisingly, one of his major aims was to harmonize the claims of science and religion. The publication of *Vestiges* provided Miller a heaven-sent opportunity—if one can use such a metaphor about so pernicious a work. Obviously so dangerous a popular book demanded a popular reply, and Miller was happy to respond, first in the columns of the *Witness,* then in 1847 in a book with the gorgeous title *Footprints of the Creator; or, The Asterolepis of Stromness.*

Chambers had argued that the fossil record gives evidence of an advance in organisms from the less complex to the more complex, just as one would expect were evolution true. Miller fully agreed that one sees such an advance—in the animal world from fish to reptiles to mammals, leading finally to man. Indeed, in a later work, *Testimony of the Rocks* (1856, p. 70) Miller went so far as to quote approvingly Agassiz's proposal that fish seem to have been most reptilian just before true reptiles appeared. However, protested Miller, certain facts in the geologic record still absolutely rule out evolution. Most important, we do not find that different kinds of organisms evolve gradually from humble beginnings. Rather, the most complex organisms appear quite suddenly in the fossil record, in finished form. Miller dwelt at length on the *Asterolepis* (a genus of fish) mentioned in his subtitle. He made much of its size and its incredibly sophisticated nature. Then he drew his triumphant conclusion: "Up to a certain point in the geologic scale we find that [fish of this kind] *are not;* and when they at length make their appearance upon the stage, they enter large in their stature and high in their organization" (Miller 1847, p. 105). Such a phenomenon, he claimed, makes evolution impossible (see fig. 12).

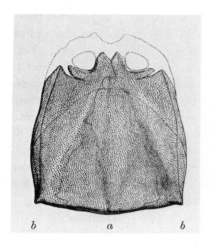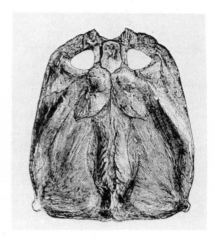

Figure 12. Miller's illustrations, designed to show the complexity of *Asterolepis*. From Miller, *Footprints of the Creator* (1847).

To this argument Miller added others: for example, the gaps in the fossil record. Why, if evolution has occurred, do we never find transitional fossils (Miller 1847, p. 105)? In any case, he argued, the evolutionist has no right to invoke the progressionism of the fossil record, for although it is undoubtedly progressionist from class to class, within classes one gets frequent degradation of characters and organisms, in a way fatal to evolutionism (Miller 1847, pp. 155–76). Thus the birds and the reptiles of today seem far less glorious than they once were—which Miller took as evidence against evolution. And so, concluded Miller and the many others who argued on related lines, the fossil record absolutely contradicts the position taken in *Vestiges*. In short, judged purely as a scientific theory, Chambers's evolutionism fails.

Philosophical Criticism of Vestiges

We come now to philosophical criticisms of Chambers's work: criticisms that dealt less with brute facts about organic evolution than with the overall scientific methodology and strategy Chambers advocated. A great deal of attention was paid to these subjects in the reviews, which is hardly surprising since so much of Chambers's argument was devoted to the claim that evolutionism follows necessarily if we take a proper scientific approach to biology. To be a good Newtonian in biology, he claimed, requires that one be an evolutionist, and he meant the theory of *Vestiges* to be the biological analogue of Newtonian astronomy.

First we have the question of the nebular hypothesis, which Chambers introduced as an analogy to biology. Many of his critics rejected the nebular hypothesis[3]—recent telescopic evidence had suggested that genuine nebulae did not exist (supposed clouds of gas resolved into distinct stars), and thus in their eyes a crucial premise of the hypothesis vanished. By the tenth edition of *Vestiges* (1853), even Chambers had modified his categorical support for the hypothesis, and thus one of his analogies—astronomical evolution, therefore biological evolution—crumbled if it did not vanish. But the critics wanted more to vanish than just the appeal to the nebular hypothesis. They disliked Chambers's whole reference to astronomy and, even more, they disliked his referring to Quetelet's findings to round out his general appeal to physics (that with planets subject to law at one end and man subject to law at the other, we must find biological evolution in the middle). For instance, Sedgwick wrote: "That man, as a moral and social being, is under law we believe true; but when it is affirmed that this law . . . is of the same order with the mechanical laws that govern the undeviating movements of the heavenly spheres, we believe the affirmation to be utterly untrue" (Sedgwick 1850, p. cxlviii). In short, the critics argued that even if Chambers could place the origin of species and evolutionary laws between Newton's laws and Quetelet's laws, which they disputed, he still could not claim that organic origins are law-bound and that his theory was genuinely scientific, because Quetelet's laws are not genuine laws like Newton's. (What kinds of laws Quetelet's are, Sedgwick did not specify.)

Next Chambers's critics turned to his use of Babbage's findings: "We think this, perhaps, the most unspeakably preposterous instance of bad reasoning in the whole volume" (Sedgwick 1845, p. 66). Again they claimed Chambers referred to laws that were not genuine (not like Newton's), this time because Babbage's laws were the results of his own machinations, whereas Newtonian laws were discovered, not made (Sedgwick 1845, p. 66). In any case, they argued, even were Chambers to prove the organic world subject to law (like the astronomical world) it would not prove biological evolution. "The laws which we study and admire, whether in the inorganic or organic world, explain the succession of phenomena, but throw no light upon the origin of the bodies in which the phenomena are observed. They tell us how things go on; they do not tell us how they began" (Gray 1846, p. 471). To be a good biological Newtonian, it is enough to pick up the laws as we now find them working—there is no need to delve into origins.

Finally, we have the philosophical criticisms of Chambers's evolutionary causal "mechanism," which suggested that when gestation is prolonged, the embryo sometimes develops from one species to another. At this point Herschel made his attack on *Vestiges*. One might think he would not have been totally unsympathetic to Chambers's views—after all, like Chambers, he thought (by analogy) that organic origins must be natural. But Chambers apparently hit a raw nerve when he had the audacity to liken his own speculations to Newton's, suggesting that his law of development was analogous to the law of gravitation. Remembering the distinction between causes and phenomenal laws, we see that for Herschel the paradigm of a cause was the law of gravitational attraction. In his *Discourse,* he had been at pains to show this as a *vera causa* at work. We have our own internal feelings of force, and when we whirl a stone on a string we have something directly analogous to, say, the moon whirling around the earth (Herschel 1831, p. 149). But at the very best Chambers had provided a phenomenal description of change. Herschel felt that no *vera causa* could be given for new species, because Chambers could not in any way point to known forces that lead to new species. (Chambers might have answered that he had tried to provide such causes, through extra light and oxygen increasing gestation length, but no doubt Herschel would have replied that he had not proved that more light and oxygen do increase gestation length or that increased gestation length causes new species to develop.)

Hence, in Herschel's eyes it was ridiculous for Chambers to claim he had found the biological analogue of gravitational attraction. At best Chambers had presented a phenomenon that needed to be explained, not an actual explanation. So we find Herschel, president of the British Association in 1845, thundering forth against writers in whose works "the idea of *law* is brought so prominently forward as not merely to throw into the background that of *cause,* but almost to thrust it out of view altogether." (Herschel 1845, pp. 675–76, his italics. Similar criticisms can be found in Explanations 1846, p. 184; Gray 1846, p. 472; and Huxley 1854.) In no way, insisted Herschel (1845, p. 676), can "the natural human craving after causes" be satisfied by such a theory as Chambers's. It gives us no more explanation of organic origins than does invoking miracles. (This reference to miracles might seem an unkind swipe at Whewell; but Whewell would have readily agreed that he had given no *scientific* explanation of organic origins).

In short, readers had a philosophical field day with *Vestiges*. Critics claimed that as good Newtonians they did not have to become evolutionists, then conversely claimed that Chambers's theory did not measure up to Newtonian standards. Chambers staked much of his case at the philosophical level. His critics answered him at this level.

Religious Criticism of Vestiges

We now come to the most sensitive subject of all—religion. Before hearing Chambers's critics, let us see just where Chambers himself stood on religion; for as even his most strenuous critics acknowledged, usually rather gracelessly, Chambers was far from being irreligious. Indeed, throughout *Vestiges* Chambers extolled his system as giving the highest conception of God—a God who does not need to work through creative interference but who has foreseen all contingencies. Having set his laws in motion, he need do nothing but sustain. There is no reason to suspect that Chambers was not sincere or to doubt that this concept of God was a major motive in his setting pen to paper.

In this context it is worth remembering that the Chambers brothers were very successful pioneers at publishing for the masses, and that in the 1830s and 1840s the British publishing trade was transformed by the industrial revolution, as steam power was applied to the printing press (Altick 1957). The Chamberses reaped the benefit of this advance, for after some difficulties in putting out their *Journal* (started in 1832) they too went over to steam, having some two hundred employees and twelve presses (Chambers 1872, p. 265). As one personally involved in the coming of the machine age, Chambers would certainly have been receptive to the notion of God as engineer. Indeed, given the crucial place of Babbage's speculations in his thought, this whole metaphor of God as machine-maker lay at the heart of Chambers's hypothesizing. "Mr. Babbage's illustration powerfully suggests that this ordinary procedure [of generation] may be subordinate to a higher law which only *permits* it for a time, and in proper season interrupts and changes it" (Chambers 1844, p. 211; his italics).

But how did Chambers deal with some of the more thorny religious questions that so exercised those who tackled the organic origins problem? As far as Genesis was concerned, Chambers openly agreed that some of his views did not mesh easily with a literal reading and quite reasonably pointed out that this was also true of much of the rest of science. "What is there in the laws of organic creation more startling to the candid theologian than in the Copernican system or the natural formation of strata?" (Chambers 1844, pp. 389–90). Well, one answer to this question would be "man"—a real difficulty for most people. Chambers's reply was that he indeed thought man had evolved from other organisms, but that he, no less than anyone, considered man the highest of organisms (1844, p. 234). He thought this was clearly illustrated by the quinary classification (1844, p. 272). And again Chambers reasonably pointed out that those violently opposed to man's coming from brutes denigrated the Creator; after all, were the brutes any less God's creations? Chambers, for all his sweet reason, was scarcely very orthodox, for he was fully prepared to allow that man might evolve into a better kind of organism. Hardly, one would think, a viable possibility for a creature made in God's own image.

But in another respect Chambers was beautifully orthodox. Think for a moment, he invited his readers, on the process by which every one of us arrives. "Were we acquainted for the first time with the circumstances attending the production of an individual of our race, we might equally think them degrading, and be eager to deny them, and exclude them from the admitted truths of nature" (1844, p. 234). But the facts are as God made them, and with practice the "healthy and natural mind" can regard them with "complacency." Hence, if God is to think up and man to act out sexual intercourse—with complacency no less—who should cavil at a little evolution? In any case, if we ourselves are at some point fish, why should not this be universally true of our ancestors?

Apart from man, one must deal with the major religious question of organic adaptation and design. With a casualness that borders on disingenuousness, Chambers fully admitted that he believed in adaptation and that adaptation did indeed prove design. He then referred the whole matter to the authorities—Paley and the Bridgewater Treatises! "It would be tiresome to present in this place even a selection of the proofs which have been adduced on this point. The Natural Theology of Paley, and the Bridgewater Treatises, place the subject in so clear a light, that the general postulate may be taken for granted" (1844, pp. 324–25). Really, however, Chambers virtually ignored the natural theology centered on God as designer, resting all on the natural theology of God as lawmaker.

Let us see how *Vestiges* fared in the hands of the critics, first in terms of revealed religion, then with respect to natural religion. One person who was open in his opposition to *Vestiges* on grounds of revealed religion was Hugh Miller: the evolutionist is faced with a stark dilemma, and neither alternative is acceptable, for he must hold "either the monstrous belief, that all the vitalities whether those of monads or of mites, of fishes or of reptiles, of birds or of beasts, are individually and inherently immortal and undying, or that human souls are *not* so" (Miller 1847, p. 13; his italics). Miller argued that an evolutionary process would rule out the possibility of a large gap between organisms without immortal souls and organisms with them. But, he contended, just such a gap must exist if some organisms do not have souls and some do. Thus an evolutionist is bound to view all organisms (including man) as on the same level with respect to souls. But to Miller it was inconceivable from a religious viewpoint either that all nonhuman organisms have souls or that humans do not. As he pointed out, somewhat horrified, if once we let go of immortal souls, from a practical viewpoint Christianity is worthless and almost anything goes. Why indeed should man "square his conduct by the requirements of the moral code, farther than a law and convenient expedience may chance to demand" (Miller 1847, p. 14)?

Two points should be made about this contention. First, what counted for Miller was man. The Bible speaks of God's creating man in his own image, breathing life into him, and giving him a soul with hope of eternal happiness. The threat to this picture frightened Miller, rather than any threat to a literalistic reading of the days of Genesis. Miller thought Genesis could be reconciled with the fossil record and believed the record shows the order of creation described by the Bible (Miller 1856). But man was central. Second, though Miller was far more open in his schemes of reconciliation than were Sedgwick and Whewell, it would be a mistake to think they and men like them were not opposed to Chambers on the same grounds of revealed religion as Miller. Sedgwick, for example, explicitly referred to the passage about God's making man in his own image as a problem for an evolutionary theory like that of *Vestiges* (Sedgwick 1845, p. 3, 12). One should be careful lest the protestations about willingness to reconcile Genesis and science blind one to the fact that revealed religion was still a powerful barrier to the acceptance of some scientific theories— even among the most educated Christian thinkers sympathetic toward science.

Next one has natural religion, centering on the question of design.

Curiously enough, Miller was tolerant toward *Vestiges* on this score. He
certainly thought that the universal adaptation he perceived in the organic
world was evidence of God's design (Miller 1856). Nevertheless, he saw
no reason why this design was incompatible with creation by law, and he
explicitly denied that the doctrine of final causes conflicts with or disproves evolutionism (Miller 1847, p. 13). But others felt differently. In
terms of science, they felt Chambers had made an appalling mistake by
not giving an adequate account of organic adaptation. In terms of religion, they thought he was ignoring—in fact, practically opposing—the
indubitable facts of final cause and of God's design. Typical of the responses to *Vestiges* was that of Whewell. We have seen that in refutation
he gathered all the explicitly antievolutionary passages in his *History* and
his *Philosophy*. Then for good measure he added everything those volumes
had said on teleology and final cause. And just in case anyone was too
naive to catch his drift, he wrote a stirring preface explicitly denying that
the evidence for organic adaptation and that for final cause and
evolution—particularly Chambers's version of evolution—can be harmonized. No natural (nonmiraculous) solution to the organic origins problem can adequately account for organic adaptation, argued Whewell. In
support of this claim he invited the reader to consider a great city, giving
all the evidence of careful planning. Could such a city, asked Whewell,
have grown by piecemeal evolution? Obviously not, came the reply
(Whewell 1846, pp. 15–16). Why then should we accept that organisms,
which show similar evidence of careful planning, had an evolutionary
origin? In other words, no miracles, no final causes. And this, thought
Whewell, was the ultimate *reductio* against *Vestiges*.

Whether Whewell was right in thinking that organic adaptation is
evidence of God's design and that it can never be the product of "blind"
unguided law is perhaps outside the scope of this book—at least the
question must wait until we have seen Darwin's contribution to the
organic origins debate. But, without unduly reading the present into the
past, we can surely acknowledge that adaptation is a major feature of the
organic world; that Chambers really had no explanation of it at all; and
that people were justified in criticizing him on this score. One might
argue that, with his reliance on Babbage's ideas, the last thing Chambers
was proposing was "blind" law. But Whewell's reply would be the one he
leveled against Lyell, that one cannot have one's metaphysical cake and eat
it too. Either one has law, in which case it is blind law; or one has
guidance, in which case one has no law. But Chambers himself seems to

have thought such criticisms as Whewell's were legitimate, for in *Explanations* and in later editions of *Vestiges* he tried in three ways to cover himself against such criticisms.

First, Chambers made clear his belief, as he perhaps should have done right at the beginning, that organic adaptation—which he himself was happy to see as evidence of God's design—could be brought about through unguided law (1845, p. 134). But, second, he denied that the world actually shows much evidence of adaptation! No one can pretend that all the animals that could have flourished in various places, past and present—and only those—actually did or do exist. For example, mammals might have existed when the coal beds were being laid down, but they did not (1845, pp. 151–52). In any case, there is not so very much adaptation that one need worry. Finally, to silence his critics, in later editions of *Vestiges* Chambers supplemented his theory with a kind of Lamarckian secondary mechanism to account for adaptations—an "impulse connected with the vital forces, tending, in the course of generations, to modify organic structures in accordance with external circumstances, as food, the nature of the habitat, and the meteoric agencies, these being the 'adaptations' of the natural theologian" (1853, pp. 155–56). It is perhaps fortunate that by this time Chambers's faith in the quinary system had somewhat collapsed, and so his new mechanism could hardly damage its perfection.

The Vertebral Archetype

Richard Owen restored sanity to the situation—at least in the eyes of Whewell and Sedgwick. In the 1840s, backed by an incredibly wide and detailed knowledge of anatomy, particularly vertebrate anatomy, Owen produced a synthesis between the ideas of two great French biologists, Georges Cuvier and Etienne Geoffroy Saint-Hilaire. The latter was perhaps best known as a sympathizer with Lamarck's evolutionism, but he is more properly remembered as one of the greatest transcendentalist thinkers (see Russell 1916, pp. 102–12; MacLeod 1965—Owen's views are expounded in Owen 1846, 1848, and 1849). Cuvier, as we know, stressed in great detail the adaptive nature of organisms, as an integral part of his doctrine of the "conditions of existence." Geoffroy, in opposition, stressed the unity of organisms and the links between them (Russell 1916, pp. 52–78). He wanted to see all organisms according to a common plan. Owen entirely agreed with Cuvier on the importance of adaptation. But in a group like the vertebrates, Owen felt that the similarities between members of different species are undeniable and are too prevalent to be left without explanation. Consider the flipper of the dugong, the forelimb of the mole, and the wing of the bat (see figs. 13, 14, 15). The bones have virtually a one-to-one correspondence. Why should this be so?

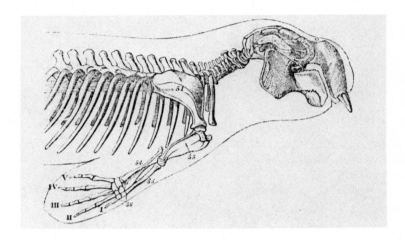

Figure 13. External form and skeleton of the pectoral fin of the dugong *(Halicore indicus)*. From Owen, *On the Nature of Limbs* (1849).

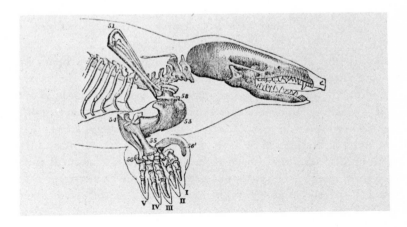

Figure 14. External form and skeleton of the forelimb of the mole *(Talpa europaea)*. From Owen, *On the Nature of Limbs* (1849).

Figure 15. External form and skeleton of the wing of the bat. From Owen, *On the Nature of Limbs* (1849).

On adaptive or teleological grounds the complications of bones in the mole and dugong are absolutely without ratiônale, for they serve no function at all (Owen 1849, pp. 13–14). Owen introduced the term "special homology" to denote a relationship like that between this flipper, limb, and wing, where the corresponding elements are known as "homologues," as opposed to "analogues," which have similar functions without similar structure. Owen argued that Cuvierian anatomy must be judged incomplete: "the attempt to explain, by the Cuvierian principles, the facts of special homology on the hypothesis of the subserviency of the parts so determined to similar ends in different animals,—to say that the same or answerable bones occur in them because they have to perform similar functions—involve many difficulties, and are opposed by numerous phaenomena" (Owen 1848, p. 73).

To escape from this dilemma Owen produced a Geoffroyan type of transcendentalist answer. For groups like vertebrates he postulated "archetypes." Such an archetype was "the basis supporting all the modifications of such part for specific powers and actions in all animals possessing it" (1849, pp. 2–3). Confining ourselves for convenience to vertebrates, we have such a plan or pattern for vertebrates, and all vertebrates are in some way modifications of it, these modifications being forced by particular adaptive needs. Thus, if we employ the term "general homology" for an organism's relationship to its archetype, special homologies can be explained as two general homologies put back to back, eliminating the middle term, the archetype. Special homology thus reflects two organisms' general homology to the same archetype.

Although we never actually see the archetype made flesh, Owen thought there were two ways the archetype could be discerned. First, in a particular group like vertebrates we see a progression from the most primitive to the most perfect (mammals and then man). There is certainly an atemporal progression, and Owen at least allowed that in time the fish were first. The most primitive forms are closest to the archetype; the highest forms like man have been subject to the greatest adaptive modification and thus are furthest from the archetype (Owen 1848, pp. 120, 132). Second, embryology gives archetypal clues, for organisms tend to be closer to the archetype when young, as is shown in man (since his special homologies with fish are stronger in the embryo; Owen 1848, p. 136). Using these guidelines one can get at the archetype, and one of Owen's most triumphant achievements was to reconstruct what he believed to be the vertebrate archetype (fig. 16).

Note how Owen's position seems to incorporate a parallelism between the fossil record and individual development. Owen had heard Agassiz speak on this subject at the Glasgow meeting, and he at once picked up the idea and endorsed it. But note that Owen's position, with its emphasis on archetypes, seems to align him more with a von Baerian type of embryology than with anything Agassiz favored, and history bears this out. Indeed, Owen had from the first endorsed and used von Baer's embryological findings (Ospovat 1976), which leads one to suspect that with their joint emphasis on basic types, his theorizing was influenced by transcendental elements in von Baer's thought as well as in Geoffroy's (about whom Owen was more open). From our viewpoint, however, it does not much matter whether Owen's transcendentalism came primarily from Geoffroy or from von Baer or from others—the important thing is

that it was there. (Remember: Although von Baer opposed the transcendentalists' key Meckel-Serres law, he was himself transcendentalist in an
important sense. There is therefore no inconsistency in tracing Owen's
transcendentalism in part to von Baer).

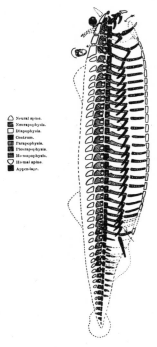

Figure 16. Owen's reconstruction of the vertebrate archetype. From Owen, *On
the Nature of Limbs* (1849). Reproduced in Russell, *Form and Function* (1916).

In the next chapter, when I examine in more detail the nature of
Owen's progressionism, more light will be thrown on his views on the
fossil-record/embryology parallel. But let me add now that though Owen
thought embryology gives archetypal clues, he was adamant that in case of
doubt the adult form takes precedence. Homological relations "are
mainly, if not wholly, determined by the relative position and connection
of the parts, and may exist independently of . . . similarity of development" (Owen 1846, p. 174). More later on the sources and consequences
of this view.

Having gone so far, influenced by the biologist Félix Vicq d'Azyr, Owen
was emboldened to take another step. Along the length of any individual
organism, one gets similarities between parts. Referring to these as

"homotypes," Owen spoke of their relationship as one of "serial homology" (1848, p. 164). In the vertebrate, the homotypes are vertebrae, and Owen argued that virtually the whole skeleton is a modification of such archetypal vertebrae (Owen 1848, p. 81), though he did allow some bones, such as those in the mammalian ear, as special adaptations (see figs. 17 and 18). Only one step remained. The transcendentalist Oken had argued that the vertebrate skull is composed of modified vertebrae.[4] For Owen this was a deduction from his theory, and he analyzed the skull into four vertebrae. Moreover, paradoxical as it may seem, he argued that "in their relation to the vertebrate archetype, the human hands and arms are parts of the head—diverging appendages of the costal or haemal arch of the occipital segment of the skull" (1848, p. 133).

First, let us consider this theory at the level of development, the ontogenetic level. Owen seems to have thought there were two forces tugging at an organism, representing the Geoffroyian or transcendentalist and the Cuvierian or functionalist elements in his thought. "There appears to be . . . during the building up of such [vertebrate] bodies, a general polarizing force, to the operation of which the similarity of forms, the repetition of parts, the signs of the unity of organization may be mainly ascribed" (1848, p. 171). Then he added, "the platonic *idea* or specific organizing principle would seem to be in antagonism with the general polarizing force, and to subdue and mould it in subservience to the exigencies of the resulting specific form" (1848, p. 171).

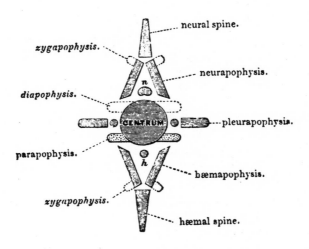

Figure 17. Ideal typical vertebra. From Owen, *On the Nature of Limbs* (1849).

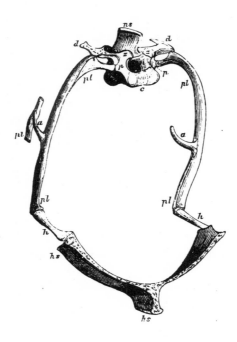

Figure 18. Natural typical vertebra: thorax of a bird. From Owen, *On the Nature of Limbs* (1849).

The archetype therefore seems to come from the "polarizing" force, which Owen also spoke of as involving "vegetative or irrelative repetition" (1848, pp. 81, 102, 132). None of this seems very clear, and one suspects that fundamentally it was not clear even to Owen himself; but the idea seems to be that throughout nature, both organic and inorganic, some force or tendency leads matter to line itself up in repetitive segments—the lower down the scale of life and being, the more the force is active or works unhampered (1848, p. 81). Owen spoke of the repetition of the elements of the spinal column as akin to the repetition involved in crystal growth (1848, p. 171), and this may be a clue to the idea's origin, for just such a concept of crystal polarization is found in Whewell's *Philosophy* (1841, 1:331–60).

A second point involves how far and in what ways Owen's theory can be called "Platonic." Owen spoke of his second force, his "specific organizing principle," which he also called the "adaptive" force (Owen 1848, pp. 108, 132, 172), as involving the Platonic *idea*. It therefore seems that

Owen's Platonism, which he appears to have been rather proud of, lay in the special adaptive modifications that make an organism a member of one species rather than another: "*ideai* of Plato . . . models, or moulds in which matter is cast, and which regularly produce the same number and diversity or species (Owen 1848, p. 172). Yet there were times when Owen spoke confusingly of the archetype as being the Platonic Form. He had a seal (for wax) made, incorporating the archetype. "It represents the archetype, . . . what Plato would have called the "Divine idea" on which the osseous frame of all vertebrate animals . . . has been constructed" (Owen 1894, 1:388). His Platonism now seems to lie not in the specific Form but in the far more abstract and general notion of the archetype. This fits in well with the way Owen often treated the archetype, for it obviously had no physical reality for him—it was not a general unifying ancestor, as it became for Darwin, but an other-worldly, yet even more real, entity.

In one sense none of this is very important. If Owen wanted to call himself a "Platonist" and got thoroughly mixed up, it was really nobody's business but his own. However, since Whewell, for one, seized gleefully (and thankfully) on Owen's ascriptions of Platonism, and since several recent commentators have argued that Platonism was a major nineteenth-century barrier to the acceptance of organic evolutionism (Mayr 1964, pp. xix–xx; Hull 1973*b*), sorting out Owen's true position becomes more pressing. One thing is clear. Insofar as Owen's Platonism centered on the archetype, it could hardly bar evolution within the archetype itself, though it might prevent transition from one archetypal form to another. Owen himself often pointed out that his vertebral archetype allowed for all kinds of forms besides those that actually exist, even suggesting that some might inhabit other planets (Owen 1848, p. 102; 1849, p. 83). Presumably, some of these might be transitions between existing forms. But even if one allows that Owen's Platonism centered primarily on the archetype, one wonders how much interarchetypal transition such Platonism might prevent. Owen allowed that cirripedes (barnacles) fit the crustacean archetype as embryos; yet because their adult forms are not like crustaceans, true to his principles he refused to classify them as Crustacea (Owen 1855, pp. 296–97). If one can switch archetypes in the course of ontogeny, one can surely do so in phylogeny.

Perhaps, therefore, Owen's Platonism was more profitable in barring evolutionism when applied to species differences, centering on adaptations. This sense of Owen's Platonism seems fairly fundamental. One might well expect a Platonist to consider the adult more basic than the embryo, and in refusing to classify cirripedes as Crustacea Owen was declaring that the adult form takes precedence. The adult member of a species is the ultimate reality, not its juvenile links. We may find that Platonism understood in this sense affected Owen's views on evolutionism, though his linking it with adaptation implies that any opposition to evolution (or mechanisms of evolution) based on adaptation would center less on Platonism than on natural theology and the general problem of how blind law can lead to adaptation. But, of course, since Plato himself propounded the argument from design, we still have Platonism in a sense!

We shall return to this point in a later chapter. Here let us look at Owen's attitude (in the 1840s) toward *Vestiges* in particular and evolutionism in general. As far as *Vestiges* was concerned, in a pattern to be repeated (subject to "that law of vegetative or irrelative repetition"?), Owen had matters both ways. To the unknown author of *Vestiges,* he wrote a very friendly letter stating that he had read the book with "pleasure and profit" (Owen 1894, 1:249). He added that "the discovery of the general secondary causes concerned in the production of organized beings upon this planet would not only be received with pleasure, but is probably the chief end which the best anatomists and physiologists have in view" (1894, 1:249–50). And he resisted entreaties by friends to condemn *Vestiges* from the pulpit of the *Quarterly Review* (1894, 1:254). But Owen fed Sedgwick and Whewell with facts contradicting *Vestiges*—for instance, he confirmed for Whewell the falsity of some of Chambers's recapitulatory speculations (1894, 1:252–53)—and, in tune with his age, he flatly stated to Whewell that no one could accept *Vestiges* because it derived man from a monkey rather than, as stated in Genesis, recognizing him as created in the image of God (unpublished letter to Whewell, in the Whewell Papers, Trinity College, Cambridge).

Notwithstanding Owen's possible involvements with Platonism, he certainly implied publicly that organic origins at least are natural (Owen 1849, p. 86). And though he professed ignorance of the actual mechanism, he tied in the appearance of species with at least some progressive elements in the fossil record (1849, p. 86). Moreover, it is clear that his archetypal speculations brought organic origins beneath law in another way, for he allowed that such origins must conform to the specifiable restrictions imposed by archetypes. But this seems to involve discernible regularity—natural law—and Owen himself spoke of "correspondences [which] flow from a higher and more general law of uniformity of type" (1848, p. 165). Although in this chapter I have opposed Owen to Chambers (primarily because of others' reactions to them), one must recognize that this is only part of the story. Owen no less than Chambers was seizing on things not previously emphasized by the British, such as homologies, and in explanation he was moving down a path not completely dissimilar from that followed by Chambers.

Whewell's Reaction to Owen

Whewell has fair claim to being the most sophisticated of the conservative thinkers on the organic origins question. Thus it is interesting to see how he reacted to Owen's theorizing, for though Owen himself favored natural causes for new species, this was hardly essential to the archetypal theory. One could continue to accept miracles and yet take over in its entirety the rest of the archetypal theory; and this is just about what Whewell did (1846; 1853; 1857, 3:553–62). Furthermore, it is not difficult to see why Whewell would make such a move. Owen's theory places plenty of emphasis on adaptations, essential for good biology and for support of natural theology, and provides one an explanation of homologies, which are biologically anomalous and threatening to natural religion (and which Whewell had therefore done his best to minimize). Moreover, it furnishes a nice tie-up between the archetype and the revealed and natural theological importance of man. It does not insist that man is the ultimate end of the progressive revelation of the archetype—I think Whewell still had trouble with progression, though he thought man came last (Whewell 1857, 3:565) and in any case was furthest from the archetype—but rather the theory emphasizes that man shows that he, of all vertebrates, received special adaptive attention from the Creator.

Then there was the question of Whewell's Platonism. Though he started as Kantian, as he grew older Whewell was more and more attracted to Platonism (Butts 1965; Ruse 1977). His fundamental ideas tended to shift from categories of the understanding to ideal forms according to which God had created the world. All this fitted in nicely with his philosophy of education, which was growing more important in his eyes, particularly now that he was master of Trinity when in the 1840s the university was coming under strong external pressure for educational reform. As a Platonist, Whewell could philosophically justify his conservative position, which put heavy emphasis on mathematics in university training. It is hardly surprising, therefore, to find that Whewell (1853) endorsed Owen's Platonism, understood at the species level.

Nevertheless, it is not clear how significant this was or became in Whewell's opposition to evolutionism. Whewell's eclectic philosophizing tended to be not merely grandiose but inconsistent. Although at times he implied that organic objects are divided into discrete kinds, as a Platonist would suppose (1860, p. 367), at other times he argued in a very non-Platonic manner that taxonomic groups such as species are very "fuzzy," with no clear boundaries (1840, 2:514–20). Since we cannot hope to delimit classes with both necessary and sufficient conditions for membership, we must be content with weaker membership criteria—for example, a list of characteristics, of which a certain number are sufficient to determine class membership, but no one of which is necessary. But a philosophy of classification like this is no bar to evolutionism. Perhaps the best thing to say about Whewell is that for the many reasons discussed in the last chapter he was an opponent of evolutionism. Having taken such a stand, he had on that score no objections to Platonism. But his Platonism on its own did not drive him to oppose evolutionism.

Finally, it is worth noting that Whewell had to pay a price for accepting Owen's solution to homologies. In the 1830s Whewell argued that as a scientist he was forced to suppose miracles to account for organic origins; at that point, such miracles seem to have been for him not only not brought about by natural causes, but in some sense entirely outside all law, including phenomenal law. By the mid-1840s we see a shift in his position, necessitated by his acceptance of archetypes. In reply to Chambers, Whewell wrote that those who do not subscribe to a transmutation of species do not thereby deny similarities between organisms (Owen's "special homologies"). "Nor do they doubt that this spectacle of analogies and resemblances implies the existence of laws in the mind of the Creator, according to which he has proceeded in the work of Creation" (Whewell 1846, p. 13). But, added Whewell, people like himself cannot follow the author "from resemblance to sequence, and from sequence to causation. They cannot venture to say that animals followed one another upon the earth in the order of their anatomical resemblances; and that their anatomical differences grew out of each other naturally by general laws" (1846, p. 13).

Whewell wanted to give away nothing at the causal level, and even at the phenomenal level he did not want to commit himself to any sequential laws. *But* it does seem now that Whewell was allowing that natural phenomenal law is in some sense applicable to organic origins. The mark of a natural phenomenal law is that we can discern it; after all, Whewell never denied that God might create by laws indiscernible to us. Just as we have seen Owen doing, Whewell now allowed that we can discern a constant plan by which God creates organisms and that, as we have seen Owen concede, consequently organic origins are in some way subject to natural law.

All in all, in the 1840s we see a slight softening in Whewell's position. Natural causes and sequential phenomenal laws are still unavailable for organic origins; however, some phenomenal laws are applicable. But Whewell was not making this move with the intent of bringing organic origins into the natural world and away from miracles. Far from it. He wanted to embrace aspects of Owen's theory and hence explain (special) homologies, one of the most troublesome phenomena for beliefs about adaptation and final cause. Yet, whatever Whewell's motives, in his reply to Chambers he was now allowing that some natural law was applicable to organic origins. Though Whewell was certainly not going as far down Chambers's path as did Owen, who wanted to establish natural origins for organisms, a critic might be forgiven for thinking that Whewell's concession was the thin end of a rather large wedge.

Vestiges *in the Broader Context*

On one thing everyone agreed. By the end of the decade, thanks to *Vestiges,* nothing was quite the same. The organic origins debate was no longer a private scientific controversy but a burning question that had been thrust upon the public eye. There are thus two final questions we must ask in this chapter. First, judged in the long run, did *Vestiges* have any significant part in effecting a resolution of the organic origins problem? Second, just why was there so much opposition, particularly from the professional scientific community? Answers to these questions may help us when we consider reactions to Darwinism.

In answer to the first question, one suspects there are two aspects. On the negative side, *Vestiges* acted as something of a lightning rod—a terrific amount of spleen and argumentation was poured out against it, to some extent exhausting the batteries of the opposing troops. Were any fresh evolutionism to come along, their arguments would look a little stale and for that reason would be less convincing. Moreover, one suspects that all the invective against *Vestiges* may have provoked a backlash. One certainly senses this with Sedgwick's contributions. If an eminent Cambridge professor declaims against a book, first for eighty-five pages in the *Edinburgh Review*, then for more than four hundred pages in a monstrously bloated preface and a three-hundred-page appendix to a rather mild little sermon on proper conduct at the university, one starts to suspect there might be something interesting afoot. Certainly it must be a book worth reading.

The positive part of the answer to whether Chambers steered people toward evolutionism is that by no means everyone rejected Chambers's work as vehemently as so many of the professional scientists. In view of its huge sale, the general public obviously found the message of *Vestiges* exciting and plausible, religious prejudices notwithstanding. Disraeli, in one of his novels of the period (1847, 1:224–26), gives a perfect sketch of how *Vestiges* was taken up in fashionable drawing rooms. Moreover, a few members (other than Owen) of the more sophisticated and informed intellectual stratum expressed sympathy for Chambers's views, although they generally did so behind the anonymity of an unsigned review or in a private letter. Baden Powell in 1848 wrote to the author in the warmest terms (Chambers 1884, p. xxx). Another who was sympathetic toward *Vestiges* was John Henry Newman's younger brother, Francis Newman (Newman 1845*a, b;* see also *Vestiges* 1845*a;* Kosmos 1845; Explanations 1848). In the next chapter we shall see how the positive influence of *Vestiges* was felt right into the 1850s and how it played a central role in converting to evolutionism the man who eventually brought Darwin into the open.

Our second question deals with the opposition to *Vestiges.* We have seen the various detailed reasons—scientific, philosophical, and religious—why people opposed *Vestiges.* But one senses that there must be something more. Sedgwick, for all his raving, was not a cruel man—indeed, by all accounts he was kind and lovable. Moreover, he had not rounded publicly on Babbage and accused him of preposterously bad reasoning, though it was from his work that Chambers had lifted his example of law by calculating machine. And Herschel's snide comments at the British Association were all very well—after all, he was on record as arguing that species origins must be natural. One gets the feeling that professional scientists took *Vestiges* very personally and that their reactions were in part a function of this personal element—though of course it is always easier to be nasty when one's opponent is anonymous. The following three suggestions may account in part for the personal tone that entered the debate.

First was the tension between professionalism and amateurism. Everyone agrees that *Vestiges* was popular with the general public—its large sale and many editions attest to this. It must have been galling to the professional scientists that a book containing so many demonstrably absurd speculations should prove so popular. This was all the more irritating because the sober scientific views of the Sedgwicks and the Whewells appealed only to the sober scientific audiences for whom they were written. In short, the author of *Vestiges* seems to have been the Velikovsky of his day, reaping all the public popularity and professional loathing that such speculators are wont to elicit.

Second, the 1840s were hard times in the British Isles—very hard times indeed (J. F. C. Harrison 1971). Not without reason were they known as the "hungry forties," for there were terrible depressions and agricultural calamities, culminating in the dreadful famines in Ireland. There was tension over the Corn Laws, Chartism, and, particularly in 1848, the frightening revolutionary examples from all over Europe. Life was tense, not least for Cambridge fellows like Whewell, who, since their college incomes were derived from rents, desperately feared the importation of cheap foreign grain (Whewell resorted to axioms to argue against free trade [Whewell 1850]; see also Checkland 1951). Against all this upheaval, Christianity, especially incorporated in a state church as it was in England, was seen to be a bulwark of morality and political stability: a guardian of the status quo. Not by sheer chance it was in 1848 that Cecil Frances Alexander published her hymn "All Things Bright and Beautiful," containing the famous lines:

> The rich man in his castle,
> The poor man at his gate,
> God made them, high or lowly,
> And order'd their estate.

In short, destroy Christianity and literally all hell might be let loose—especially in a society in which most of the population could not vote, lived near or in poverty, and (if they were lucky) toiled incessantly so that a small minority might live in luxury. Hence, insofar as *Vestiges* was seen as an attack on Christianity, and we know it was—"sapping the foundations of religion" (Brewster 1844)—it was seen as an attack on morality and ultimately on the political status quo. Finish Christianity and anything goes. And, as Hugh Miller for one pointed out (1847, p. 16), this was just what *Vestiges* was trying to do. Hence, *Vestiges* was seen as not just atheistic: it was dangerous.

For more enthusiastic Christians like Sedgwick and Brewster, *Vestiges* was particularly painful. These men were treading a very fine line. As good Protestants they put their faith in the Bible, and as good scientists they put their faith in science. But their science had to be very careful not to threaten the Bible. Just as Chambers had his critics, so Sedgwick had his, particularly one Dean Cockburn of York, who argued (virtually) that any truck with science, certainly any truck with modern geological science, went against the teaching of the Bible. Consequently, for all their talk about harmony between science and religion, the Sedgwicks were particularly sensitive to scientific works that went too far and spoiled things for everyone, tainting and dragging down serious scientists in the eyes of the majority who were unable to separate scientific dross from gold. Thus, as Brewster stated openly: "in forming a theory of creation for the study and reception of a community, either really or professedly religious, the theorist is not entitled to the privilege which we concede to the original inquirer" (Brewster 1844, p. 474).

And *Vestiges* did go too far. Hence, on the one hand it was open to critics to decry it and to argue that all scientific work was dangerous— "The work before us cannot fail to strengthen such prejudices" (Brewster 1844, p. 505). On the other hand, it was open to critics of a different stripe to argue that, since science contradicts a literal reading of the Old Testament, it is futile to put one's ultimate faith in a book (the Bible); faith must be centered in something else. Newman and his followers, as they turned to Rome right in the middle of the decade, thought that in the continuity of the Catholic church they found that something else, and they were not slow to point out the frailty of a faith based exclusively on the Bible (Willey 1949). In short, *Vestiges* upset the science/religion harmony and threatened to make men choose among some frightening alternatives—atheism, popery, or abjuring science. *Vestiges* simply had to be proved bad science.

A third reason *Vestiges* was so disliked centered on woman. It is generally agreed that *Vestiges* was very popular with women. I am not sure precisely why. Miller (1856, p. 251) was not beyond suggesting that women—upper-class English women, that is—represent the highest form of humanity, with Tierra del Fuegians and the Irish at the bottom. But Chambers does not claim that women have evolved beyond men, and, given his rather prim views on queen bees, one might even infer that he depreciated the roles into which so many Victorian matrons (from the queen down) were being thrust. Perhaps *Vestiges'* popularity with women lay in its understandability: for once, "science" was not cloaked in abstruse language known only by men. Also, the conclusions of *Vestiges* undoubtedly were appealing because they shocked staid members of society. But popularity with women there certainly was. Disraeli's parody illustrates this, and several reviewers, particularly Sedgwick and Brewster, harped incessantly on this point, in a grossly sexist manner. Sedgwick, having speculated that only a woman could have written such a work, then taken back this speculation once the idea was sown, warned against the influence of such books on the female mind: "the ascent up the hill of science is rugged and thorny, and ill-fitted for the drapery of a petticoat" (1845, p. 4). Brewster too bemoaned the way quasi-scientific ideas like phrenology and evolution were finding favor in women's eyes. "It would augur ill for the rising generation, if the mothers of England were infected with the errors of Phrenology: it would auger worse were they tainted with Materialism" (1844, p. 503).

It is tempting to dismiss these outbursts as nineteenth-century bigotry. But it may be more profitable to suggest that *Vestiges* was taken as a threat to the sentimental, idealized role of wife and mother into which Victorians were casting women. Both Brewster and Sedgwick openly endorsed this vision. "The mould in which Providence has cast the female mind, does not present to us those rough phases of masculine strength which can sound depths, and grasp syllogisms, and cross-examine nature" (Brewster 1844, p. 503). Rather, woman has a "soft and gentle temperament," "quick appreciation of character," and "instinctive knowledge of what is right and good" (Sedgwick 1845, p. 4). But women's liking for *Vestiges* augured the end of Christianity and morality in the home, and it meant having to argue at home about whether men are descended from monkeys. Apart from its intrinsic unpleasantness, this was all most frustrating because women are constitutionally incapable of telling right from wrong in these matters. Let the gentler sex therefore stick to drawing and painting fossils, as Brewster was happy to praise one Lady Cumming Gordon of Altyre for doing (1844, p. 488 n). And let us qualified male scientists go after *Vestiges* with renewed vigor for having so poisoned the temple of the hearth. Little wonder therefore that *Vestiges* was not merely opposed. It was hated.

6 *On the Eve of the "Origin"*

The 1840s were bad years for Britons, particularly those who, like Sedgwick, were both sincere Christians and committed scientists. But God tempers the wind to the shorn geologist, for a time at least, and from the publication of Sedgwick's phantasmagoric reply to *Vestiges* in 1850 until the *Origin of Species* was published in 1859, the organic origins debate simmered rather than boiled over. There were no more scientific thunderbolts such as Chambers had hurled, and life in general was much less threatening (Briggs 1973; Clark 1963). But this does not mean time stood still. One of the fascinating aspects of the Darwinian Revolution is that in 1844 Chambers's evolutionary speculations drew such derision, whereas only fifteen years later Darwin's evolutionism could convert a significant number of people. Though this was in large part due to the respective merit of the contributions, it seems likely that Darwin's success was furthered because by 1859 people were "readier" for evolutionism than they had been in 1844. But Chambers himself was one of the main reasons people were readier. For all its inadequacies, *Vestiges* reduced the shock value of evolutionism, and it converted to evolutionism some of the men I will discuss in this chapter.

The Fossil Record Again

Our old friends Lyell and Owen had differed over the fossil record, specifically its allegedly progressive nature. Through the previous decade it had become almost inevitable that Chambers's critics, while denying his interpretation of the fossil record, nevertheless strongly emphasized its progression, from humble organisms up to mammals and finally to man. In Britain this was particularly the theme of Sedgwick and Miller and, in a limited sense, of Owen too. Lyell, who twenty years earlier had flatly denied such progressionism, seized his opportunity as president of the Geological Society to deliver an address in 1851 in which he made one final effort to stem the tide and to refurbish as an alternative a nonprogressive interpretation of the record.

Marshaling all his skills, Lyell argued at length and in detail that the organic world is in a steady state and pointed out the anomalies in a progressive reading. So, he contended, it is no wonder the earliest known plants are very primitive; they are sea plants, and "the contemporaneous Silurian land may very probably have been inhabited by plants more highly organized" (1851, p. xxxviii). Moreover, the earliest known terrestrial plants include palms, which are among the most highly developed vegetable organisms. There are also anomalies in the animal world. Referring to the deposits at Stonesfield, as he had in the first edition of the *Principles,* Lyell noted that among the bones of pterodactyls we find a bird bone, something not to be expected were there progression.[1] Moreover, fossils of placental and marsupial mammals appear in these same deposits—again, progressionist anomalies. Referring to Owen's great authority, Lyell rather smugly suggested that such organisms probably imply other quadrupeds, which, being carnivorous, "keep down the numbers of the Phascolotheres and Amphitheres, which were probably, like the quadrupeds now most nearly allied to them, quick breeders" (1851, p. lxv, quoting Owen 1846). Lyell also made much of the progressionist claim that Cetacea (whales) and other marine mammals do not appear before a certain period in earth history, and because one fossil shell was identified as having housed a whale parasite, he pushed back the appearance of whales themselves beyond what progressionists wanted.[2]

Throughout this attack on progressionism Lyell repeatedly implied that even if we ignore the anomalies the fossil record is too fragmentary to support a progressionist interpretation. Thus he helpfully suggested that we find few birds earlier because their carcasses were invariably eaten by predators. And throughout Lyell's address the discerning listener found many hints that his fears for man's dignity had in no way abated since he wrote the *Principles*. Lyell emphasized that a progressionist chain links man most closely with other organisms; then, like everyone else trying to have matters both ways, he assured his listeners that even if progression is proved it in no way threatens man (1851, p. xxxix, lxxiii). Though Lyell, unlike Sedgwick, refused to see man as necessarily being God's final creation, apparently he found the absence of human remains in earlier strata sufficient proof that man is of recent origin (1851, pp. lxxiii, lxxi).

Owen was not one to take contradiction lightly, and before the year was out he vigorously struck back. Only because he was under "the influence of those uniformitarian views that have chiefly guided his labours in the field of science" (Owen 1851, p. 424) could Lyell have made the claims he did. The fossil record was indeed progressive in a sense, Lyell's anomalies were nothing like as anomalous as he thought, and his referring all difficulties to gaps in the record was ridiculous. For instance, what right had Lyell to suggest that the terrestrial contemporaries of the earliest plants (which are, as a matter of fact, marine) would be highly organized? They might just as well be low-grade plants like lichens and mosses (1851, p. 420). In any case, Lyell's reference to the first terrestrial plants was irrelevant—the progressionist sees plants in general as low compared with animals; so their level of development has little to do with animal progression. Not that Lyell's reference to plants saved his case, for even in the plant world we get progression.

Concerning the Stonesfield deposits, Owen was inclined to interpret his own words rather differently from Lyell. The pterodactyls and the like could have kept the Stonesfield mammals in check—there was no need to call in "the aid of larger hypothetical mammals" (Owen 1851, p. 442). As for the parasite from which Lyell deduced whales, Owen grew so heavy-handed with his sarcasm that one feels he must have thanked Lyell for so thoroughly exposing himself. He considered Lyell's interpretation of the shell highly tenuous, particularly since the period to which Lyell ascribed his Cetacea yielded absolutely no fossil evidence of them. A little pressure and Lyell's feet would slip right out from beneath him (Owen 1851, p. 440). Finally, Owen demolished Lyell's case against the fossilization of birds. Considering the rate at which birds are known to breed, multiply, and be killed, it is madness to pretend that they existed in early times but were not fossilized: "We can only express our regret that the Philosopher should have been suffered to subscribe so far into the Advocate" (1851, p. 438). But, for all his overbearing manner, Owen's attack contained at least one lacuna. He could hardly deny the existence or the placing of the Stonesfield mammals. This was just as well, for during the 1850s a great many more such mammals were discovered (Wilson 1971). But, as we can see from a chart Owen presented in 1860 illustrating progressionism (fig. 19), the Stonesfield mammals hardly wrecked progressionism—they merely pushed back the earliest date of mammals. And as Murchison (1854), another progressionist, noted, in the overall scheme the Stonesfield mammals are not all that significant.

This was Owen's direct attack on Lyell. From the tone of Owen's argument, as well as from his later chart, one might think that (with a few slight variations) he was merely endorsing the kind of progressionism worked out in the 1840s by people like Agassiz and Miller—a more-or-less linear progression from the primitive up to man. But if we look carefully at Owen's reply to Lyell (as well as at his later work), we see that, for all his trumpeting of progressionist orthodoxy in opposition to Lyell (to the extent of quoting Sedgwick against Lyell), Owen was in fact endorsing a picture significantly different from that proposed by progressionists (perhaps including himself) in the 1840s.

For a start, Owen was not proposing a linear progression, even with few side diversions. The notion of branching, or divergence, was fundamental to his picture. Accepting von Baer's central embryological claim that in development we see a gradual change from the general to the special—indeed, practically claiming it as his own invention—Owen (like Agassiz) read embryological development into the fossil record: It "is a principle which is illustrated in a remarkable degree by the succession of animal forms on our planet" (Owen 1851, p. 430 n). In other words, Owen believed that all organisms started with basic archetypal forms, but from then on we see divergence and progressive adaptation and specialization to particular ecological niches (the effects of his adaptive force). There is therefore no linear development. Although by 1860 Owen was apparently prepared to order the appearance of organic forms on earth as invertebrates-fish-reptiles, in his reply to Lyell he envisioned all four *embranchements* coming together and even considered reptiles to be as ancient as fish (1851, pp. 419, 422).

Second, Owen moved away from Miller's claim (also his own earlier claim) that there is no succession within groups like fish. In line both with a fossil modeling on von Baer's embryology and with his own archetypal

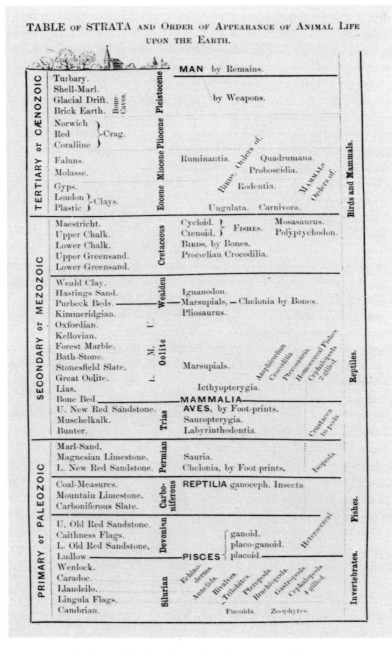

Figure 19. Geological table from Owen's *Paleontology* (1861).

theory, he argued that we start with primitive fish and move to more complex (progressed) kinds. "Palaeontology demonstrates that there has been not only a *successive* development in this class, but, as regards their vertebrate skeleton, a progressive one" (1851, p. 426). Take the strata of the Silurian and Devonian periods. These huge strata have been extensively investigated, but no completely ossified vertebra of a fish has been found. One has only remains of primitive fish; not until later strata does one get fish with proper backbones. And some sort of progression within the group also holds for reptiles, birds, and mammals. All along we get progression from the general to the specialized.

All this adds up to a fundamental (though virtually unacknowledged) transformation of progressionism as previously understood, as going from the simple to the complex, from the most primitive to the most sophisticated—man. Although the status of man was always at the center of Owen's own thought, he had now knocked sideways the anthropocentrism of previous progressionism. No longer do we have a progression up to man, possibly with occasional sideline diversions. Divergence is now fundamental, and each line must be considered in its own right. Man is no longer the measure of all things. The measure now is the efficiency with which organisms specialize and adapt themselves to their own particular niches. Because efficiency often implies complexity and sophistication, one might still expect to see elements of the old progressionism within and between groups. But essentially Owen smoothing the way for a theory that sees a great deal of adaptive divergence, sees gradual change from the primitive to the specialized (and hence considers adaptation more of a process than a fixed state), and does not perceive the whole fossil record as exclusively directed toward man (Bowler 1976*a*).

Whatever the differences between the progressionists, Lyell's antiprogressionism was a minority position. But some supported him, whether their motives were pro-Lyell or anti-Owen. More light will be thrown on this as we welcome two newcomers who were in Lyell's corner.

Species and Skulls

J. D. Hooker and T. H. Huxley were to become, above all others, the champions of Darwin and his theory. Let us discuss their careers up to 1859 and see how their work helped prepare the way for evolutionism.

Joseph Dalton Hooker (1817–1911) was born to the scientific purple (L. Huxley 1918). His father, William Jackson Hooker, was one of Britain's leading botanists, first being regius professor of botany at Glasgow, then for twenty-five years the highly successful director of the Kew botanical gardens. Joseph Hooker was trained as a doctor but started his career as a naturalist on board ship, the H.M.S. *Erebus*, which spent from 1839 to 1843 in Australia and the Antarctic. On his return to England Hooker began analyzing and publishing the massive results of his botanical collecting. This took him fifteen years, as he covered the flora of the Antarctic, New Zealand, and Tasmania. Hooker continued his travels throughout his very long life: his most adventurous excursion undoubtedly was his trip to the Indian subcontinent in 1848–51, when he traveled through the Himalayas, visiting parts of Tibet and Nepal that were not again entered by Europeans until the next century.

In 1855 Hooker, whose scientific talents and achievements were by then unquestioned, was appointed assistant to his father at Kew, and upon his father's death ten years later he became director. Since Hooker grew up in Scotland, he did not come in contact with the scientific network through the English universities. But there were family friendships with members of the group, particularly with Lyell and Henslow, and Hooker married Henslow's daughter. He became close friends with Charles Darwin soon after the *Erebus* trip, and in 1844 he was the first to be let in on the great secret of natural selection, though he was not then converted. To round out his portrait, let me add that Hooker was incredibly hardworking, rather peppery in temper, but warm and a loyal and lovable friend.

Although Thomas Henry Huxley (1825–95) came from a far more modest background than did Hooker, in important respects their early careers ran parallel (L. Huxley 1900). Supporting himself on scholarships, Huxley made a truly brilliant performance as a medical student at Charing Cross Hospital. Then, joining the navy, he was appointed to H.M.S. *Rattlesnake*, just about to start for the South Seas. He spent from 1846 to 1850 on this trip, and during this time he laid the foundation for a highly successful career as a comparative anatomist. In particular, Huxley studied and wrote important memoirs on organisms that float on the surface of the sea—hydrozoans, tunicates, and mollusks. Since they are so delicate, they could be studied properly only where they naturally occurred (Winsor 1976). Leaving England as an unknown naval officer, Huxley returned to find that his work had brought him professional recognition. But until middle age money was always a pressing problem for him,and it was not until 1854 that he gained the scientific post of lecturer in natural history to the Royal School of Mines and (in 1855) naturalist to the Geological Survey. From then on Huxley's life was a success story of the kind so beloved by Victorians. Although tortured with chronic indigestion and fits of crushing depression, his public self-confidence and energy made him outstanding in Victorian Britain—a time and place when men were publicly self-confident and energetic as never before or since—and by the end of the 1850s he had risen toward the top of his profession.

I confess I find it difficult to warm to Huxley, though in important respects he was a fine man. Although Huxley was a good friend, he lacked generosity toward opponents. He wanted not merely to defeat the other side, but to attack their moral and intellectual credentials. We shall see evidence of this after the publication of the *Origin;* but the attitude also came through forcefully in the 1850s. Huxley's tendency to turn his disputes into personal quarrels may have been the mark of an inner insecurity—the poor scholarship boy always trying to prove himself better than the next man. Such great outward self-confidence compensating for a basic insecurity was very much a mid-Victorian trait (Houghton 1957), and I suspect Huxley (a bit like Whewell) was a paradigm. Though he had a powerful personality, one misses the inborn aristocratic belief in one's own self-worth that was Darwin's—perhaps reflecting their differing financial states. But however one feels about Huxley, no one could deny that he was grand company, a first-class public speaker with a beautiful grasp of English, and just the man to appoint to a boring but crucial committee: excellent potential for a college dean, which is just what he became.

Until the end of the decade, Hooker and Huxley were not evolutionists, but they give fascinating evidence of how the scientific community was moving toward evolutionism. Hooker's "Introductory Essay" was published in 1853 at the beginning of his work on the flora of New Zealand. Hooker felt the need to explain at some length why he could not be an evolutionist; he was not going to let the organic origins question fade away through neglect. But first he was careful to point out that his antievolutionism was essentially methodological, since he was required to give some permanence to his specific descriptions. He said he recognized that one could not solve the organic origins problem by considering New Zealand alone and noted that he did not put forward his views "intending it to be interpreted into an avowal of the adoption of a fixed and unalterable opinion on my part" (Hooker 1853, p. vii). Helpfully, he listed alternatives to his position. Then Hooker ran through the usual list of reasons against transmutation—the difficulty or impossibility of producing specific changes artificially, the lack of significant change when organisms move to different climates, and so on.

Nevertheless, for all his antievolutionism, Hooker brought forward facts and interpretations calculated to make an evolutionist feel good indeed. In modern taxonomic parlance, Hooker was a "lumper" rather than a "splitter" (1853, p. xiii). His inclination was to put different populations of similar organisms in the same species rather than in different ones. He thus tolerated a high degree of intraspecific variation: just what one must have to postulate a theory of evolution that depends on all sorts of variations coming into populations. Second, in flat contradiction to what he had said earlier, he argued that a population isolated from the main body of its species (say on an island) may well be changed by the new conditions, to the extent that it is very difficult to tell if it came from the original species. Moreover, "To connect those dissevered members is often a work of great difficulty, for individuals of such races frequently retain their character even when they have been under cultivation for many years" (1853, p. xv). Third, Hooker pointed out experiments at Kew that showed how plants from different climates and different parts of the world can be transported and survive elsewhere (1853, p. xvi). This strikes at the idea that organisms cannot have been distributed around the globe by natural means and that each organism implies creation because of the peculiar adaptations necessary for its particular location.

Fourth, Hooker mentioned that individual islands, despite similarity of climate and so on, tend to have their own peculiar species (1853, p. xx). Although he said nothing explicit, this fact of geographical distribution is hard to explain in terms of special creations but would be welcomed by an evolutionist. Fifth, admitting his indebtedness to the speculations of those well-known uniformitarians Charles Lyell and Charles Darwin, but outdoing them, Hooker argued that the existing distribution of plants, though inexplicable by the methods of distribution known today, is readily understood if we presuppose former land masses that have since disappeared through the regular uniformitarian (actualistic, noncatastrophic) geological processes (1853, p. xxvi; Darwin thought existing methods of transportation sufficient, and Hooker later accepted this). Finally, Hooker obliquely hinted at, if not a transmutationary origin for species, at least an ongoing natural one. Having pointed out that this is the case for extinction, he suggested that from this "established premiss the speculator may draw his own conclusions" (1853, p. xxvi).

With friends like Hooker, the special creationists did not need enemies. But, though at this time Hooker did know of Darwin's theory, he was not an evolutionist. Indeed, to support Lyell he came out with an attack on fossil plant progression (Hooker 1856). The confusion of his essay was genuine and reflects the confusion of many scientists at this time—particularly those interested in geographic distribution.

Huxley too joined in the cry against evolutionism, with his attack in 1853 on Chambers's *Vestiges* ("the product of coarse feeling operating in a crude intellect"; 1854, p. 425). But Huxley's researches into comparative anatomy concern us here, both insofar as they counter Owen's archetypal theory and insofar as they could be seen very practically as countering this theory. Intimations of what was to come surfaced in a talk Huxley gave at the Royal Institution in 1854 (Huxley 1851–54), but his major excursion into comparative vertebrate anatomy came in 1858 in his Croonian lecture to the Royal Society (Huxley 1857–59). Taking as his topic the vertebrate skull, Huxley distinguished two lines of inquiry. First, "Are all vertebrate skulls constructed upon the same plan?" and, second, "Is such plan, supposing it to exist, identical with that of the vertebral column?" Huxley 1857–59, 1:540). Letting himself be guided both by comparisons of adult skeletons and by comparisons of developmental relations, he considered the skulls of a sheep, a bird, a turtle, and a carp and answered his first question with a resounding yes. The second question, again considering developmental phenomena, Huxley answered with an equally resounding no. The supposed isomorphism between bones of the skull and bones of the backbone simply does not exist.

On their own terms, Huxley's conclusions were of obvious importance to the evolutionary debate, though their full force was not visible until later. They preserved and reaffirmed those parts of Owen's archetypal theory that an evolutionist like Darwin would want and rejected those parts that such an evolutionist would find irrelevant or even unwanted. Using Owen's terminology, Huxley supported "special homology" between the bones of different organisms, the very thing any evolutionist would seize on as evidence of common ancestry. Huxley rejected as false "serial homology" between bones of the same body. Certainly he rejected homologies between *all* the bones of the body, which regarded all bones as modifications of the vertebrae, something at best inexplicable or explicable only ad hoc by a Darwinian evolutionist. (An evolutionist might accept—even expect—*some* serial homology.) Finally, Huxley showed that one could study comparative anatomy yet regard "general homology" as meaningless: that one can entirely ignore any relations of an organism to its supposed transcendental archetype. Again, this favors a naturalistic evolutionary theory, though nothing Huxley said prevented one from interpreting the archetype (or something akin to it) not as a Platonic Form but as an ancestor.

In a sense, the scientific importance of Huxley's paper lay less in its support of a position than in its criticism of Owen's archetypal theory. The vertebral theory of the skull was not just an appendage to Owen's position but was meant to be deducible right from its heart. Hence, Huxley was trying to destroy perhaps the most sophisticated and generally acceptable British analysis, before Darwin, of the ultimate nature of organisms. But Huxley was able to draw his conclusions only by embracing a methodological rule that, at least in the context of his lecture, went unjustified—that the key to homology lies in development rather than in adult forms. Presumably Huxley got his rule from reading German embryologists—he cited them by name (1857–59, 1:541) and he was one of the few Englishmen who could read German (in a letter he acknowledged von Baer's influence; L. Huxley 1900, 1:175). It is possible, though, that he was also inspired by the newly developing science of linguistics and its success in tracing links through origins and roots.

But, despite Huxley's approach, it seems that a supporter of Owen's archetypes might argue that adult forms cannot be demoted this way. Indeed, a Platonist might argue that adult forms are the truest reflection of reality; and Owen both claimed to be a Platonist and consistently argued that ultimately the adult form must take precedence. I am not denying that a nonevolutionist might want to work from embryological homologies rather than adult links—as von Baer did. But even if one can justify such a procedure because one's classifications thus contain fewer anomalies and point to more subtle links, Huxley himself offered no such justification. Moreover, a Platonist might feel that the advantages of working from embryological homology do not outweigh the danger of ignoring the deeper ontological reality revealed by adults. When we come to Owen's disagreement with Darwin over barnacle classification, I will suggest that Owen's moves stem from just such a Platonism. One suspects that Huxley was able to draw conclusions comforting to Owen's scientific opponents primarily because he chose the right methodological rules; but there was far more to Huxley's lecture than disagreement about scientific claims. There was a personal side that had important ramifications for our story.

The Owen-Huxley Quarrel

When young Huxley first returned to England, Owen—by far the most important of English biologists—took a friendly interest in his career, even helping him get extended paid leave from the navy (L. Huxley 1900,

1:65, 103). But almost from the first Huxley hounded the older man, trying to hold him up to scientific ridicule at every turn. Despite his initial friendliness to Huxley, Owen was not a pleasant man, and it seems clear that Huxley had not been back in London long before Owen felt threatened by his success and drive, to the point of trying to block publication of one of his papers (L. Huxley 1900, 1:105). But Huxley himself showed a thoroughly mean streak. His review of *Vestiges*, for example, was as much an attack on Owen as on Chambers. Pretending not to know who had replied to Lyell—though he knew very well it was Owen (1900, 1:101)—Huxley deliberately introduced Owen's reply, linking its progressionism to Chambers's, referring to its "ludicrous classification" of plants, and in refutation quoting Owen's own earlier position on the advanced nature of early forms (Huxley 1854, p. 7). Prudence, at least, might have dictated a little charity, for Owen was not the only one with scientific skeletons in his cupboard—in the early 1850s Huxley had only just emerged from a romance with Macleay's quinary system (Winsor 1976).

Huxley's baiting of Owen continued throughout the decade. In 1855 he labeled Owen's classification of the Radiata "one of the most thoroughly retrograde steps ever taken since zoology has been a science" (Huxley 1856, p. 484), and in 1858 he dismantled Owen's views on parthenogenesis and refuted them to the Royal Institution and the Linnaean Society (Huxley 1858). Finally came the Croonian lecture. By this time there had apparently been a personal falling-out, involving protocol at the School of Mines. Owen, a visiting lecturer, took to styling himself "professor," which Huxley took as a slight to his own status (L. Huxley 1900, 1:153). Huxley seized the opportunity of the Croonian lecture not merely to counter Owen's ideas, but to make a bitter personal attack on Owen—though in the grand style of William Whewell he did it without naming Owen. In particular, Huxley spoke of "later writers on the Theory of the Skull [who] have given a retrograde impulse to inquiry, and [who] have thrown obscurity and confusion upon that which twenty years ago had been made plain and clear" (Huxley 1857–59, 1:542). He spoke of the word "archetype" as being "fundamentally opposed to the spirit of modern science" (1:571). And he spoke of his opponents (Owen) as having attempted "to introduce the phraseology and mode of thought of an obsolete and scholastic realism into biology" (1:585). In short, Huxley treated Owen's ideas in such a way that Owen could not fail to take personal umbrage, as he did.

We see here the public sealing of a division between Owen and Huxley that led to a polarization of the two men and everything they stood for (Macleod 1965). At the political level, for example, in 1860 Owen deliberately supported a particular candidate for the Linacre professorship in anatomy at Oxford because Huxley was supporting another candidate.[3] At the scientific level Owen was determined to oppose Huxley, and if Huxley was to support Darwinism then so be it. Owen thus pushed himself into an opposition to Darwinism that may have gone beyond what he really felt. And since Huxley and his son outlived Owen and wrote the history, Owen has been portrayed as more opposed to Darwin than he apparently felt or acted.[4]

I am not suggesting that Huxley alone was Machiavellian enough to bait Owen into an unfelt opposition to Darwinism. Owen had religious and scientific reasons for rejecting or at least modifying anything that impinged on man or on final causes (as Darwin was to do). Besides this, Owen was touchy enough in his own right to oppose, without any prodding, a speculative theory like Darwin's that drew the light away from his own sober theorizing. But the Huxley-Owen clash, climaxed by Huxley's Croonian lecture, helped create a rather artificial opposition to Darwinism, which the Darwinians, particularly Huxley, portrayed as more implacable and formidable than it really was. After all, Owen had given some friendly response to *Vestiges;* when he reviewed the *Origin* in 1860 he by no means denied evolutionism (he even praised *Vestiges!*). And at least until 1853—when Huxley started attacking Owen and when Darwin and Huxley became close—there was warm feeling between Darwin and Owen (Darwin and Seward 1903, 1:75). Moreover, Owen's first public reaction to Darwin's ideas (probably the first of all such reactions), was guarded but by no means unsympathetic (Owen 1858*a*, pp. xci–xciii). But this was before Huxley took up Darwin's cause.

In short, the Owen-Huxley clash provided not so much preparation for Darwin's ideas as preparation for opposition to Darwin's ideas—opposition that, because it was based on hatred rather than sober scientific commitment, the Darwinians would relish.

Philosophy in the 1850s

The philosophy of science of the 1850s shows no radical innovations or major syntheses. But, as with science, we see the ground being prepared for Darwinian evolutionism, both by those who were to welcome Darwin's theory and by those who were less than enthusiastic. In particular, we see reiteration and extension of the claim that all the universe, including man, is subject to law—unbreakable natural regularity. Though one could hold to unbroken law and still deny organic evolutionism, we have also seen that appeal to law was a powerful factor in favor of evolution. And it almost goes without saying that insofar as man can be shown to be subject to law, the argument against evolutionism based on man's peculiarities is diminished. A major factor in spreading the acceptance of unbroken law, even as applied to man, was undoubtedly the popularity and widespread use of John Stuart Mill's *System of Logic,* first published in 1843. Mill was not a practicing scientist like Herschel and Whewell, and his work does not convey the same immediacy of philosophical problems arising from science—indeed, Mill admitted that he got much of his science from Whewell's *History of the Inductive Sciences* (Mill 1873, pp. 145–46). Unlike Herschel and Whewell, who used philosophy to understand science, Mill tended to use science to understand philosophy. Although Mill was a greater philosopher, Herschel and Whewell spoke directly to the scientific community in a way that Mill did not. This is probably why he plays far less of a direct role in our story (quite apart from his not publishing until 1843, when for Darwin, at least, the important scientific moves had already been made). However, Mill's importance in influencing general opinion cannot be denied. By the 1850s his *Logic* was in use as a textbook at Oxford. Students not only were being exposed to his famous methods of inquiry, but also were learning that everything is subject to the law of universal causation, that "it is a law that every event depends on some law" (Mill 1872, 1:376). Moreover, Mill was not loath to push his inquiries into the social sciences. Asking "Are the actions of human beings, like all other natural events, subject to invariable laws?" (1872, vol. 2, bk. 4), Mill was happy to reassure his readers that they were.

But Mill was not alone in wanting to explore man's subjection to the rule of law. Although Sedgwick dismissed with contempt Chambers's appeal to Quetelet's findings about statistical regularities among men, on the grounds that such regularities were not genuine laws, in a review of Quetelet's work Herschel (1850, p. 17) took a far more sympathetic approach. Such statistical regularities, he argued, disclose "tendencies, working through opportunities," and these in turn point to law-governed causes of which we are at present ignorant. Referring to Quetelet's findings about humans en masse—constancies in the ratios of male to female births, illegitimate to legitimate births, stillbirths to live births, marriages to the total population, first to second marriages, marriages of widows to bachelors, marriages of widowers to spinsters, and so on— Herschel drew the blunt conclusion that "taken in the mass, and in reference both to the physical and moral laws of his existence, the boasted freedom of man disappears" (1850, p. 22). Man, no less than anything else, is subject to law. (One might of course take a cue from Chambers's critics and argue that man's now being subject to law proves nothing about origins. This was no doubt Herschel's own position. But, as our story has already shown, arguing that things run by law tempts people to suppose they originate by law.)

Baden Powell (1855, p. 108) was another who joined in the hymn to the unbroken rule of law. As we shall see shortly, Powell was one man who was prepared to draw evolutionary consequences from his position even before the *Origin* appeared. But let us conclude this section by referring briefly to H. T. Buckle. Buckle was a historian, not a philosopher; his very popular *History of Civilization in England* started to appear in 1857. Partly for this reason, Buckle illustrates perfectly that by the eve of the *Origin* the average Briton had been thoroughly exposed to the idea of unbroken law that applies even to man. Buckle aimed to show a pattern of improvement and progress in society, and to this end he was eager to show that the development of human groups is a function not of mysterious entities, chance, or divine interventions, but of discoverable natural causes governed by fixed laws. These causes included climate, food, and the general aspect of nature, the last being something that excites the imagination and suggests "those innumerable superstitions which are the great obstacles to advancing knowledge" (Buckle 1890, 1:29). To prepare the way, Buckle opened his case by positing that man, no less than anything else, is subject to law, and once again we find reference to Quetelet's statistics. Everything about man is statistically regular, even down to the annual number of letters incorrectly addressed! (1890, 1:24).

No less than science, philosophy was helping to clear the way to a law-governed solution to the origins problem.

Religion in the 1850s

Granting that a revealed religion based on the Bible can block the acceptance of scientific ideas (including evolutionary ideas), we know that this opposition can be weakened as a by-product of "higher criticism," where the Bible is examined for internal consistency and is related to its sources and to our knowledge of history. Clearly, insofar as such criticism tends to question the literal truth of the Bible, it eases the way for scientific claims that clash with Scripture. Even by 1830 German influence on this score was being felt in England, though this influence was offensive to many, particularly to John Henry Newman. Nevertheless, not even Newman could stop the course of history, particularly the course of German scholarship. Germany continued its analysis of the Bible, and the greatest work produced undoubtedly was David Strauss's *Life of Jesus,* in which Strauss argued his "mythic theory," suggesting that the historical Jesus of Nazareth is given to us in the Bible as viewed through the lens of the messianic idea, constructed before Jesus' birth on the basis of Old Testament Prophecies.

There was no wholesale rush by British believers to embrace the most avant garde elements of German thought[5] (or in Germany either—Strauss lost his job). But biblical criticism did continue to seep across the Channel. George Eliot, for example, translated Strauss. Charles Hennell, a Coventry Unitarian minister, produced a rather unsophisticated, but perhaps all the more convincing, home-grown analysis of the Bible. Hennell's work seemed to run parallel to German thought rather than to be directly inspired by it. And, surely proving that God moves in a mysterious way, John Henry Newman's younger brother Francis wrote a series of works as he moved away from orthodox faith—works in which the tenets of Christianity were examined and found dreadfully wanting (Benn 1906, 2:17–36).

It was not until the 1860s that the whole question of biblical criticism blew wide open in Britain, and the row it engendered probably helped spread evolutionary ideas. But even before the *Origin* German-inspired criticism was bringing many to realize that only a drastically modified Christianity had any hope of keeping a place in the minds of intelligent, well-informed men. Thus Benjamin Jowett, the future master of Balliol, took a very liberal line toward certain Christian doctrines in his 1855 edition of Paul's epistles to the Thessalonians, Galatians, and Romans. His ideas were so offensive to some of his fellow Oxonians that they blocked his full payment as regius professor of Greek (Abbott and Campbell 1897). But by the 1850s men of Jowett's intellectual caliber and influence were feeling the need to promote a substantially transformed Christianity—one far less threatening to science. And we should remember that Francis Newman was one of the most enthusiastic recipients of the doctrines of *Vestiges* (Newman 1845*a, b*).

The more traditionally British way of handling conflict between science and revealed religion was not just to destroy religion, but to interpret both so that they no longer seem to conflict. Although he was far from being an evolutionist, Hugh Miller doggedly pursued this route to the end of his life. In his posthumous set of essays, *Testimony of the Rocks* (1856), he revived the "era" theory of Creation, arguing that the geological record supports Genesis as long as the Bible's "days" are understood as referring to long periods of time. Miller's importance to us is not the intrinsic worth of his ideas, for the thesis was not original with him (Millhauser 1954). But his interpretation led him to reemphasize to his huge audience a progressive nature for the fossil record like that Agassiz propounded, which, despite all he said, could only please an evolutionist. As Lyell worried in 1856, "it is strange how much the 6-day theories have led towards C. Darwin. Hugh Miller is more of an advocate for the evolution of Man out of preexisting inferior grades than I have been" (Wilson 1970, pp. 88–89). Evolutionism still had a long way to go, but even in revealed religious thought there was something to cheer the ardent proponent.

Let us turn next to natural theology. In his brilliant pioneering work, *Form and Function,* E. S. Russell (1916, p. 78) wrote as follows: "The contrast between the teleological attitude, with its insistence upon the priority of function to structure, and the morphological attitude, with its conviction of the priority of structure to function, is one of the most fundamental in biology." We have encountered these two attitudes as they are expressed in natural theology. We have the argument from design centering on adaptation, the "utilitarian" design argument. And in contrast we have the argument from order, harmony, symmetry, and some regularized pattern. In Britain in the 1830s the former argument was ascendant. This was the main argument of the Bridgewater Treatises and was crucial to antievolutionism. But the argument from morphology started to flood in during the 1840s (Bowler 1977). This was in part Continental influence, particularly various strands of transcendentalism, and in part it was home-grown. For example, Macleay's quinary system (taken up by Chambers) sees design in the harmonious patterns repeated throughout the animal world. Yet, whatever its sources, the morphological argument poses far less of a barrier to evolutionism of some kind. Of course, many strong supporters of this argument were violent antievolutionists—Agassiz, for example—and if one insists on seeing absolutely fixed patterns in the world one is going to have trouble explaining them through blind law. But the key evidence for the argument, Owen's

special homologies, almost cries out for evolutionary interpretation—
though this does not exclude a natural theological interpretation.

Paradoxically, in the 1850s Whewell was doing his part to spread the
morphological argument and variants because he had embroiled himself in
a dispute over extraterrestrial life ("plurality of worlds"). Desperate to
deny humanlike life elsewhere in the universe, since he thought it
threatened man's unique relationship with God (Todhunter 1876, 2:292),
Whewell found himself forced to emphasize all the things that do not
show direct purpose so that he might argue that other worlds need not
have inhabitants and thus exhibit direct purpose (Whewell 1853).
Whewell himself was still a staunch advocate of the utilitarian argument
and a violent opponent of evolutionism; nevertheless, he was now prom-
ulgating the morphological argument (and conceptual siblings like the
argument from law) in its own right, not just as an escape clause. Nor
were Whewell's opponents in the dispute hurting the cause of
evolutionism. Whewell's chief critic, David Brewster (1854), though also
an antievolutionist, was showing that the notion of direct purpose carried
to the extreme leads to ridiculous consequences (by 1850 standards). He
himself was arguing for humanlike inhabitants of the sun!

In short, natural religion, like revealed religion, was also preparing the
way for evolutionism. Owen himself, the major influence in making the
morphological argument plausible to Britons, was edging very close to
evolutionism. Furthermore, returning to the adaptation design argument,
remember that (perhaps in spite of himself) Owen was forcing it toward
evolution. With his picture of a gradually changing, adaptively
specializing fossil record, he was getting away from the idea of adaptation
as some static phenomena, fixed by God once and for all—or at least until
the next catastrophe. For him adaptation almost necessarily was becoming
dynamic, changing to stay on top of circumstances. This, as we shall see,
was a vital move.

It is most instructive to look once again at Huxley. Although he later
called himself an "agnostic"—indeed, he invented the term (L. Huxley
1900, 1:343–44)—Huxley strikes one as being, by temperament, intensely religious. He really cared about ultimate questions. In another age
he would have been pope, or at least archbishop of Canterbury. How could
such a man, in the 1850s, synthesize a religious position that would leave
unfettered his overwhelming passion for science? Concerning revealed religion we find that he drew from German thought through Carlyle: "*Sartor
Resartus* led me to know that a deep sense of religion was compatible

with the entire absence of theology" (L. Huxley 1900, 1:237). Concerning natural religion, obviously linked to Carlyle's natural supernaturalism, Huxley showed that his sympathies lay less with God as direct designer than with God as creator of law, symmetry, harmony, and beauty (Huxley 1854–58, p. 311). This reflected his scientific interests—as an invertebrate taxonomist his task had been to discern homologies beneath the confusing special adaptations—as well as the influence of Macleay's system. But, whatever the sources, we see how current religious thought had enabled at least one scientist to ready himself for an entirely non-theological attack on the organic origins question—to such an extent that it somewhat backfired.

We come next to the evolutionists of the 1850s—first men who supported the idea of evolution, but not the mechanism of natural selection, then the man who finally grasped Darwin's mechanism and used it in an evolutionary sense.

The Evolutionists

On 1 June 1850, Alfred Tennyson's *In Memoriam* was published. So great and so instant was the poem's success that on 5 November Tennyson was made poet laureate (Ross 1973, pp. 114–15). That most Victorian of Victorians, the queen herself, said all that was needed when, after the death of the prince consort, she told the poet: "Next to the Bible 'In Memoriam' is my comfort" (Ross 1973, p. 93). But perhaps Sedgwick had been right. Before the ink was dry on his response to *Vestiges,* the gullible public was avidly reading a poem that not only was directly inspired by Chambers's evolutionism, but followed the doctrine into a blasphemous perversion of Christianity.[6]

In Memoriam, as is well known, was Tennyson's tribute to the memory of his friend Arthur Hallam, who died in 1833 at the age of twenty-two. As is also well known, Tennyson was fascinated by science (Whewell had been his tutor), and a major theme of the poem involves the poet's reactions to various scientific works, spread over several years, since the poem was begun in 1833 but not finished until 1849. About the middle of the poem the poet's sense of purpose, his hopes for himself and for Hallam, sinks to the lowest depth of despair, faced with what he took to be the meaningless lack of direction of Lyellian geology.

> Are God and Nature then at strife,
>   That Nature lends such evil dreams?

> So careful of the type she seems,
> So careless of the single life;
>
> . . . .
>
> 'So careful of the type?' but no.
>   From scarped cliff and quarried stone
>   She cries, 'A thousand types are gone:
> I care for nothing, all shall go.'[7]

Given Nature, "red in tooth and claw" (Ross 1973, p. 36, from sec. 56), going nowhere, all seems pointless.

But Tennyson, toward the end of his poem, regains his faith through an evolutionism like Chambers's with its theme of progress—a progress that might lead to a higher being than man, of which Hallam was a premature example.

> A soul shall draw from out the vast
> And strike his being into bounds,
>
> And moved thro' life of lower phase,
>   Result in man, be born and think,
>   And act and love, a closer link
> Betwixt us and the crowning race
>
> . . . .
>
> Whereof the man, that with me trod
>   This planet, was a noble type
>   Appearing ere the times were ripe,
> That friend of mine who lives in God.[8]

In an important way this poem tells us more than all the sober passages of science and philosophy and religion. It was sincerely loved by the Victorians, was quoted endlessly, and provided a very genuine source of comfort, from the widowed queen on down. Yet here is the paradox. Its hope comes from evolutionism and its prospect of progress to a race of supermen like Hallam. If this is not a travesty of Christianity, I do not know what is. Part of the explanation of this paradox probably lies in the fact that poetry is by its very nature open to interpretation. Tennyson could say things hinting at progress, and the reader could take this essential message, dressing it in orthodox or unorthodox garb to suit his own prejudices. But there is more than this. One suspects that deep down many Victorians did not much care whether organic evolutionism was true. They did not indeed care much about the truth of the doctrinal niceties of conventional Christianity. What they did care about was the frightening rapidity of change in their lifetimes and the essential lack of security of their society (Houghton 1957)—a society supported and surrounded by the vast underprivileged, frequently starving masses. When Tennyson held out the hand of hope and of progress to a better state, they grasped it thankfully, not bothering about details.

If this thesis is true, we should expect to see echoes of it after the publication of Darwin's *Origin.* I suggest that this is precisely what we do find, particularly surrounding the evolutionism of Herbert Spencer. But this lies in the future. Here, noting that Tennyson does not really say much about a mechanism for evolution (other than hinting at some sort of recapitulation), let us turn to two other evolutionists of the 1850s: first Spencer himself, then the irrepressible Baden Powell.

Herbert Spencer (1820–1903) was born in Derby, of a father with Quaker sympathies (Spencer 1904; Duncan 1908; Greene 1962; Burrow 1966; Peel 1971). His education was almost the complete reverse of that undergone by poor John Stuart Mill, who was forced to start learning Greek at the age of three. Young Spencer was allowed to dabble in anything he pleased, and dabble is just what he did. So although he picked up a little science and mathematics, his education was more remarkable for what he did not learn than for what he did. He started adult life as a railway engineer, and through this came a lifetime's devotion to the cause of evolutionism. Attracted to the fossils exposed by the railway cuttings, in 1840 Spencer read Lyell's *Principles of Geology,* with precisely the opposite effect from what the author intended, for he was promptly converted to a Lamarckian evolutionism (Spencer 1904, 1:176). It was not until 1850 that Spencer began publishing on the subject, by which time he had found a post as subeditor of the *Economist.* Once he started, however, a plethora of evolutionary writings streamed from his pen, and he filled volume after volume as only nineteenth-century writers could.

Selecting drastically, as one must when dealing with Spencer, we have first a little article from 1852 in the *Leader* (Spencer 1852a), in which Spencer faced squarely what he took to be the essential dichotomy, "law *versus* miracle." Either we must believe that species were specially created or we must be transmutationists (he made no mention of the third option that people like Herschel and Lyell tried to maintain), and with the question put this way Spencer had no qualms about opting for transmutationism. Then in 1855 Spencer published his *Principles of Psychology,* in which he applied his evolutionism not just to man as physical being but to man as mental being. Finally, in 1857, in "Progress: Its Law and

Cause,"[9] Spencer started to put everything together and argue for an overall evolutionary world view: in the inorganic, the organic, and the purely human worlds we see the same pattern manifested. In particular, we see a progress wherein relatively homogeneous beginnings are transformed into relatively heterogeneous ends. Causally, Spencer explained his phenomenal law thus: "Every active force produces more than one change—every cause produces more than one effect" (Spencer 1857, 1:32). His link between cause and effect was obvious: one starts with one or a few causes, these multiply, and thus from homogeneity develops heterogeneity. This idea of going from the general to the specialized bears strong resemblance to von Baer's theories, and though he felt that he himself had earlier grasped something akin to the idea, Spencer was quite open about his debt. In 1851 he had read in his textbook an account by Carpenter of von Baer's embryological findings (Spencer 1904). Although at one point Spencer was careful to note that he was trying not to link such progress directly to the teleological end of human happiness (1857, 1:2), he was nevertheless prepared to speak of it as a "beneficent necessity" (1857, 1:58). And in illustrating the development of human societies, Spencer started with homogeneous savage groups and ended with something that, for all its heterogeneity, sounded very much like nineteenth-century Britain. He went so far as to say explicitly that "it is more especially in virtue of having carried [the] subdivision of function to a greater extent and completeness, that the English language is superior to all others" (1857. 1:17). Despite his protests, Spencerian progress took an old-fashioned, jingoistic direction; especially since "the civilized European departs more widely from the vertebrate archetype than does the savage" (1857, 1:50).

Finally, Spencer published a fascinating essay somewhat earlier, in 1852. In this essay, "A Theory of Population, Deduced from the General Law of Animal Fertility," Spencer was concerned with Malthus's claims about population growth, particularly his gloomy conclusions about near-inevitable human struggles for existence. To such an optimist as Spencer (1852b), seeing everywhere an "inherent tendency of things towards good" and "at work an essential beneficence," a doctrine like this was a challenge, not a refutation, especially since politically Spencer was a champion of laissez-faire economics, with its central theme that human happiness is maximized by man and that the state should not tamper with the natural laws of political economy.

Rising to the Malthusian challenge, Spencer averred that the ability to

maintain individual life is inversely proportional to the ability to reproduce (1852*b*, p. 498). Although we learn that this is an a priori claim,
Spencer thought it politic to offer lengthy empirical justification. Apparently it all revolves around the fact that what one puts into reproduction
takes away from the individual. For example, undue production of sperm
cells in man leads first to headaches: "this is followed by stupidity; should
the disorder continue, imbecility supervenes, ending occasionally in insanity" (1852*b*, p. 493). Though this is much what one might expect
from a Victorian bachelor, later we shall see that this line of thought had
an influence on Darwin, for Spencer's claim stemmed directly from his
belief (shared by Darwin) that the germ cells are produced from the whole
body, especially the brain, not just from the sex cells. At this point
Spencer was content to rest his case on the similarity of chemical content
between the sperm cells and the brain.

Next, Spencer pointed to the apparent link between physiological and
social evolution in mankind. Members of advanced civilizations, particularly Englishmen, have much larger brains than savages. How has this
come about? Undoubtedly, through the pressure of population. England
offers fewer resources without effort than do the habitats of savages; hence
past generations of Englishmen striving to combat population pressure
have applied their intelligence and moral sense more strenuously than
others; this has brought on enlargement of intelligence and moral sense;
and thus, through a Lamarckian inheritance of acquired characteristics, the
English have progressed toward a higher form.

For good measure, Spencer threw in an anticipation of natural selection.
"All mankind in turn subject themselves more or less to the discipline
described; . . . but . . . only those who *do* advance under it eventually survive" (1852*b,* p. 499). Hence, "as those prematurely carried off must, in
the average of cases, be those in whom the power of self-preservation is
least, it unavoidably follows, that those left behind to continue the race
are those in whom the power of self-preservation is the greatest—are the
select of their generation" (1852*b*, p. 500). And, showing he was Victorian in more things than sex, Spencer cited the dreadful example of the
Irish, who had failed to progress.

His conclusion was now easy to draw. As we progress up the evolutionary scale our fertility drops, and finally we achieve an equilibrium, caused
by but escaping from Malthusianism. Having brought culture and intellect to their highest points, having brought "all processes for the satisfaction of human wants to the greatest perfection," then "the pressure of
population . . . must gradually bring itself to an end" (1852*b*, p. 501). In

short, with the coming of something much like *Homo britannicus,* Spencer saw evolution reach a glorious culmination.

Let us conclude with Baden Powell. By the end of the 1840s Baden Powell was an evolutionist, and in 1855 he proclaimed his position publicly. Perhaps Baden Powell's position as an Anglican divine and Oxford professor carried more weight than what he wrote, for in fact there was little in his work that had not been said before. Acknowledging that *Vestiges* was riddled with problems, Powell openly supported its central transmutationary message. Moreover, he made little secret of why he supported such a message. God, we must assume, works only through unbroken law, and Powell gave as strong an endorsement to God as industrialist as can be found anywhere. "Precisely in proportion as a fabric manufactured by machinery affords a higher proof of intellect than one produced by hand; so a world evolved by a long train of orderly disposed physical causes is a higher proof of Supreme intelligence than one in whose structure we can trace no indications of such progressive action" (Powell 1855, p. 272).

It is clear that Powell went beyond Lyell, since he became a transmutationist, but it is both interesting and highly significant that he did not think he was repudiating Lyell's approach to the organic world. Far from it—Baden Powell quoted Lyell's *Principles* to support his attitude toward the organic world. As Lyell had shown, in the inorganic world we always have the same laws governing the same kinds of causes, in a steady way. We must therefore suppose that the same holds for the organic world. Hence, new species come "not by the agency of imaginary convulsive paroxysms," but by transmutation (Powell 1855, p. 362). We repeatedly see this partial acceptance of Lyell, as we saw it with Spencer. Lyell's geological strategy is adopted entirely; his antievolutionary arguments are countered or ignored; and it is claimed that a Lyellian must be an evolutionist!

We see therefore that through the 1850s organic evolutionism refused to vanish. This was bound to lead to something eventually, and, as we now learn, indeed it did!

Alfred Russel Wallace

Alfred Russel Wallace (1823–1913) was born in Monmouthshire (Wallace 1905; Marchant 1916; George 1964; McKinney 1972; Smith 1972). Coming from a humble background, he received a rather weak education. Wallace left school at fourteen and spent most of the next ten years as a surveyor. During this time he developed an interest in natural history and began to collect specimens, first of plants, then of beetles. In 1848, with his friend and fellow naturalist Henry Walter Bates, he set off to the Amazon on a collecting trip. He spent four years there, but when he returned to England his ship burned and nearly all of his collection was destroyed. Undaunted, in 1854 Wallace set off again, this time to the Malay Archipelago (modern Indonesia), where he spent some eight years. By the time he returned he had become a well-known and respected naturalist, particularly because of his evolutionary discoveries. But he was never very successful at gaining a post that paid a salary, and since he was constantly in need of money he was forced to write prolifically on a variety of subjects to supplement the income from the sale of his collections and (later) a small pension. Always rather solitary, Wallace became enthusiastic about a variety of topics that in both the nineteenth century and the twentieth (though perhaps for different reasons) labeled him as something of a crank—spiritualism, phrenology, opposition to vaccination, and even socialism! One of the fascinations of Wallace is the almost total contrast between him and Darwin. Wallace was always on the outside looking in, whereas Darwin was in every respect part of the establishment.

By 1845 Wallace was an evolutionist or at least was very close to being one. He had read *Vestiges* and found Chambers's arguments highly plausible. "Although I saw that it really offered no explanation of the process of change of species, yet in the view that change was effected, not through any unimaginable process, but through the known laws of reproduction commended itself to me as perfectly satisfactory, and as affording the first step towards a more complete and explanatory theory" (Wallace 1898, pp. 137–38). There is not much evidence on just why Wallace was so influenced by *Vestiges,* though we do know that in these early years he had very little religious faith, based on either revealed or natural religion, so that was no barrier. But an important clue is that Wallace, particularly in his youth, was *not* a member of the professional scientific community. He had not grown up with the argument from design and years of antievolutionary diatribes, nor did he have good friends like Sedgwick declaiming into his ear the inequities of *Vestiges.* He was a rather ignorant young man with an interest in natural history—one of the many members of the general public who were enthralled by Chambers's sweeping and engaging hypotheses. What distinguished Wallace from the rest is that he did something about his enthusiasm, for it is clear that he had the organic origins problem in mind when he went to Brazil. It is also clear that in

South America he was mightily impressed by the facts of geographical distribution—for example, the way species on different sides of natural barriers like rivers are often very similar, naturally suggesting to an evolutionist that a parent species became differentiated on the two sides of the barrier.

By about 1854, therefore, we find Wallace fascinated by the organic origins question—a convinced evolutionist seeking a mechanism and feeling that geographical distribution is a key to the whole question. As he planned a book on the problem, it was natural that Wallace should turn to Lyell's *Principles* for guidance. Lyell had written one of the definitive treatments of the organic origins and species questions, and he had much to say about geographical distribution and methods of dispersal. Like others, Wallace sympathized entirely with Lyell's geological strategy, then turned round and claimed that to be a consistent Lyellian one must become some kind of transmutationist in the organic world.

> While the in-organic world has been strictly shown to be the result of a series of changes from the earliest periods produced by causes still acting, it would be most unphilosophical to conclude without the strongest evidence that the organic world so intimately connected with it had been subject to other laws which have now ceased to act, and that the extinctions and productions of species and genera had at some late period suddenly ceased. The change is so perfectly gradual from the latest Geological to the modern epoch that we cannot help believing the present condition of the earth and its inhabitants to be the natural result of its immediately preceding state modified by causes which have always been and still continue in action. [Quoted by McKinney 1972, pp. 32–33, from Wallace's notebook]

In his own mind, Wallace debated at some length with Lyell's antievolutionism. At the same time, however, Lyellian geology convinced Wallace of the correctness of his own approach, emphasized the importance of geographical distribution, and taught him something about the geological succession of species. This great influence, combined with his own experience and general reading (particularly of paleontological information from the Swiss scientist François Jules Pictet), was enough to spark Wallace's first important independent evolutionary move. In 1855 Wallace published a paper entitled "On the Law Which Has Regulated the Introduction of New Species," this law being: "Every species has come into existence coincident both in space and time with a pre-existing closely allied species" (Wallace 1855, p. 82).

This law ostensibly was intended to counter the "polarity theory" of the British scientist Edward Forbes,[10] a theory claiming that organic creation occurred more frequently at some periods than at others. Wallace turned to questions like geographical distribution, showing that the law was confirmed by the facts and in turn threw light on these facts. Thus, for example, he made reference to the Galápagos archipelago, a group of islands off South America, where Darwin had discovered similar but different endemic species of finches and tortoises on various islands (mentioned in his *Journal of Researches*). Wallace gave an entirely Lyellian account of how the islands would initially have been populated by dispersion from the mainland. Then, "we can account for the separate islands having each their peculiar species, either on the supposition that the same original emigration peopled the whole of the islands with the same species from which differently modified prototypes were created, or that the islands were successively peopled from each other, but that new species have been created in each on the plan of the pre-existing ones" (Wallace 1855, p. 74). Particularly pertinent, Wallace thought, was the observation that geologically ancient islands seem to show more diversity from an original form than do more recent islands.

Although at the end of his paper, employing the familiar astronomical metaphor, Wallace likened his law to gravitation explaining facts akin to Kepler's laws, it is clear that Wallace had produced, in the philosopher's terminology, a phenomenal law. Causes had still to be sought, and Wallace's paper was sufficiently ambiguous that one could almost read it in a creationist sense. Three years later, however, he hit on his causal mechanism for evolutionary change. While sick with fever early in 1858, Wallace thought about the checks keeping savage populations from expanding drastically. This sparked recollection of a book he had read some thirteen years earlier, Malthus's *Essay on the Principle of Population* (Wallace 1905, 1:361), the same Malthus and the same ideas about the struggle for existence that had engaged Spencer. Knowing full well from Lyell's work, if from nowhere else, how prevalent struggle is in nature, Wallace connected Malthus's ideas with the animal world and moved right on to the idea of natural selection—the idea that not all organisms will survive to reproduce, that survival depends on an organism's peculiar characteristics, and that there will therefore be a "natural selection" for such characteristics, culminating in time in full-blown evolutionary change (Wallace 1858; he did not use the term "natural selection").

As soon as his fever subsided, Wallace quickly wrote up his inspired idea in a short paper, intending to send it off to England for publication. Because Wallace's ideas in many respects mirrored Darwin's so exactly, and because Darwin dealt with the matter so much more thoroughly, we can leave detailed exposition of these ideas to the next chapter. But one move Wallace made was *not* made by Darwin. In response to Lyellian objections to evolutionary theories based on analogies with the domestic world, Wallace contended strongly that one can draw no true analogy between wild and domestic organisms. He argued that domestic forms have been selected for abnormalities, and that were they released in the wild one would expect them to be eliminated or to rapidly revert to type—as one finds does happen. Hence, Wallace explained, the instability of domestic change does not, as Lyellians think, prove that all selective change is unstable. It points to the intense selective pressures of the wild, which can lead to thoroughly stable changes. That the domestic world does not act like the wild world supports rather than undermines the selective mechanism. Wallace's failure to exploit the artificial selection analogy may perhaps be linked with an insensitivity in his paper to the true importance of selection within groups, for he placed most emphasis on struggle and selection between varieties (Bowler 1976c). But Wallace certainly does discuss selection between individuals, so his not using the analogy in no way destroys his legitimate claim to being a codiscoverer of natural selection as a mechanism for evolution.

Having written his paper, Wallace had to decide to whom it should be sent. After his earlier paper was published in 1855, Wallace had been disappointed that more notice was not taken of it. As we shall see, notice had been taken. But in any case, Wallace had begun to correspond with other men interested in the organic origins question. One of these was Charles Darwin, who was rumored to hold heretical views on the problem. It was therefore not surprising that Wallace sent his paper on selection to Darwin, suggesting that if it seemed worthy, he send it on for publication.

So Darwin comes back into our story. Many thinkers through the 1850s had cleared the way, and the next moves were up to him. Why he made these moves, and what they were, we shall learn in the next chapter.

7 Charles Darwin and the "Origin of Species"

So far we have met the public Darwin—the rich, well-connected, ambitious, personable young man who was showing himself one of Britain's brightest new students of geology. We must now turn to the private Darwin: the man who, unbeknown to the scientific community, was successfully cracking the conundrum Herschel had so aptly named "the mystery of mysteries"—the question of organic origins.

Charles Darwin, born in 1809, attended university first at Edinburgh and then (from 1828 to 1831) at Cambridge and spent the years from 1831 to 1836 circumnavigating the globe as naturalist on the H.M.S. *Beagle*. Very soon after returning to Britain, probably in the early spring of 1837, Darwin became an evolutionist, and in the fall of 1838 he hit on the mechanism of natural selection brought on by the struggle for existence. In 1842 he wrote a thirty-five-page sketch of his theory (hereafter referred to as the *Sketch*), and in 1844 this was expanded to a 230-page essay (the *Essay*) (both are included in Darwin and Wallace 1958). None of this was made public, though he showed the *Essay* to Hooker. Darwin spent practically the whole of the next ten years concealing his evolutionism, as he labored to produce tomes on barnacle systematics. Only when this was done did he return full time to his evolutionary work, and in the mid-1850s he began a massive work on natural selection and evolution. He was interrupted by the arrival of Wallace's essay, and after this and short evolutionary extracts from Darwin's earlier writings had been published by the Linnaean Society (Darwin and Wallace 1958), Darwin rapidly wrote an "abstract" of his ideas. This abstract, *On the Origin of Species by Means of Natural Selection; or, The Preservation of Favoured Races in the Struggle for Life,* was published in November 1859. The private Darwin had joined the public Darwin.

First to Evolution

Before Charles Darwin was born, the name of Darwin was already famous, or notorious, for the idea of organic evolution. At the end of the eighteenth century, Charles's grandfather, Erasmus Darwin, had extolled in verse and prose the virtues and beauties of such a developmental world picture (Greene 1959; J. Harrison 1971). Although he predated Lamarck in his evolutionism, many of Erasmus Darwin's speculations anticipate those of the French biologist. In his best-known work, *Zoönomia* (1794–96), Erasmus Darwin postulated something very much like the so-called Lamarckian inheritance of acquired characteristics. And, like Lamarck, Erasmus Darwin visualized a rather vague principle of progress pulling along the whole evolutionary process. One difference, however, is that whereas Lamarck postulated a continuous process of spontaneous generation, Darwin suggested that "all warm-blooded animals have arisen from one living filament" (Darwin 1794–96).

As a teen-ager, Charles Darwin read *Zoönomia,* (Darwin 1969, p. 49). Moreover, at Edinburgh he became friendly with just about the only Lamarckian in Britain, Robert Grant, who later became a professor in London. Darwin recorded that Grant expounded Lamarckian evolutionism to him with enthusiasm.[1] So, coupled with his own admission that he greatly admired his grandfather's *Zoönomia,* young Darwin probably had a more sympathetic introduction to organic evolutionism than any other person in Britain at the time (on the other hand, he attended lectures on zoology and geology by Robert Jameson, Cuvier's British editor). However, there was certainly no immediate conversion. Darwin himself, no doubt with truth, later wrote that "it is probable that the hearing rather early in life such [evolutionary] views maintained and praised may have favoured my upholding them under a different form in my *Origin of Species*" (Darwin 1969, p. 49). But when he went to Cambridge, to become a clergyman, he did "not then in the least doubt the strict and literal truth of every word in the Bible" (1969, p. 57).

At the Cambridge of Whewell and Sedgwick, Darwin stood in no danger of forgetting about the organic origins question, and one can feel certain his elders made sure the grandson of "the celebrated Dr. Darwin" knew all the good reasons why his grandfather's wild hypotheses were just that. On the other hand, Darwin was plunged into a community that was feeling the tension between science and religion—a community in which leading members like Sedgwick and Whewell felt unable to take portions of the Old Testament absolutely literally. And toward the end of his undergraduate studies Darwin felt the stimulating effect of the empiricist philosophy of John Herschel, who took an even more liberal line on biblical truth. Perhaps partly because of Herschel's influence, on the *Beagle* voyage Darwin soon started to become a Lyellian geologist. Then in 1832 he received the second volume of Lyell's *Principles* (Darwin 1969, p. 77 n), with its detailed discussions of Lamarckian evolutionism, organic struggles for existence, and the ways organisms might be distributed naturally around the world, its veiled hints about the naturalness of species creations, and its overwhelming evidence of organic extinction. As a Lyellian geologist Darwin was forced to think hard about the organic world and to think about it differently—for all that Lyell himself sometimes wished otherwise—from his catastrophist teachers at Cambridge, or from Lamarck (Limoges 1970; Egerton 1968).

The Lyellian sees inevitable gradual change within a context of overall stability. Struggle and the failure to adapt result in the need to migrate or, eventually in any case, the death of the species. The whole question of adaptation, therefore, becomes crucial and *dynamic* (or relativized) as it is not for Lamarck or for a catastrophist. For Lamarck, as we saw, it is self-evident that organisms adapt. Moreover, their tendency to reproduce at great speed guarantees their safety as a species. If things do get too threatening, there is always the escape of progression up the chain of being. For the catastrophist also, adaptation is a ubiquitous fact of organic life, and although there may be some extinction at all times, organisms generally are safe in their adaptation. Major changes and extinctions come only through catastrophes. By the end of the 1830s, Whewell was really trying to accept constant extinction. But, when it came to major extinction, he still posited something like a worldwide freeze: a phenomenon close to a catastrophe (Whewell 1837, 3:591).

Certainly the catastrophists recognized Malthusian population pressures and saw that throughout the organic world organisms (or species) compete for resources. But, for all that, it was a gelded struggle: in the tradition of Paley himself, the struggle was seen as evidence of God's goodness, that animals not die lingering deaths (Buckland 1836). It was less struggle for existence than "balance of nature." For a Lyellian, however, despite the paradox that Lyell's own motives were as religious as any, the threat of extinction was far more constant. Adaptation was not a suit one put on to guarantee safety until the next catastrophe. The battle was constant—stay on top, stay adapted, leave, or perish. The picture is a dynamic one fraught with force and tension. In this respect it is the picture toward which Owen's paleontology headed, although that happened twenty years later and in a context that was in other respects anti-Lyellian.

Darwin was therefore looking at the organic world in the Lyellian way. In a field notebook of February 1835 we find him saying to himself: "With respect then to the *death* of species of Terrestrial mammalia in the S. Part of South America I am strongly inclined to reject the action of any sudden debacle. —Indeed the very numbers of the remains render it to me more probable that they are owing to a succession of deaths after the ordinary course of nature" (Darwin MSS, 42; quoted in Herbert 1974, p. 236 n). And the very next sentence shows Lyell is master here. "As Mr. Lyell supposes species may perish as well as individuals; to the arguments he adduces I hope the Caria of B. Blanca will be one more small instance of at least a relation to certain genera with certain districts of the earth. This co-relation to my mind renders the gradual birth and death of species more probable." (Herbert 1974, p. 236 n. By "gradual birth" in this context, Darwin is not hinting at evolutionism. He is referring to Lyell's constant creation of new species). In fact, Darwin was so far Lyellian in his thinking about the organic world that we find him making the ultra-Lyellian appeal to God's design in keeping the organic world in a steady state. If species are not being successively created to compensate for continual successive deaths of species, we must allow that on earth "the number of its inhabitants has varied exceedingly at different periods. —A supposition in contradiction to the fitness which the Author of Nature has now established" (Herbert 1974, p. 233 n).

But, as a Lyellian, in looking at the organic world Darwin was presented with the nagging question any Lyellian faced. If not by evolution, if not by miraculous special creations, then how on earth did one get a constant natural supply of new species? Moreover, Darwin had his own Lyellian ax to grind, which concentrated his attention even more on organisms, their predecessors, their distributions, and their origins. Darwin wanted to find some universal law of elevation and subsidence to explain the Lyellian steady-state picture. But before he could inquire more deeply into causes, he needed phenomenal data about areas of elevation and subsidence. This led him directly to questions of inorganic geographical distribution. Since Lyell dealt with questions of geographical distribution by twining almost inextricably the inorganic and organic worlds, Darwin was practically pitchforked into the organic origins question. To a Lyellian, the whole question of the age of certain areas—the times they had been at certain heights and so on—was intimately connected with the nature of the flora and fauna of these areas and with the fossil record. (Think back to Darwin's evidence for his claim that the Plain of Coquimbo had recently been elevated. As figure 6 shows clearly, Darwin, like Lyell, was fusing the organic with the inorganic.)

With his attention directed this way, Darwin apparently found three phenomena particularly striking and felt they were not adequately accounted for within the Lyellian world picture (Darwin 1969, p. 118). First, the Pampean formation contains fossils of armadillo species, now extinct, that resemble existing armadillos; second, very similar organisms replace each other going southward over the South American continent; and, third and most important, as Wallace was to recognize some twenty years later, organisms are distributed in a peculiar way on the Galápagos archipelago (see fig. 20). On different islands one gets different species of finches and tortoises, all very similar to each other and to South American forms. As a Lyellian, Darwin knew these species had to be naturally caused. From Lyell's *Principles* he could find no reason why they were distributed as they were. The problem was compounded because the Galápagos archipelago is of fairly recent geological origin and thus the organisms of the various islands seem more recent than the mainland forms.

In a sense, as Whewell pointed out in his *History,* (1837, 3:589) not many options are open to the Lyellian faced with the problem of organic origins. Moreover, if he is honest with himself, one option recommends itself above all others. The Lyellian must believe in some kind of lawbound organic creation from nothing; or from something, organic or inorganic, widely different from that which is created; or he must believe in natural creation from similar forms. By any reasonable principle of parsimony the last choice is far preferable to the first two. But if this

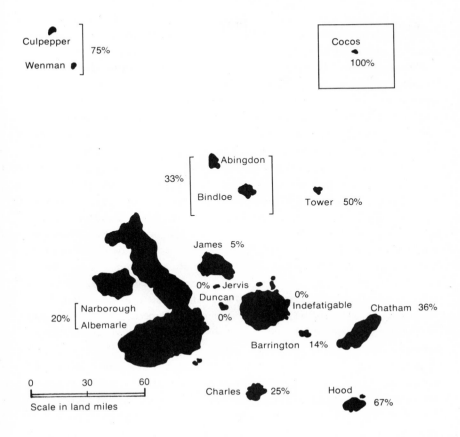

Figure 20. Percentage of endemic forms of finch on the various Galápagos islands. Darwin, of course, did not have these percentages. From David Lack's *Darwin's Finches* (1947). By permission.

option is taken, though one may be loath to admit it, one is accepting a saltatory evolutionary theory. Although Lyell himself avoided posing this choice, Darwin was more forthright. When he returned to England he set about writing up his diary into a travel book, published as the *Journal of Researches*. This set him thinking again about the things he had seen, particularly in the Galápagos islands. And so in March or April of 1837 Darwin asked himself the question Lyell had refused to pose and, answering it in the obvious manner, moved across the divide and became an evolutionist. The Galápagos finches had to be explained as the natural law-bound product of one parent stock (Herbert 1974). Significantly, it

was not until after the voyage, early in 1837, that the ornithologist John
Gould convinced Darwin that the finches formed real species, not just
varieties (Grinnell 1974, p. 262).

Breaking with his mentor on this question of evolution did not mean
Darwin ceased to be a Lyellian. Though Darwin's evolutionism went
against Lyell's steady-state system, Darwin was pushed toward organic
evolutionism precisely because he was so committed to the system in the
inorganic world. And in another respect Darwin perhaps became more
Lyellian than ever before. Lyell had argued that gradual change would be
impossible because the transitional forms would be at too great a disadvantage in the struggle for existence: given his dynamic concept of adaptation, Lyell felt that a changing species was bound to succumb. Darwin's
first written evolutionary ideas mirrored this concern and led him to
speculate about one-step, saltatory species transformations. About this
time, he wrote in his notebook: "not *gradual* change or degeneration from
circumstances: if one species does change into another it must be per
saltum—or species may perish."[2] Apparently Darwin first had in mind
that species, like organisms, might have definite life spans and then would
(or at least could) change into new species: "Tempted to believe animals
created for definite time: —not extinguished by change of circumstances"
(Herbert 1974, p. 247 n).

It is plain that Darwin was thinking frantically and that nothing was
very stable in his mind. In the summer of 1837, about three months after
he became an evolutionist, he decided he could think more systematically
if he kept notebooks devoted to the organic origins question. He kept
these "species notebooks" for two years, right through the time when he
discovered natural selection, and they are invaluable guides to tracing the
minutiae of his thought. Let us examine these notebooks and see how,
about eighteen months after his first rudimentary speculations, Darwin
came upon the mechanism for which he is so famous.[3]

. . . and Then to Natural Selection

From the first notebook it seems clear that Darwin's earliest speculations
on the nature and causes of evolution did not last long. By midsummer
1837 the Lyellian worries about the possibility of gradually changing
species had diminished. The Galápagos experience not only turned Darwin toward evolutionism, it influenced his thinking about the causes of
evolution (Grinnell 1974). In particular, Darwin had in mind the model of

a group of organisms, *isolated* from all others, evolving into a new species. "Let a pair be introduced and increase slowly, from many enemies, so as often to intermarry—who will dare say what result. According to this view animals on separate islands, ought to become different if kept long enough apart, with slightly differ[ent] circumstances. —Now Galapagos tortoises, mocking birds, Falkland fox, Chiloe fox. —English and Irish Hare—" (de Beer et al. 1960–67, B, p. 7). In such a model, as this passage shows Darwin recognized, the threat from external competition is eliminated; thus Lyellian fears for the adaptedness of changing species vanish. Darwin therefore felt free to posit gradual organic evolution based on small changes rather than sudden leaps. And he did switch to such minute changes, though we shall see remnants of saltatory changes in his thought for some time.

Having blocked off one Lyellian worry, Darwin continued to synthesize his position from Lyellian elements. Given our knowledge of Darwin as a geologist, this is what we would expect. First, Darwin thought that the constantly changing Lyellian world would create new isolated areas—the rising of the Galápagos archipelago, for example. But, second, he realized that isolation in itself would not lead to change. To affect organisms, one needs constant inorganic change in the isolated places. Once again Lyellian geology comes to the rescue. "We *know* world subject to cycle of change, temperature and all circumstances, which influence living beings" (B, pp. 2–3). Third, the Lyellian world view requires a whole new perspective on adaptation. It is no longer static, but is dynamic, with death the constant penalty for lack of success. "With respect to extinction we can easily see the variety of ostrich Petise may not be well adapted, and thus perish out, or on other hand like Orpheus being favourable, many might be produced. This requires principle that the permanent varieties, produced by confined breeding and changing circumstances are continued and produce according to the adaptation of such circumstances, and therefore that death of species is a consequence (contrary to what would appear from America) of non-adaptation of circumstances" (B, pp. 37–39). Darwin drew a diagram (fig. 21) showing just how extinction would lead to the distributions of species we find today, with big gaps between some genera but not between others.

Notice that even at this early point Darwin was thinking of irregularly branching, not linear, evolution. Indeed, he speculated that in the animal world we might get three main branches (for land, sea, and air), then have subbranches. "Organized beings represent a tree, *irregularly branched*" (B,

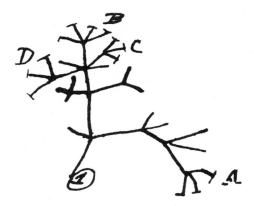

Figure 21. How species become separated through divergence and extinction. From Darwin's *Species Notebook* B.

p. 21), although "the tree of life should perhaps be called the coral of life, base of branches dead, so that passages cannot be seen" (B, p. 25; and see fig. 22). We know that a major implication of branching (considered as fundamental) is that man is no longer a measure of all things, and Darwin is aware of this. "It is absurd to talk of one animal being higher than another—we consider those where the { cerebral structure / intellectual faculties } most developed, as highest. —A bee doubtless would where the instincts were" (B, p. 74). More on this later.

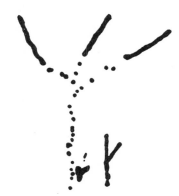

Figure 22. The "coral of life." From Darwin's *Species Notebook* B.

Returning to the Lyellian influence, we find that Darwin followed Lyell in believing that an evolutionary theory must attempt to explain the fossil record (B, p. 14). Lamarck did not see matters this way, and Lyell misinterpreted him. Before he had finished with his notebooks, Darwin was worrying about the way the Stonesfield mammals wrecked the progressive nature of the fossil record.[4] Lyell was not so naive in his moves after all! Finally, there is the question of migration. Natural means of migration were essential to Lyell, and they were equally so for Darwin. Having emphasized isolation, he had to explain how organisms got to their isolated spots in the first place, before they started changing. We therefore find an intense interest in means of travel, and what an actualist/uniformitarian moving of continents, throwing up of land bridges, then destroying them, and so on (Grinnell 1974)!

These were the speculations Darwin took with him through his first year as an evolutionist. He was doing just what Lyell could not do, and hence— as with the status of man—challenging just what Lyell could not challenge. Yet, paradoxically, in many respects his thought was Lyellian through and through. Darwin never entirely relinquished this model of evolution through isolation, though it began to play a less and less prominent role in his thought, and though he certainly came to believe one could have evolution from species to species without geographical isolation. But, as he started on his second species notebook in February 1838, other concerns started to capture his attention—in particular, the causes of new variation and their relationship to adaptive advantage.

In the first species notebook, Darwin had tended to let variation look after itself. It occurred, and Darwin thought it was in some way a function of environmental conditions. But he had been concentrating on organisms evolving in isolation and thinking of the threat to an organism as coming primarily from outside its own species; though, as a passage quoted above shows, he was very conscious that adaptive failure spells extinction. Thus he had been able to shelve the question of adaptation, at least for newly evolving organisms. His first concern was to permit change—any kind of change. But about the beginning of 1838 Darwin began to worry in earnest. After all, isolation simply sets aside the problem of adaptation; it does not eliminate it. Even if they are protected from outside threats, isolated organisms like those on the Galápagos islands are building up adaptations, and as a good Lyellian—as well as the protégé of those rabid natural theologians Sedgwick and Whewell—Darwin knew this problem had to be solved. "With belief of transmutation and geographical group-ing we are led to endeavour to discover *causes* of changes—the manner of adaptation (wish of parents??)" (de Beer et al. 1960–67, B, p. 227). Just how does one link new variation with the fact of organic adaptation? "Can the wishing of the Parent produce any character in offspring? Does the mind produce any change in offspring? If so, adaptation of species by *generation* explained?" (B, p. 219). But this is just wild guessing, and so Darwin turned to the only possible source of information about new variations, the way they are passed on, and their relation to adaptation. He turned to the domestic world, the world of the animal and plant breeder.

At this point we must tread carefully, for from here on the thought processes we find in the notebooks and the account Darwin later gave of his road to natural selection do not quite coincide (Limoges 1970; Herbert 1971). In one of many similar accounts of his discovery, Darwin wrote: "I came to the conclusion that selection was the principle of change from the study of domesticated productions; and then, reading Malthus, I saw at once how to apply this principle" (Darwin and Seward 1903, 1:118). This seems fairly clear. In the domestic world, Darwin saw that the selection of minute useful variations could lead to overall adaptive advantages; and though he could not at first understand how it happened in the wild, he could see by analogy that such a process was possible.

But in the notebooks we do not get this picture. Darwin quickly realized (if he had not realized it before) that the art of breeding consists in picking out favorable variations, and that the variations one normally needs will be small ones. "All these facts clearly point out two kinds of varieties. —One approaching to nature of monster, hereditary, other adaptation" (C, p. 4). At one point he even speculated about a counter-part where, because new variations give organisms vigor, they are likely to be the ones to breed and pass on variations. "Whether species may not be made by a little more vigour being given to the chance offspring who have any slight peculiarity of structure. hence seals take victorious seals, hence deer victorious deer, hence males armed and pugnacious all order" (C, p. 61). Hence we have selection, albeit not sparked by population pressure or involving the adaptive advantage of the new variations. But at other times Darwin could see no worthwhile analogy. "The changes in species must be very slow owing to physical changes slow and offspring not picked.—as man do when making varieties" (C, p. 17). This doubt of a significant analogy between the domestic and wild worlds continued in Darwin's mind right up to his reading of Malthus at the end of September 1838. Earlier that month Darwin wrote: "It certainly appears in domesticated animals that the amount of variation is soon reached—as in pidgeons no new races" (D, p. 104).

Darwin's later accounts of his discovery were therefore somewhat misleading. Having turned to the domestic world, he did not suddenly and unequivocally parade the analogy between artificial and natural selection, shelving all else in his search for the force behind natural selection. But he did, through the spring and summer of 1838, start to explore in some depth another way organisms might get heritable adaptive variations— that perennial favorite, the inheritance of acquired characteristics, particularly those acquired through habit. Thus, "All structures either direct effect of habit, or hereditary and combined effect of habit" (C, p. 63), and "According to my views, habits give structure, ∴ habits precede structure, ∴ habitual instincts precede structure" (C, p. 199). Somehow, he thought, we might explain organic adaptation this way. (This all sounds a bit Lamarckian and may indeed reflect a direct influence.)

Once again Darwin had developed something that would always be, for him, a piece of the overall evolutionary picture. Like isolation, the inheritance of acquired characteristics remained a Darwinian doctrine to the end. But he was not really satisfied. The wax in one's ear, its bitterness a perfect adaptation for discouraging insects, hardly seems a function of habit (C, p. 174). And the eye! "We never may be able to trace the steps by which the organization of the eye passed from simpler stages to more perfect preserving its relations. —the wonderful power of adaptation given to organization. —This really perhaps greatest difficulty to whole theory" (C, p. 175).

During the summer of 1838 Darwin in no way actively promoted the analogy between the artificial and the natural worlds; but it is clear that his reading was not letting him forget that selection is the means of change in the domestic world or that analogies might be drawn with the natural world. Trying to understand heredity, variation, and adaptation, Darwin was reading deeply on breeding, and he encountered two pamphlets that described in some detail the process of artificial selection, its great effects, and the possibilities that such effects might be permanent—the very thing Lyell, and everyone else, was denying in support of the impossibility of evolution. Darwin responded enthusiastically to these pamphlets, by John Wilkinson (1820) and Sir John Sebright (1809). (Darwin's response is in C, p. 133; I discuss the pamphlets in detail in Ruse 1975c.) Although they still did not convince him that artificial selection is the clue to the natural mechanism of species change, it is plain that they were influencing him in that direction. In the margin of one (Sebright 1809, pp. 14–15) he wrote: "In plants man presents mixtures, varies conditions, and destroys, the unfavourable kind—could he do this last effectively and keep on the same exact conditions for many generations he would make species, which would be infertile with other species." Moreover, in a passage Darwin noted, the idea of natural selection was presented and the analogy drawn between it and artificial selection: "A severe winter, or a scarcity of food, by destroying the weak and the unhealthy, has all the good effects of the most skilful selection" (Sebright 1809, pp. 15–16). Passages like this, containing an idea of natural selection, occurred several times in the early nineteenth century (Wells 1818; Matthew 1831; Blyth 1837; see Limoges 1970 for a discussion). It goes without saying that the passage in itself does not mean a great deal. Sebright was in no way using it as a mechanism of evolutionary change.

We now come to September 1838. By then Darwin knew all about artificial selection and had even come across an idea of natural selection. But, contrary to his later recollections, he was not thinking strongly of the analogy between them and probably did not consider natural selection a significant candidate as an evolutionary mechanism. He thought the key to adaptation lay in use and disuse, and he was ambivalent about selection and analogies from the domestic world. It would be a great thing if one could find a natural kind of selection, but how would one do this, and what guarantee would there be that it could cause indefinite evolution? But Darwin's study of the breeders was starting to push him in one very important aspect: it was showing him that organisms of the same species are not identical and that in the differences can lie the key to change. Lamarck's mechanism and Chambers's mechanism could work solely at the level of the individual because both evolutionists presupposed with their nonevolutionary opponents that all instances of a particular form of organism would be the same and that changes in a group would thus involve a simple summation of identical changes in individuals. For Darwin's natural selection, however, the group or population was vital: change involved the alteration of differences between individuals, which collectively compose the group. Certainly, Darwin's study of breeders was priming him for this conceptual shift. (On this point, see Ghiselin 1969; Hull 1973b; Mayr 1972b, 1977.) Then, at the end of the month, Darwin read Malthus (1826) on population—the tendency to increase in number and the limited food and space all leading to an inevitable struggle for existence. Although he already thought of adaptation in a dynamic way, never before had Darwin realized what pressure there is on organisms or how critical and dynamic is the concept of adaptation: "Population is increase at geometrical ration in FAR SHORTER time than 25 years— yet until the one sentence of Malthus no one clearly perceived the great check amongst men" (de Beer et al. 1960–67, D, p. 135). Consequently, "One may say there is a force like a hundred thousand wedges trying [to] force every kind of adapted structure into the gaps in the oeconomy of nature. or rather forming gaps by thrusting out weaker ones" (D, p. 135).

In some manner (to be discussed shortly), this reading of Malthus was enough. Darwin had all he needed. Malthus showed that the survival and reproductive differentials come from the desperate struggle for existence. To survive, an organism must have an adaptive edge not only over members of other species, but over members of its own species (Vorzimmer 1969). Darwin was able to connect this with what he knew about selection, which demands the very differential reproduction the struggle provides; and he saw that an organism's adaptive edge over other organisms could not be defined independent of the struggle. Adaptation was a function of the peculiarities of organisms that won out. Winning, and only winning, was what counted. Thus was born the idea of natural selection as an evolutionary mechanism. Some organisms win, because they have helpful characteristics the losers lack; and in the long run these characteristics add up to full-blown adaptations and significant evolutionary change. As Darwin later wrote, "Here, then, I had at last got a theory by which to work" (Darwin 1969, p. 120). And it is not long (27 November 1838) before we find him referring to his mechanism to explain change: "An habitual action must some way affect the brain in a manner which can be transmitted. —this is analogous to a blacksmith having children with strong arms. —The other principle of those children which *chance* produced with strong arms, outliving the weaker ones, may be applicable to the formation of instincts, independently of habits" (Gruber and Barrett 1974, N, p. 42).

But, despite what Darwin said, a mechanism is not a fully developed theory, and he still had much work to do before he could claim to have one. Let us turn now to the period from 1838 to 1844, the year the *Essay* was written.

Darwin and Philosophy

Whatever their differences, for both Herschel and Whewell Newtonian astronomy was the paradigm of a scientific theory. Throughout the species notebooks, up to and including the *Origin,* Darwin constantly followed the philosophers on this, claiming that any adequate solution to the organic origins question must satisfy the canons of Newtonian astronomy—but adding the corollary that Darwinian evolutionism will satisfy such canons whereas special creation will not! Thus, right in the middle of the first species notebook (summer 1837), Darwin wrote:

> Astronomers might formerly have said that God ordered each planet to move in its particular destiny. In same manner God orders each animal created with certain form in certain country, but how much more simple and sublime power let attraction act according to certain law, such are inevitable consequences—let animal be created, then by the fixed laws of generation, such will be their successors. Let the powers of transportal be such, and so will be the forms of one country to another. —Let geological changes go at such a rate, so will be the number and distribution of the species!! [B, pp. 101–2]

Of course at this time everyone was pushing Newton and the rule of law, so we ought not ascribe this general influence exclusively to the philosophers, though Darwin was impressed by them and their work. For instance, shortly before conceiving of natural selection as an evolutionary mechanism (August 1838), Darwin read with avid interest a review by Brewster of Comte's *Cours de philosophie positive.* What he would have got from this is that the aim of science is the "positive" stage, that "the fundamental character of *Positive Philosophy* is to regard all phenomena as subjected to invariable natural laws," and that the best of all laws is the Newtonian law of gravitational attraction (Brewster 1837*b*). This must surely have been a strong and (given the timing) crucial confirmation for Darwin of the importance of his "Newtonianism."[5] Nevertheless, the philosophers Herschel and Whewell went into far more detail than most about the precise implications of Newtonianism, and considering the extent to which Darwin followed their prescriptions—as he did in geology—direct influence seems plausible. Let us see how this is so.

There has long been a question about why Darwin reacted so enthusiastically to Malthus's work. That Darwin used Malthus's ideas in an essential way must be granted. But why was the actual reading of Malthus so important, since Darwin already knew all about the struggle for existence from the very detailed analysis in Lyell's *Principles* (Vorzimmer 1969)? Moreover, though he accepted Malthus's premises, Darwin wanted to stand Malthus's conclusion on its head: Whereas Darwin saw struggle as leading to change, Malthus essentially saw struggle as ruling out change!

Malthus, by concentrating on man, helped Darwin see struggle as intraspecific, not just as between groups or between a group and its environment. But we must qualify this debt by pointing out that Malthus himself was more interested in man versus his environment, restricting his account of bloody intragroup struggle primarily to primitive man (Bowler 1976*b*). Hence, Darwin probably got from Malthus what he needed to get. As I pointed out above, Darwin's key move to populational thinking—to recognizing critical variation within species—was probably much aided by his study of breeders' methods and results. We must therefore seek elsewhere for Darwin's major debt to Malthus. In light of Darwin's comment that his mechanism became clear as soon as he read Malthus (stressing the geometric ratio of population increase), and the way he presented his theory, undoubtedly the debt springs from Malthus's quasi-mathematical presentation of his subject. Right at the beginning of his *Essay* Malthus laid things out starkly: geometric population increases outstrip arithmetic increases in food supply. This leads to a crunch unless one invokes some sort of restraint (Malthus 1826, chaps. 1 and 2; see Ruse 1973*a*). Helpfully, Malthus argued analogically from the nonhuman to the human world; so all Darwin had to do was argue back, dropping restraint and concluding that an organism struggles with everything, including its fellows. But the method of presentation was what counted, and the Herschel-Whewell influence puts this whole matter in context and makes Malthus's importance obvious. According to Herschel and Whewell's Newtonian philosophy, the best laws are quantitative, like the law of gravitational attraction. These laws do not occur in isolated splendor but are bound together in hypothetico-deductive systems. Darwin found Malthus's *Essay* new and striking not because of the truth of his premises—nearly everyone in the 1830s knew these and accepted them as indubitable—but because Malthus presented his ideas in lawlike, quantitative form, with a deductive approach. This was just what Darwin, soaked in the contemporary philosophy of science, was looking for. Not until Darwin read Malthus could he fit the struggle for existence into his mechanism of organic change—a mechanism that had to be scientific as he understood the word. Because of the way Malthus presented the struggle,

Darwin realized it was sufficiently universal, inevitable, and powerful to cause indefinite organic change, unlike artificial selection (as he then thought).

As soon as Darwin had read Malthus, he started to think in terms of forces and pressures that were pushing organisms into the available and not-so-available gaps in the economy of nature, and this whole notion of force was central to the way the philosophers interpreted Newtonian physics. One must explain through causes, preferably *verae causae*. But the paradigm of a *vera causa* (for either Herschel or Whewell) was force. Hence, insofar as Malthus was able to show him something akin to a force at work on organisms, Darwin was becoming aware of something that would be a prime candidate for a scientific evolutionary mechanism. Because of the philosophical lens through which he viewed matters, he was highly receptive to the way Malthus presented the struggle for existence—because of his viewpoint, Darwin knew it had to be critical.

But this still takes us only to the mechanism. Toward the end of 1838, Darwin had to start thinking about ways to develop his mechanism into a full theory of evolution. Once again we find the philosophical influence critical. Darwin was a highly professional young scientist. He felt it essential to put his theory, controversial as it would undoubtedly be, into a form that professional scientists would respect. People might reject his theory, but they would not reject it because of poor structure. What were the guidelines for a good theory and where would one find them? They were the ones specified by those arbiters of science Herschel and Whewell, and one would find them in the philosopher's conversation (at this time Darwin was mixing frequently with these men) and in their writings. In particular one must look at Herschel's *Discourse* and Whewell's recently published *History of the Inductive Sciences*.

At the end of 1838, this is just what Darwin was doing. He reread the *Discourse,* and went carefully through Whewell's *History* for the second time.[6] Judging from his comments in the margin of the *History,* he was most interested in those things Whewell particularly admired in Newtonian astronomy and the arguments Whewell brought against organic evolutionism. Darwin wanted to make his theory as Newtonian as possible and to anticipate the worst attacks the critics could make. But how was he to do this? We know that Herschel and Whewell, though very close methodologically, came to differ over the concept of a *vera causa:* for Herschel, the empiricist, analogical argument from experience was what really counted; for Whewell, the rationalist, it was a consilience of inductions pointing upward. Darwin plainly strove to satisfy both these conceptions of *vera causa*.

Take first Herschel's analogical notion. In the context of discovery Darwin probably was most influenced, as he later said, by the artificial/ natural selection analogy, though it did not play the overwhelming role he later implied it did. This is as may be. The important point is that once we have the notion of natural selection, we enter the context of justification; there was no need for Darwin to mention the analogy again. Indeed, there were good reasons he might refrain from mentioning it, as Wallace refrained twenty years later. For Lyell, Whewell, and everyone else, the *reductio* of evolutionism was that artificial selection does not lead to sustained, permanent change. But, though Darwin apparently shared this view just before discovering natural selection, no sooner had he read Herschel again than he suddenly started to emphasize the analogy. "It is a beautiful part of my theory, that domesticated races of organics are made by precisely same means as species" (E, p. 71; I take it that these "means" included, as they always did for Darwin, the inheritance of acquired characteristics). Moreover, Darwin soon thought of using the analogy as a way to present his theory publicly.

> Varieties are made in two ways—local varieties when whole mass of species are subjected to same influence, & this would take place from changing country: but greyhound race-horse & poulter Pidgeon have not been thus produced, but by training, & crossing & keeping breed pure—& so in plants *effectually* the offspring are picked & not allowed to cross.—Has nature any process analogous—if so she can produce great ends—But how—even if placed on Isl^d. if &c &c—make the difficulty apparent by cross-questioning—Here give my theory.—excellently true theory. [E, p. 118]

This paradoxical about-face is paradoxical no longer when we realize that Darwin was trying to present natural selection as a Herschelian *vera causa*. We argue by analogy from actually experienced ways of changing organisms to presumed ways. Moreover, Herschel's discussion of *verae causae* in the *Discourse* shows added reason for Darwin's joy that so close an analogy can be drawn between artificial and natural ways of producing change. Herschel (1831, p. 149) argued that "if the analogy of two phenomena be very close and striking, while, at the same time, the cause of one is very obvious, it becomes scarcely possible to refuse to admit the

action of an analogous cause in the other, though not so obvious in itself."
He then went on to explain that the force we feel and cause when
whirling a stone at the end of a string points undeniably to the existence of
an analogous force keeping the moon in orbit as it whirls around the earth.
This is just Darwin's situation. A causal force, artificial selection, is
directly perceived and caused by us. Therefore we have the best evidence
for the analogous causal force of natural selection. (Although Darwin was
going against Lyell here, Herschel and Lyell shared the same *vera causa*
concept, and so in another sense Darwin was being thoroughly Lyellian:
indeed, a Lyellian actualist. More generally, everything in this section
complements and reinforces the Lyellian influence; though in the *Principles*
Lyell was not hypothetico-deductive, and so we must go beyond Lyell for
philosophical/methodological influences on Darwin.)

By the beginning of 1839, then, Darwin was convinced that he should
exploit to the full the analogy between the domestic and natural worlds.
From his reading lists (a notebook in the Darwin collection labeled "Books
to Be Read"), we find that in the next few years he set to this task with
gusto, delving into the world of breeders and reading, for example, the
classic works on breeding by Youatt.[7] Undoubtedly Darwin was convinced he should exploit the domestic/natural analogy before he did much
research. But, once convinced of its importance, Darwin was in a peculiarly favored position, which helped confirm his conviction that the opponents of evolutionism were mistaken when they trotted out examples from
the domestic world to disprove evolution. Darwin was working at a time
when scientific animal and plant breeding was being developed and
refined as never before, as agriculturalists tried to keep up with the
voracious demands of an exploding and increasingly urban population.
Coming as he did from Shropshire, the heart of agricultural England,
Darwin was able to draw extensively on local and family connections. For
instance, his uncle and (by early 1839) father-in-law Josiah Wedgwood,
with whom Darwin had a very close relationship, was one of England's
leading sheep breeders,[8] as well as an officer of the Society for the Diffusion of Useful Knowledge, which was then actively promoting the principles of scientific agriculture.[9] And the Darwin family itself had long been
famous in Shrewsbury as pigeon fanciers, a tradition Charles Darwin
continued (Meteyard 1871).

Once convinced of the importance of the domestic/natural analogy,
Darwin developed it to the hilt. In the *Sketch,* the *Essay,* and the *Origin,*
Darwin began by discussing change in the domestic world, pointing to

the importance of selection. Then, after noting variation in the wild and presenting the struggle for existence as a counterpart of man's desire for selection, Darwin argued by analogy to natural selection. Moreover, he repeatedly used the analogy to illustrate important points of his theory. Developing this analogy in the context of justification probably led Darwin back to the context of discovery. It seems likely that sometime between 1839 and 1842, when he wrote the *Sketch,* he was led to one of his subsidiary evolutionary mechanisms, sexual selection. Man selects not only for qualities that will aid his livelihood—heavier cows, shaggier sheep, bigger vegetables—but on occasion also for qualities that give him pleasure. These qualities tend to be of two kinds—combative strength, as when one breeds a fiercer bulldog or cock, and beauty, as when one breeds a fancier pigeon. Darwin mirrored these qualities in his analysis of selection. Natural selection corresponds to selecting for things that help man survive. Sexual selection corresponds to selecting for things that give man pleasure: Darwin in turn divides sexual selection into selection through male combat, where the stronger male gets the female(s), and selection through female choice, where the more attractive male gets the female(s) (Darwin and Wallace 1958, pp. 48–49, 120–21).

Although there certainly were hints of sexual selection in the species notebooks even before Darwin read Malthus,[10] as well as in Erasmus Darwin's *Zoönomia* (J. Harrison 1971) and elsewhere in Darwin's early reading, the way Darwin tied sexual selection so tightly to these kinds of artificial selection in the *Origin* makes it highly probable that the analogy played an important role in his path of discovery—certainly in his decision that the natural/sexual selection dichotomy was genuine and worth putting into the theory. Wallace, who did not use the artificial/natural analogy, did not mention sexual selection in his 1858 essay; and, as we shall see, Wallace later had trouble with sexual selection through female choice.

So far we have considered Darwin's attention to Herschel's notion of a *vera causa.* Whewell's notion of a *vera causa* tied to a consilience was equally significant for Darwin,[11] though this will become plainer when we have laid out Darwin's completed theory of evolution. But, even before discovering natural selection as an evolutionary mechanism, Darwin showed that a consilient theory was his ideal. "Absolute knowledge that species die and other replace them. —Two hypotheses: fresh creations is mere assumption, it explains nothing further; points gained if any facts are connected" (de Beer et al. 1960–67, B, p. 104). Then as he labored toward the writing of the preliminary *Sketch* in 1842, Darwin worked hard to show how natural selection (and his other mechanisms) could be applied to explanations in behavior, paleontology, biogeography (geographical distribution), anatomy, systematics, embryology, and so on, all these areas being tied together chiefly by one mechanism. Just the thing a consilient theory should do, and just the thing that Darwin, sharing Whewell's theory ideals, thought a theory should do.

In 1842 and 1844, Darwin wrote out the preliminary versions of his theory, which in essence were very little changed in the *Origin*. We have a discussion of domestic selection. We get the struggle, then the analogical counterpart of domestic selection, natural selection. Then, along with discussions of the nature of sterility and the like, we get the mechanism of selection applied to all the problem areas mentioned above. In short, we have something Darwin himself could accept as a properly structured theory. Before we can continue chronologically, however, we must pause and go back. For Darwin's contemporaries, his teachers and his seniors in the scientific network, religious questions were a most important barrier to acceptance of an organic evolutionary theory. Why were they not a barrier for Darwin?

Darwin and Religion

At the center of revealed religion, based on faith and revelation, is the Bible, and we know that when he went up to Cambridge Darwin took the Bible literally. When he left Cambridge to join the *Beagle,* his faith was fairly orthodox, but during the voyage it started to crumble. Undoubtedly the major reason for this was Darwin's growing conviction that the Bible, particularly the Old Testament, was incompatible with science, particularly uniformitarian geology (Darwin 1969, p. 185). As Darwin became committed to science, he became more and more committed to the rule of law, which in turn excluded miracles. But for Darwin, Christianity without miracles was nothing (at least as a divinely inspired religion), and so his adherence to Christianity faded away (Darwin 1969, pp. 86–87). That he rested the case for Christianity so thoroughly on miracles was perfectly natural, for he was brought up on Paley's *Evidences of Christianity* (Darwin 1969, p. 59). As we have seen, in the traditional English empiricist manner, Paley made the whole truth of the Christian revelation entirely dependent on the genuineness of the biblical miracles. So when miracles went for Darwin, Christianity went too.

After the *Origin* was published, Darwin became something of an agnostic about the existence of God (Darwin 1969, p. 94; Mandelbaum 1958). Until that point it is probable that, although no Christian, he was neither atheist nor even agnostic. He was a deist of a kind—believing in an unmoved Creator who worked entirely through unbroken, unchanging law. He therefore accepted a natural religion based on reason and sense. This is the language Darwin used in the *Origin* (1859, p. 488), and it seems improbable that he was hypocritically using such a vocabulary for tactical reasons—he was going to ruffle the Christian feathers anyway. Certainly Darwin was a deist while he was discovering his theory, for he constantly used such language in his notebooks, when he was talking only to himself (Gruber and Barrett 1974, M, p. 154). But one suspects there is more to the story. Lyell was a deist in the sense ascribed to Darwin, but religion continued to be a major antievolutionary stumbling block for Lyell. Why did Darwin remain unawed by the question of man, feeling none of the emotions Lyell felt about man's peculiarity and special status? In the first species notebook (late 1837), man was firmly put in his place: "People often talk of the wonderful event of intellectual man appearing. The appearance of insects with other senses is more wonderful" (de Beer et al. 1960–67, B, p. 207). And no sooner had Darwin grasped natural selection than he was speculating on how it might apply to man (M, p. 42). Man was just not the barrier for Darwin that he was for so many others.

Darwin took this attitude for a number of reasons, feeling sure that man and his origin, like everything else, must come beneath the rule of unbroken law. First, unlike others of his group, Darwin had firsthand experience of savages in the wild—the natives of Tierra del Fuego. Their primitive life-style clearly impressed on Darwin how very nonpeculiar and close to the brutes man is. "Compare the Fuegian and the Ourang-outang, and dare to say differences so great" (M, p. 153). Second was Darwin's family background. Although Darwin himself at one point believed the Bible literally and intended to become a clergyman, his grandfather Erasmus was at best a weak deist, quite able to believe in evolution, whom Coleridge (1895. 1:152) thought an atheist; his father Robert, who had an overwhelming influence on Darwin, was an unbeliever (Darwin 1969, p. 87); his uncle Josiah Wedgwood was a Unitarian (he supported Coleridge for a number of years; Meteyard 1871); and, most important of all, Charles's older brother Erasmus had become an unbeliever by the time Charles returned from the *Beagle* voyage. One suspects that all of these things must have influenced Charles, making him fairly unimpressed on the subject of man. There is no doubt that Erasmus directed or confirmed Charles's thought, for Emma wrote to Charles just after their marriage, tactfully bemoaning Erasmus's bad influence (Darwin 1969, p. 236).

Third was the intellectual community with which Darwin was mixing in London while he was formulating his ideas. Not only was Darwin exposed to believing scientists like Lyell and Whewell, he was also intimate with men whose religious ideas were—or could be taken to be— much more tolerant of evolutionism. He was good friends with Babbage (Darwin 1969, p. 108) and read the *Ninth Bridgewater Treatise* (de Beer et al. 1960–67, E, p. 59), which ascribed everything to the rule of law. Also, Darwin was much in the company of Carlyle, a close friend of his brother Erasmus, and was indeed greatly taken with him, for he wrote to Emma just before they were married (January 1839) that "to my mind Carlyle is the best worth listening to of any man I know" (Litchfield 1915, 2:21). There apparently was little scientific sympathy between Darwin and Carlyle, but Carlyle's unorthodox religious beliefs may have fallen on fertile ground. In particular, his natural supernaturalism, with its deliberate refusal to see any part of the creation as more or less marvelous or miraculous than any other, may well have influenced Darwin, as it undoubtedly influenced Huxley. Be this as it may, Darwin was certainly mixing in company that was not as ready to exalt man's status as were Whewell and Lyell.

Finally, it seems that Darwin simply cared less about religion than many other men. He admitted this himself (1969, p. 91). Darwin thought about the subject as much as most men, if not more—his studies forced him to do this. But essentially he just wanted to get on with his science, whatever the consequences. From Darwin one never gets the burning religious zeal to be found in, say, Sedgwick or, in a different sense, Huxley (Herbert 1977).

In that aspect of natural religion centering on design, Paley was once again a starting point, for at Cambridge Darwin read the *Natural Theology* and was much impressed (Darwin 1969, p. 59). Darwin always accepted the major premise of the natural theologians of the 1830s, that the organic world must be understood in terms of its designlike appearance, its adaptedness, its relation to ends: we have seen this throughout Darwin's discovery of natural selection, a mechanism that itself focuses on the concept of adaptation. When discovering his theory Darwin took such adaptation as evidence of some designer (Gruber and Barrett 1974, M, p.

136), and at that time he was ready to allow that a purpose of earth's creation was man (de Beer et al. 1963, E, p. 49). In later years, however, Darwin seems to have felt that natural selection made the argument from design redundant to the point of untenability (Darwin 1969, p. 87), though until the end of his life he continued to have flashes that all this adaptation must spell design (F. Darwin 1887, 1:316 n).

Of course, even in the 1830s Darwin was not particularly conventional about natural religion, especially as it centered on adaptation. First, Darwin thought adaptation, designed or not, could be produced by normal, unbroken law. And since views like this were often linked to the picture of God as industrialist, it is interesting that, true to his rural connections, Darwin invoked its country cousin—God as farmer. In both the *Sketch* and the *Essay* Darwin explicitly likened natural selection to the work of a superbeing (Darwin and Wallace 1958, pp. 114–16). (Wallace went even further, arguing that selection is just like a self-regulating machine; Darwin and Wallace 1958, p. 278.) Second, Darwin showed his unconventionality by refusing to accept that organisms are essentially perfectly adapted to their environments. It was necessary to his organic world picture that organic adaptive failure be a regular phenomenon—not something that waits on catastrophes. Indeed, Darwin pushed to the limit the revolution about adaptation Lyell had started: From the absolute static property of the catastrophists, through the fallible property of Lyell, to his own thoroughly relativized dynamic phenomenon, where in any species some organisms must be adaptively inadequate at virtually any time (B, pp. 37–38, 90). The third point is minor but had an interesting consequence. Darwin veered from conventional thought on adaptation in that he could not consider anything in one species as provided solely for the benefit of some other species. He explicitly acknowledged that such an adaptation would refute his theory (Darwin 1859, p. 211).

Nevertheless, though he was putting a strain on adaptation, particularly as a bond between God and his creation (a fact he fully recognized; M, p. 70), this does not deny the great importance of adaptation for Darwin. He could not accept such adaptation as the product of divine intervention; but he knew from his scientific/religious background that any adequate biological theory must meet the problem head on, and he tried to do this with his mechanism of natural selection. Indeed, Darwin's sensitivity about adaptation had a somewhat paradoxical result. Darwin was educated and was doing his great creative work in the 1820s and 1830s, when the concept of design through adaptation was at its peak. By the time he published his work, as we have seen, this strand of natural theology had been much buffeted. Hence, though his nontheological theory is often portrayed as taking the teleology out of biology, if anything Darwin was bringing it back in! Adaptation, with its orientation towards ends, was a more significant facet of the organic world for Darwin than it was for Huxley. Because the creative Darwin was a man of the 1830s rather than the 1850s, after the *Origin* biology was in a sense brought back to a teleology based on adaptation, from which it had started to stray.

The Long Wait

We come now to the major puzzle in the Darwinian story. By the middle of 1844 Darwin had completed a version of his theory in the 230-page *Essay.* Although he made some changes when he wrote the *Origin,* they were comparatively minor. Why then did Darwin not publish his work at once, as soon as he had got something down on paper, rather than finish some geological work and then plunge into a massive, eight-year systematic study of barnacles?

The usual reason given for the delay is some version of that offered by Huxley. The only person to whom Darwin showed his *Essay* (when it was just finished) was Hooker, with whom a friendship was rapidly ripening (Darwin and Wallace 1958, p. 257). Unconverted, Hooker suggested that, before publishing, Darwin might well deepen his understanding of biology. Darwin saw the value of this and turned to barnacles with just such an aim. Thus, as Huxley wrote, "Like the rest of us, he had no proper training in biological science, and it has always struck me as a remarkable instance of his scientific insight, that he saw the necessity of giving himself such training, and of his courage, that he did not shirk the labour of obtaining it" (F. Darwin 1887, 1:347).

There are two reasons why this sanitized version of the cause of the delay is unconvincing. First, there is no great difference between the *Essay* and the *Origin.* Darwin became more convinced that the smallest of new variations ("individual differences") are the building blocks of evolution, and this was probably influenced by his barnacle work. Isolation as a factor in speciation decreased in his eyes, and something he called the "principle of divergence" came into being. But, overall, content and structure are virtually unchanged. Darwin did not need nearly a decade of invertebrate systematics to write the *Origin.* Second, he was not the man to let a little ignorance stand in the way of the publication of a bold and sweeping hypothesis. I say this not sarcastically but literally—Darwin could not wait to get into print on the subjects of coral reefs and Glen Roy. He was ambitious and wanted to make his mark in the scientific community. A few blank spots were not going to stop him from publishing his solution to what his group took as the major scientific problem of the day. He was realistic enough to know how good his solution was.

The true answer has to be sought in Darwin's professionalism, just as his success at finding and developing his theory must be sought there. Darwin was not amateur outsider like Chambers. He was part of the scientific network, a product of Cambridge and a close friend of Lyell, and he knew well the dread and the hatred most of the network had for evolutionism. If he had any doubts, the publication of *Vestiges* in the same year as he wrote his *Essay* confirmed them. Sedgwick raged against *Vestiges* for eighty-five pages in the *Edinburgh Review;* Whewell denounced it in *Indications;* and Herschel condemned it from the presidential chair of the British Association. Darwin knew his theory was much better than Chambers's—"better" in that it more adequately answered the problems as then understood—but it was evolutionary and materialistic nonetheless, and it was certainly not going to make its author very popular. When telling Hooker of his evolutionism, Darwin confessed that it was like admitting to a murder (F. Darwin 1887, 2:23). It was a murder—the purported murder of Christianity, and Darwin was not keen to be cast in this role. Hence the *Essay* went unpublished.

Was Darwin a coward? Not really. He could not have accomplished what he did without his background, firmly rooted in the scientific community. But because of this background, he could not do more. Later we will see that a central element in the story is the way Darwin, before the *Origin* was published, built around himself a new scientific community from which his evolutionism could be launched. Also, Darwin had no idea the delay would be so long. Stimulated by a strange barnacle he discovered while on the *Beagle,* Darwin decided to do a little work on barnacles (Darwin 1969, pp. 117–18). This project exploded into a full-length study, taking a great deal of time, and was further dragged out by the constant, severe illness that crippled him. Days, weeks, and months were lost when he was unable to work. From the pushy, vibrant young man of the 1830s, Darwin was reduced to an invalid. And so year after year the *Essay* on species lay untouched—with strict instructions that it be published in the event of his death (Darwin and Wallace 1958, pp. 35–36). Darwin had no desire to be ignored by posterity.

The work on barnacles provides a fascinating interlude (Darwin 1851a, *b;* 1854*a, b;* Ghiselin 1969). Darwin dared not let the reader know explicitly why he thought barnacles are as they are. Yet the work was pregnant with veiled hints. We see Darwin consciously preparing the way for the *Origin*, just as most of the authors we have considered were unconsciously preparing the way.

Not unexpectedly, we get much information about adaptation. In sessile cirripedes, for example, we learn that "the action of the cirri is really beautiful" (Darwin 1854*a*), since they are "beautifully adapted to catch any object floating or swimming in the water." Similarly, "we have the structure of the shell extremely complicated, yet beautifully adapted for strength, and for the protection of the included body" (Darwin 1854*a*, p. 152). And in a particular genus of pendunculated cirripedes "the teeth on the valves and scales are sharp, and fit for wearing soft stone, at the very period when the animal has to increase in size" (Darwin 1851*a*, p. 345). And so on, and on.

Second, we find that there are many new facts about the variation between individual barnacles—which, as Darwin privately noted to Hooker, made life wretched for the systematist but brought joy to the evolutionist advocating natural selection (F. Darwin 1887, 2:37). In sessile cirripedes, hardly a single external character does not vary widely. Moreover, "as whole groups of specimens often vary in exactly the same manner, it is not easy to exaggerate the difficulty of discriminating species and varieties" (Darwin 1854*a*, p. 3). In the *Essay* Darwin spoke hesitantly about the amount of variation in the wild. In the *Origin* he was far more confident. His barnacle work must have been important here.

Third, Darwin put strong emphasis on embryology in classification: "embryonic parts, as is well known, possess the highest classificatory value" (Darwin 1851*a*, p. 285). This led to an interesting and significant difference between Darwin and Owen (Winsor 1969). Cirripedes metamorphose drastically. Following the discoverer of this fact, John Vaughan Thompson, on the basis of their embryonic forms Darwin placed cirripedes in the class Crustacea. In 1843 Owen had refused to do this in his *Lectures on the Comparative Anatomy and Physiology of the Invertebrate Animals.* He knew of Thompson's discovery, but because the *adults* lose their motility, he felt they could not be Crustacea. And Owen held to this position in the second edition of his *Lectures,* published in 1855, though he doubled his section on cirripedes through augmentation with Darwin's findings. Darwin was keen on embryological classification because he thought embryos more significant than adults for evolutionary connections. He thought natural selection had less of a tendency to separate embryos morphologically. Since Owen's Platonic metaphysics led him to emphasize adults, he and Darwin classified cirripedes differently.

Continuing the interaction with Owen, we find Darwin working out the archetypal cirripede and interested in homologies. This was no Platonism, of course. Darwin thought the archetype was the actual ancestral form, not some metaphysical nonsensory reality. We find some depreciating remarks about Owen's notion of vegetative repetition (which leads to serial homology). The Balanidae, we learn, are the most complicated of cirripedes, but not "by mere vegetative repetition" (Darwin 1851a, p. 152). As we know, vegetative repetition was hardly the critical phenomenon for a Darwinian evolutionist that it was for Owen, though, not so surprisingly, both in his early notebooks and in his *Essay* Darwin accepted the vertebral theory of the skull (de Beer et al. 1960–67, E, p. 89; Darwin and Wallace 1958, p. 221). In the *Origin*, written after Huxley's Croonian lecture, he obviously did not; but long before this he was putting pressure on Owen's speculations.

There were two barnacle phenomena that Darwin thought cried out for an evolutionary explanation, and he came very close to telling the reader so. First, he was convinced that the apparatus by which barnacles cement themselves to their hosts or to the substrate is a modified ovarian tube, the cement itself being modified ova. He offered detailed hypotheses on how this might have happened, were one one of "those naturalists who believe that all gaps in the chain of nature would be filled up, if the structure of every extinct and existing creature were known" (Darwin 1854a, p. 151; Darwin was in error here—what he described as ovaries were really salivary glands; see F. Darwin 1887, 2:345). Second, Darwin made a brilliant discovery about barnacle sexuality. He found that not all barnacles, as had hitherto been believed, are hermaphrodites. Some species include very tiny males, like embryos; and in turn some of these species have true females and others have hermaphrodites, some also with small male organs. Darwin was, with justification, terribly excited about this discovery, and he presented it in a sequential form, virtually demanding an evolutionary interpretation, with true hermaphrodites sliding into hermaphrodites with males ("complemental" males), and on to true unisexual organisms (Darwin 1854a, p. 23). It did not take a very imaginative reader to see Darwin postulating (as he thought) hermaphroditism turning into male-female sexuality, and he gave all kinds of hints about how the males, being so embryolike, might be a function of arrested development, while the females were caused by disuse of the male organs (Darwin 1851a, p. 282). This discovery was particularly satisfying for Darwin because back in 1838, just at the moment when he was reading Malthus, he had been speculating about sex and had decided that unisexual organisms are transformed hermaphrodites. (De Beer et al. 1960–67, C, p. 162. Today we would put the evolution as from separate sexuality to hermaphroditism.) Here was confirmation. And, in turn, here was a problem in barnacle systematics for which his evolutionary theory had prepared him.

A fascinating sideline to Darwin's work on barnacle systematics ties in with a point made earlier about his position on adaptation. To convince the reader that the parasites on the females are indeed males, Darwin pointed out that the females have special places to accommodate the males. He added, "Now there is a strong and manifest improbability in an animal being specially modified to favour the parasitism of another" (Darwin 1851a, p. 285). To a Darwinian evolutionist, as we have seen, there is an impossibility. To a special creationist, there is no difficulty. We are all God's creatures. In short, Darwin was presupposing evolution to make his case.

Finally, there were a few not-so-guarded hints about the order of appearance of fossil barnacles (Darwin 1851b, p. 5). A special creation progressionist might welcome them, but they would certainly bring little comfort to a Lyellian. Once again, Darwin did his part to pave the way for his own theory.

The barnacle study finished, Darwin returned to the organic origins problem. At the urging of his friends, he started to write a colossal work on evolution (Stauffer 1975; Hodge 1977)—a kind of science by filibuster that would overwhelm his opponents by fact and footnote. This project was (luckily) interrupted in 1858 by the arrival of Wallace's paper. Dropping everything, in just over a year Darwin wrote the *Origin*. So we finally come to Darwin's major public pronouncement on organic evolution.

On the Origin of Species

Darwin's first chapter, "Variation under Domestication," deals with animal and plant breeding. He argues that organisms with many different forms, like pigeons, have common ancestors and that the main reason for these diverse forms, besides use and disuse, is man's power of selection. The second chapter, "Variation under Nature" establishes the widespread variation in the wild. Darwin was concerned with very small variations, "individual differences," as opposed to larger changes, "single variations" (Darwin 1859, pp. 44–45). He showed more confidence in their ubiquitous existence in the *Origin* than in the *Essay*, perhaps in part because of his work on barnacles. In the *Essay* he explicitly allowed that natural evolution might occasionally occur through a saltation from one form to another (certainly not, however, across a species; Darwin and Wallace 1958, p. 150; Vorzimmer 1963). In the *Origin* all natural changes are smooth, fueled by individual differences.

The next two chapters are the crucial ones. First, we get the derivation of the "Struggle for Existence":

> A struggle for existence inevitably follows from the high rate at which all organic beings tend to increase. Every being, which during its natural lifetime produces several eggs or seeds, must suffer destruction . . . otherwise, on the principle of geometrical increase, its numbers would quickly become so inordinately great that no country could support the product. Hence, as more individuals are produced than can possibly survive, there must in every case be a struggle for existence. . . . It is the doctrine of Malthus applied with manifold force to the whole animal and vegetable kingdoms. . . . Although some species may be now increasing, more or less rapidly, in numbers, all cannot do so, for the world would not hold them. [1859, pp. 63–64]

Then in the fourth chapter, "Natural Selection," Darwin goes on to derive his key mechanism:

> How will the struggle for existence . . . act in regard to variation? . . . Let it be borne in mind in what an endless number of strange peculiarities our domestic productions, and, in a lesser degree, those under nature, vary; and how strong the hereditary tendency is. . . . Can it, then, be thought improbable, seeing that variations useful to man have undoubtedly occurred, that other variations useful in some way to each being in the great and complex battle of life, should sometimes occur in the course of thousands of generations? If such do occur, can we doubt (remembering that many more individuals are born than can possibly survive) that individuals having any advantage, however slight, over others, would have the best chance of surviving and of procreating their kind? On the other hand, we may feel sure that any variation in the lest degree injurious would be

rigidly destroyed. This preservation of favourable variations and
the rejection of injurious variations, I call Natural Selection.
[1859, pp. 80–81]

These arguments are not presented in absolutely formal fashion; but one
can see the philosophers' influence (Ruse 1971, 1973a, 1975a, d). The
argumentation is certainly much closer to the hypothetico-deductive ideal
than to anything, say, in Lyell. In the first argument he begins with the
Malthusian premises that organic beings tend to increase at a geometric
rate and that such increase outstrips supplies of food, which increase only
arithmetically (not to mention space, which increases not at all). This
leads to the struggle for existence. In the second argument he starts with
this struggle as premise, adds in claims, based on analogy from the
domestic world, that some of the variation in nature will help in the
struggle and some will hinder, and thus neatly implies natural selection:
organisms with useful heritable variations have a better chance of surviving and reproducing than organisms with injurious heritable variations.

Along with natural selection we get sexual selection, divided into
selection through male combat and selection through female choice (Darwin 1859, pp. 87–90). Again Darwin justifies the claims by analogy to
the domestic world, as he exploits to the full the *vera causa* justification of
his evolutionary mechanisms. Then he illustrates and expands on his
mechanisms, in particular natural selection. We are shown, for example,
how scarcity of food might lead to competition among wolves, with
victory going to the fastest. "I can under such circumstances see no reason
to doubt that the swiftest and slimmest wolves would have the best chance
of surviving, and so be preserved or selected . . . I can see no more reason to
doubt this, than that man can improve the fleetness of his greyhounds by
careful and methodical selection" (1859, p. 90).

This chapter contains two important subsidiary discussions. First,
Darwin shows that he has definitely relinquished geographical isolation as
an essential element in evolutionary speciation. Relying entirely on the
smallest of variations, Darwin reveals his fear that isolated populations,
tending by their very nature to be small, might not be injected with
enough new variation to cause significant change. He therefore gives up
exclusive reliance on isolation in return for being able to argue that
(splitting) speciation might occur between subgroups of large populations, where there definitely would be ample new variation. But even in
large populations he seems to favor some kind of ecological isolation as
helping speciation—"haunting different stations" or "breeding at slightly
different seasons" (1859, p. 103).

Second is the "Principle of Divergence." Arguing analogically from pigeons, Darwin suggests that "the more diversified the descendants from any one species become in structure, constitution, and habits, by so much will they be better enabled to seize on many and widely diversified places in the polity of nature, and so be enabled to increase in numbers" (1859, p. 112). In other words, the reason there are so many kinds of species and so much splitting is that it confers a selective advantage. Darwin likened the principle of divergence to the "physiological division of labor" of the French biologist Henri Milne-Edwards, which claims that the more specialized the parts of the body (e.g., the stomach fitted for vegetable food alone), the more efficient the body is. In his *Autobiography* (1969, pp. 120–21) Darwin claimed that this principle of divergence came to him several years after the writing of the *Essay* (probably 1852), and while there is no reason to doubt that he did not consciously realize the problem (and the solution) till then, there nevertheless are strong intimations of both problem and solution in his earliest evolutionary writings (de Beer et al. 1960–67, E, pp. 95–97). This principle of divergence allowed Darwin to introduce his famous metaphor likening fossil and contemporary species to the dead and living branches of the Tree of Life (see fig. 23). Of course, the actual phenomenon of splitting and divergence (without cause or recognition of a need for cause) was not new for Darwin: it had been an integral part of his evolutionism from the beginning.

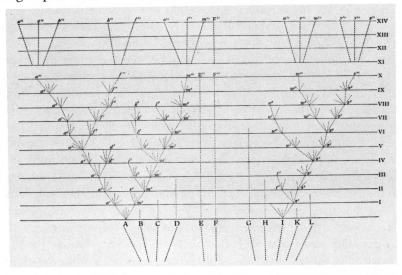

Figure 23. Darwin's diagram of descent (time moves upward). From the *Origin of Species*.

Next we get the chapter "Laws of Variation," a subject about which Darwin candidly admitted "our ignorance...is profound" (1859, p. 167). He seemed to think there were two basic kinds of variation. One type was more or less random (with respect to needs), and Darwin suspected these variations might be due to conditions impinging in some way on the reproductive system. Those of the other type do seem to involve a direct response to the environment: "I believe that the nearly wingless condition of several birds, which now inhabit or have lately inhabited several oceanic islands, tenanted by no beast of prey, has been caused by disuse" (1859, p. 134). In later chapters we shall learn that Darwin had philosophical reasons for feeling dissatisfied with his treatment of heredity in the *Origin* and see how he strove to remedy matters.

The sixth chapter deals with "Difficulties on Theory." One problem Darwin tackles in this chapter is the absence of intermediate forms between species. Why is it that we do not "everywhere see innumerable transitional forms?" (1859, p. 171). Darwin posited that, excluding cases where speciation forms through isolation and where one would thus expect no intermediates, a group intermediate between two diverging varieties will tend to be smaller than the main groups because the intermediate groups will be in the rather narrow zones between the larger areas in which the main groups are adaptively diverging. These intermediate groups, because they are smaller, are more likely to be wiped out than the main groups, chiefly because they have less new variation on which selection can act and thus respond less readily to external challenge. Hence, generally we should not expect transitional forms.

This chapter also takes up the problem of organs that are highly perfected, like the eye. Although we cannot trace the eye's evolution, Darwin averred that the possiblilty of such evolution through natural selection is shown because in the Articulata we can set up a scale of living beings with eyes ranging from the simplest to the most complex (1859, p. 187). Hence, evolution could certainly have happened—the idea of transitory forms is not impossible—and, given the supporting evidence, there is no reason to believe such evolution did not occur. In any case, "The correction for the aberration of light is said, on high authority, not to be perfect even in that most perfect organ, the eye" (1859, p. 202).

A chapter titled "Instinct" follows, in which Darwin explains that instincts, like structures, vary; that they can be of great adaptive value; and that it is reasonable to think of them as subject to and produced by natural selection. Following this comes the chapter "Hybridism." Here he argues, as one might expect given the Darwinian position, that there is a very gradual gradation between perfect fertility and perfect sterility (p. 248), thus blurring the distinction between variety and species (p. 276). Darwin also suggests that sterility is a by-product of the laws of growth and heredity—it is "incidental on other differences, and not a specially endowed quality" (p. 261). These reproductive barriers between species are not deliberately fashioned by selection.

This discussion concludes just about every element of Darwin's solution to the problem of the origin of species—the problem of how and why organisms split into distinct groups as opposed to the problem of organic origins *simpliciter*. Darwin presented his solution piecemeal through the first half of the *Origin* rather than offering it all in one unit. Speciation occurs because organisms change in response to changing environments and because the more diversified and specialized they become the more efficiently they can exploit their environments. This can happen in isolation from other groups, but more often it involves geographically linked varieties of the same group responding to the different challenges of different zones. In these latter cases, intermediates disappear because their groups have less variation to draw on than the main groups. Sterility develops between groups as a result of other differences, not directly through selection.

We come next to the geological chapters: "On the Imperfection of the Geological Record" and "On the Geological Succession of Organic Beings." The first of these chapters is very Lyellian—explicitly so (p. 310). Darwin admits that, prima facie, the geological record raises major problems for an evolutionary theory like his own: he recognizes the objection that there was inadequate time for the slow process of natural selection and that the abrupt transition from one species to another (as revealed by the record) disproves evolution, as does the first appearance of life in all its sophistication. In reply to the first objection, Darwin counters that there was indeed enough time. One of the detailed examples on which Darwin stakes his case is the denudation of the Weald (in Sussex), which he calculates, estimating the time required for the sea to wear away a cliff, must have taken some three hundred million years (pp. 285–87; Burchfield 1974). This was the absolute time required for a phenomenon that (judging from the strata) had begun only in the latter part of the Secondary. One can therefore imagine, or perhaps cannot imagine, the total time available for evolution. In reply to the second objection, Darwin argues that the gaps in the fossil record can be explained by the inadequacy of the record—fossils were not deposited, we have failed to find the fossils, and so on. In any case, speciation occurs most frequently when the ground is being elevated, thus causing new stations (as on islands). But since this is not a time of deposition, no transitional fossils are being laid down (p. 292). Third, Why did life appear full-blown at the beginning of the Silurian? This problem, Darwin admits, worries him. It is not because the older the rock, the more metamorphosed it automatically becomes, and that all pre-Silurian rocks are so metamorphosed that they can show no fossils. The Silurian is too rich in fossils to allow that aging necessarily metamorphoses. Ever ready with a hypothesis, Darwin suggested that pre-Silurian organisms might have flourished on continents where oceans are now. And he guarded his hypothesis against contrary evidence like deep-sea borings by suggesting that the fossils of these organisms may now be metamorphosed by the great weight of the ocean above them! (pp. 306–10).

The second geological chapter was much more positive. For instance, it is necessary to Darwin's theory, unlike Lamarck's (and in certain respects Chambers's too), that there be no escalatorlike evolution, where the loss of a particular species is compensated by the evolution of the same species at a later time. For Darwin, a species has only one chance, and he was pleased that the fossil record confirms this (p. 313). Similarly, he seized on the fact that throughout the world the fossil record shows parallel developments. New kinds of species, perfected by natural selection, would have an advantage over the old forms all over the world and thus would spread and leave similar fossil deposits worldwide (p. 327). Also, the fact that "the more ancient a form is, by so much the more it tends to connect by some of its characters groups now widely separated from each other" (p. 330) can easily be explained by descent with modification, as can the fact that "fossils from two consecutive formations are far more closely related to each other, than are the fossils from two remote formations" (p. 335). But what about the key question? What about progression? From background and previous hints we might expect that Darwin would take a position not unlike (the later) Owen, and this is what we find. Divergence was fundamental for Darwin, and since man is not ontologically different from the rest of creation, there could be no unilinear man-directed progression. We have divergent evolution leading to ever greater adaptive specialization. But within this framework Darwin was prepared to concede some vague subjective sense of progression: he too, as Lyell had always feared, linked progression and evolution. "The inhabitants of each successive period in the world's history have beaten their predecessors in the race for life, and are, in so far, higher in the scale of nature; and this may account for that vague yet ill-defined sentiment, felt by many paleontologists, that organisation on the whole has progressed" (p. 345).

In Owen's case we have the dual influence of von Baer and Agassiz, which led to the belief that early fossil forms are like modern embryonic forms. These two influences were at work on Darwin in the same way. Müller, Carpenter, and Owen himself convinced Darwin of the truth of von Baer's embryology (reinforced by Darwin's reliance on his principle of divergence, by his debt to Milne-Edwards, and by the fact that Milne-Edwards proposed an embryology just like von Baer's; Ospovat 1976). On the other hand, Darwin knew Agassiz's work and, like many others, was inclined to accept his links between embryology and the fossil record, though not his transcendental recapitulationism (Darwin 1859, p. 449). By the time the *Origin* was written, Huxley had attacked Owen, and so Darwin had to tread carefully. But he wrote, "If it should hereafter be proved that ancient animals resemble to a certain extent the embryos of more recent animals of the same class, the fact will be intelligible" (p. 345). We shall see in a moment, Huxley notwithstanding, why Darwin thought the fact would be intelligible.

The two chapters following the geological discussion, on "Geographical Distribution," contain some of Darwin's strongest cards, as our knowledge of his path to discovery might lead us to suspect. Here he gives some of the most striking facts of geographical distribution. For instance, it is well known that the Old and New Worlds have some very similar stations: "There is hardly a climate or condition in the Old World which cannot be paralleled in the New" (p. 346); yet the organisms of the two land masses are very different. Conversely, going down or up a continent like South America we find closely related organisms in different stations. "The plains near the straits of Magellan are inhabited by one species of Rhea (American ostrich), and northward the plains of La Plata by another species of the same genus" (p. 349). On top of this, where there are barriers we find different forms, though the absolute distances may be slight, whereas where there are no barriers we find similar forms, though they may be far apart.

Having set up his case, Darwin points out that all these phenomena could arise through Darwinian evolutionism and natural modes of transportation. And to support his general case that each new species was formed only once, Darwin provides detailed discussions of how organisms might or might not be transported around the globe, along with speculations on such things as the effects of the glacial period. As might be expected, the geographical distribution of the organisms of the Galápagos archipelago receives special attention, as he shows that Darwinian evolutionism, and only Darwinian evolutionism, can provide a satisfactory analysis. "We can clearly see why all the inhabitants of an archipelago, though specifically distinct on the several islets, should be closely related to each other, and likewise be related, but less closely, to those of the nearest continent or other source whence immigrants were probably derived" (p. 409).

Penultimately, we get a grab-bag chapter: "Mutual Affinities of Organic Beings: Morphology: Embryology: Rudimentary Organs." The natural system is explained as simply a function of common descent. Morphological problems dissolve in the same way. Take the classic homology between the hand of man, paw of the mole, leg of the horse, flipper of the porpoise, and wing of the bat. Naming Owen to support his position, Darwin argues that this problem is too much for the doctrine of final causes (p. 435; Owen had hardly adopted Darwin's solution!). But it can all be easily explained by a theory of descent with modification owing to natural selection. The same is true for such serial homologies as do exist. As I have mentioned, Darwin sided with Huxley over the vertebrate theory of the skull (p. 438). However one decides on this, insofar as organisms, through the rather useless homologies of their different body parts, show a metamorphosis from a more primitive, repetitive structure, one has evidence for transmutation. This whole question of serial homology is something of a one-edged sword. Its existence was critical to Owen's position; but, while it could be used, it was not critical to Darwin. He therefore used it where it existed but felt free to point out that there seems to be little repetition and serial homology in the Mollusca (p. 438; of course, were serial homology universal, one might consider it anomalous in Darwin's theory).

Turning to embryology, Darwin argues that the differences between embryos and adults are easy to explain in terms of the different selective pressures to which they are subject (pp. 439–50). He contends that new characteristics make themselves felt only in the adult form because embryos of different species do not feel different selective pressures as the adults do, and that this is why the embryos of species with widely different adults are often very similar. Selection forces adults apart, leaving embryos together. Darwin justifies this argument by detailed reference to related phenomena in the domestic world, and he points out that his explanation shows why the key to classification lies in embryology. Classification has its ultimate rationale in ancestry, with organisms classed according to the closeness of their lines of descent. But the embryos of modern forms tend to be like the embryos of ancestral forms; hence, two extant organisms with similar embryos probably have common ancestors, even though the adult forms are widely different (p. 449). Darwin is thus able to offer theoretical underpinning for the methodology Huxley extolled.

Furthermore, assuming that much evolution does involve adding characters to the adult, leaving the embryo untouched, we have a reason why modern embryos may well be like ancestral adults. The ancient form remained unchanged through ontogeny; it is the modern adult that is changed. Agassiz's parallelism is thus supported, although, mindful of Huxley, Darwin is careful to add that he hoped only "to see the law hereafter proved true" (p. 449). But Darwin was not a transcendental recapitualist. He followed von Baer in embryology and was thus close to Owen's position in paleontology.

Rounding out this chapter we find a brief discussion of "rudimentary, atrophied, or aborted organs." Again Darwin contends that only a theory of descent with modification can provide a satisfactory explanation, though it seems that he thought the main agency responsible for rudimentary organs was disuse rather than selection (p. 454). Finally comes a chapter titled "Recapitulation and Conclusion," noteworthy primarily for containing what must be the understatement of the nineteenth century: "Light will be thrown on the origin of man and his history" (p. 488). Virtually throughout the *Origin* Darwin had carefully avoided reference to that controversial being, man. But lest he be thought to have dishonorably concealed his views (1969, p. 130)[12] right at the end he left little doubt that evolution through natural selection was intended to apply, without exception, to all organisms.

Note three final things about Darwin's work as a whole. First, Darwin intertwined the question of evolution itself with the mechanism for evolution. His cases for evolution and for natural selection are presented as a whole. To deal with them singly, one must separate them out himself. Second, speculation about spontaneous generation is conspicuously absent. Darwin was well aware of Lyell's comments about Lamarck; so he assumed the initial fact of life as a given and wisely stayed silent about its cause. Third, confirming earlier discussion, the overall nature of Darwin's theory was consilient. He once wrote of the *Origin* as "one long argument from the beginning to the end" (1969, p. 140), and in an important sense it is. Having derived his mechanism, Darwin applied it to many subdisciplines like behavior, paleontology, and biogeography. The theory as a whole has a fanlike appearance, with natural selection functioning as a Whewellian *vera causa* (see fig. 24).

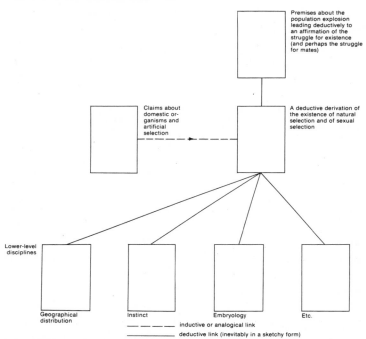

Figure 24. The three key structural elements of Darwin's theory: (1) the empiricist *vera causa*, from artificial selection; (2) the deductive core, leading to natural selection; (3) the rationalist *vera causa*, unifying the lower-level disciplines beneath natural selection. The diagram simplifies Darwin's theory: for instance, all the subsidiary mechanisms are ignored, as are some subsidiary disciplines. Also, the exact links between the core and the rest of the theory were to be questioned.

Darwin's Predecessors

In the final chapter I shall present an overall analysis of the Darwinian Revolution. But, with Darwin's theory fresh in our minds and before our view has been clouded by contemporary reactions to the *Origin*, let us briefly compare Darwin with his predecessors. We can already draw a major conclusion from our study. Without denying that earlier public evolutionists, particularly Chambers, may have prepared the way for Darwinism, we can see it would be a mistake to think that the most significant intellectual continuity in the Darwinian Revolution existed among those one would label "evolutionist."

One suspects that Darwin was eased toward evolutionism by the ideas of his grandfather and of Lamarck. Darwin shared causal speculations with them and with the later evolutionists Chambers and Spencer, specifically about the inheritance of acquired characteristics: there may well have been a direct link between Darwin and the first two on this. Generally speaking, all the evolutionists including Darwin were united by a burning urge to bring organic origins under normal, unbroken law. But after this there is almost total difference between Darwin (and Wallace) and those who publicly accepted evolution before them.[13] In no sense can Darwin be seen as the natural climax of a chain of evolutionists, even if we consider his seminal work as coming in the late 1850s rather than fifteen to twenty years earlier.

Consider Lamarck for a moment. He supported continuous spontaneous generation. He saw a constantly repeating procession up the chain of being, with—allowing for a few side trips—a progression to man. He saw a balance in nature. He was relatively uninterested in the fossil record. He had no notion of selection, and his mechanism of evolution concentrated entirely on the individual. He found species something of an embarrassment. On all these absolutely critical matters Darwin differed drastically. He had no place for spontaneous generation. He saw evolution as a non-repeatable process. He saw divergence as fundamental. He did not see progress to man. He saw a struggle for existence. He was fascinated by the fossil record. He discovered selection, and his mechanism of evolution necessarily concentrated on the group and on intrapopulation differences. And he had a carefully developed theory of the origin of species.

Much the same tale of difference can be told for the other evolutionists like Chambers and Spencer. (I exclude Wallace, who really does stand with Darwin.) Perhaps the extent to which Darwin differed from his fellows can be most dramatically illustrated by the ways he and Chambers dealt with the Galápagos fauna. Darwin, we know, explained the finches as common stock from mainland founders, molded on the various islands by different selective forces. He explained the absence of mammals in terms of difficulty in reaching the islands (Darwin 1859, pp. 393–95). For Chambers, however, who learned about the Galápagos fauna from Darwin's writings, the finches represented separate lines that had evolved independently, though in a parallel way, from different ancestors up from the sea. Absence of mammals showed that there had not yet been time for them to evolve from the present inhabitants (Chambers 1844, pp. 161–63; Hodge 1972). The two explanations could not be more different. Not only were there no sequential links from Chambers to Darwin (the reverse was true, since Chambers got his facts from Darwin), conceptually they were worlds apart. Even when someone like Spencer grasped the significance of some concept important to Darwin, such as calculations of Malthus, he did so only to show how we are all evolving into Englishmen, at which happy point Malthus will become irrelevant!

Darwin and the *Origin* therefore were not the natural culmination of a long line of evolutionists and their writings. Yet Darwin and his work did not spring from nowhere. For direct links, as well as intellectual sympathies, we must look primarily to the scientific group from which he came. Above all, we see in Darwin the influence of Lyell. We know that Darwin's geology was Lyellian through and through, and the same is largely true of his attack on the organic origins problem, even though he was doing just what Lyell could not do. Darwin as geologist satisfied all three of our divisions of Lyellianism, and as might be expected we find that the evolutionist Darwin was a complete actualist. He wanted to explain the origin of organisms by causes of a kind we see about us at present, in both the domestic and the natural worlds. Similarly, Darwin was a uniformitarian. He wanted no causes of an unknown intensity. Thus, after some initial speculating (itself a function of Lyellian worries), he wanted no supersaltations creating new species at a leap. His link to steady-statism was more tenuous. In the inorganic world he accepted it entirely and believed it ultimately fueled evolution. In the organic world Darwin was in one sense going against Lyellian steady-statism, but in other senses he was not. He was not a normal progressionist: for him, progression was almost incidental and evolution could (and sometimes did) go against it. Also, Darwin, like Lyell, saw new species appearing and old disappearing on a fairly regular, continuous basis. Even here he did not break completely with his mentor.

Darwin was Lyellian in other respects too; for example, in his suppositions about the imperfections of the fossil record, in seeing the fossil record as pertinent to evolution, and in recognizing that speculations about spontaneous generation do the evolutionist no good. (Reactions to Chambers must have convinced Darwin of his own wisdom on this score;

but his intention not to touch spontaneous generation predates Chambers.) And, Lyell apart, Darwin drew from others in his group: from the philosophers of science he learned methodology; from all the members he learned the importance of adaptation. And he also drew on those the group admired and championed, as in embryology and in understanding the significance of the fossil record. Of course there were also nonscientific influences on Darwin, particularly from his family (Gruber and Barrett 1974). But all this influence-tracing is not an attempt to deny Darwin's own genius. Even in 1859 the *Origin* was a fine achievement. That Darwin did most of the work nearly twenty years earlier is remarkable. But Darwin had antecedents and influences, and we can find them if we look in the right direction.

But enough of the *Origin*'s past. Let us turn to its future.

8 *After the "Origin": Science*

On 29 April 1856, Charles Lyell wrote in his notebook: "After conversation with Mill, Huxley, Hooker, Carpenter and Busk at Philos. Club, conclude that the belief in species as permanent, fixed and invariable, and as comprehending individuals descending from single pairs or protoplasts is growing fainter—no very clear creed to substitute."[1] This confirms our conclusions about the three decades before the *Origin*. The question of organic origins was an open sore; to many in the scientific community, none of the answers hitherto provided seemed adequate. But no one had anything better to offer.

Now, however, at the end of the 1850s, Darwin made public his theorizing on the problem. For those who wanted one, it supplied a detailed and sophisticated evolutionary answer to inquiries about organic origins. Here they found no wild speculations like Chambers's, but a careful and systematic attack on the problem by one of England's prominent men of science. This attack, moreover, at once became a focus of attention both within and outside the scientific community. Owen made brief reference to the Darwin-Wallace papers in his presidential address to the British Association at Leeds in September 1858 (Owen 1858a, pp. xci–xciii), and as soon as the *Origin* was published it was discussed at length in leading newspapers, reviews, and other media. Clergymen and other speakers denounced its vile doctrines from the pulpit and the podium, and Darwin's friends, particularly Huxley and Hooker, praised Darwin and the *Origin* in equally fervent terms. Whether the conflict pitted religion and the religious against science and the scientists to quite the extent often claimed will have to be explored. But it cannot be denied that generally the *Origin* sparked an explosion, as controversy swirled around what came to be known as "Darwinism" or, less reverently, the "ape theory." (For background, see Ellegård 1958; Vorzimmer 1970; Hull 1973*b;* and Hodge 1974.) "Darwinism," as a theory including natural selection, will be distinguished here from the more limited "evolutionism." A "Darwinian" is someone who identified with Darwin, but not necessarily someone who accepted all of Darwin's ideas.

In scientific reactions to the *Origin,* we find that there was a great deal of repetition, and some of the more popular examples and arguments against Darwin's position appeared over and over throughout the period being discussed. Sacrificing a strict historical ordering for the sake of systematic clarity, I shall follow Darwin's order of topics in the *Origin,* dealing with elaborations, criticisms, and rebuttals. Though this plan may not always yield the very first instance of a criticism, I shall try to give representative and influential examples. Since we want an overall feeling for scientific reaction to Darwinism, I shall also consider work that might be counted in Darwin's favor.

Artificial Selection

The *Origin* begins with Darwin's argument from the domestic world, based both on the changes man has wrought on plants and animals and also on the mechanism of artificial selection. In using these to support his position, Darwin reversed the position of the antievolutionists, who had always used the impermanence and limitation of changes in domestic forms as an argument against evolution. Not surprisingly, therefore, right at the beginning Darwin's work faced criticism. One who felt that Darwin's argument from the domestic world did not do all one might have hoped was Huxley. Huxley became an evolutionist, and he accepted that natural selection probably played a significant part in evolutionary changes. But because breeders never achieve full-blown physiological reproductive separation—where two groups cannot interbreed, though descended from the same known parents—Huxley always felt that natural

biologists generally, who gave way before the overall thrust of Darwin's arguments throughout the *Origin,* insofar as these proved evolution per se. From now on they allowed not just that organic origins were natural, but also that they were evolutionary. But these critics argued that natural selection could not be the only or even the main cause of evolutionary change, and they wanted to supplement it with other mechanisms. Although not necessarily for the same reasons as Huxley, supporters of other major mechanisms also often invoked some class of large variations (what Darwin called "single variations" as opposed to "individual differences") that would allow for "saltatory" evolutionary changes.[2] Thus William H. Harvey, professor of botany at Trinity College, Dublin, wrote to Darwin that though he could allow small variations for specific changes, "where a generic limit has to be passed, bearing in mind how *persistent* generic differences are, I think we require a *saltus* (it may be a small one) or a real break in the chain, namely, a sudden divarication."[3] To illustrate this position Harvey drew an analogy with that favorite Victorian toy, the kaleidoscope: as we turn it we get small changes, then suddenly everything goes and we get a whole new pattern.

The appeal to large variations was certainly made more plausible because artificial selection working on small variations failed to produce specific changes, not to mention that no one had any direct evidence that natural selection in the wild changes one species into another. On the other hand, for many of the natural selection supplementers, no less than for those who rejected evolutionism outright, it is impossible to separate scientific and religious motives completely.

Turning back to Darwin, we find that he did not budge from his position that the key to evolutionary change lies in the smallest of variations: individual differences. From the beginning he had distinguished large and small variations, looking upon the large changes as monstrosities, and through the years before the *Origin* he had become convinced that these large variations play no significant role in evolution. Of course Darwin, like his critics, always believed that natural selection is only one of a number of evolutionary mechanisms; but, whereas they wanted to demote it to a lesser role, he wanted to keep natural selection of small differences as the major cause of evolution. In reply to Harvey and similar critics, Darwin reiterated his position, claiming that, in any case, monstrosities are usually sterile or inadequate in other ways (Darwin 1959, p. 121). Most particularly, Darwin again noted that large variations entirely fail to meet the challenge of adaptation. There is no reason a

saltation should be adaptive: usually they are not. This is precisely why
one needs a selective process and why large variations are therefore redundant. And as we shall see shortly, another of Jenkin's criticisms reinforced
Darwin's belief that he should have nothing to do with any but the
smallest of heritable differences.

By this point one might wonder whether anything in the reaction to the
Origin involved new empirical evidence. Darwin could not provide direct
proof of significant evolutionary change through the selection of minute
differences; on the other hand, his critics could not provide direct proof of
significant evolutionary change through saltations (or "sports") or through
any other mechanism. But there was indeed some new evidence that
Darwin counted in his favor. Though not the direct evidence of selection
actually seen changing one species to another, it was nevertheless strongly
relevant. In particular, soon after the *Origin* was published, Henry Walter
Bates, Wallace's traveling companion in the Amazon region, produced a
brilliant piece of science supporting not merely the efficacy of natural
selection but also the untenability of large variations in species formations
and evolution generally.

Bates (1862) set himself the problem of explaining why many species of
butterflies of the Amazon Valley mimic the forms of other species—in
some cases, species not closely related. Bates found his answer within the
context of Darwinian selection. The mimicking is a form of adaptation.
For example, when Heliconidae are mimicked by members of other
families, the adaptive advantage probably is that Heliconidae are unpalatable to enemies such as birds. Hence the mimicking forms deceive their
enemies into thinking they too are unpalatable. But exactly how are these
mimicking adaptations formed? "The explanation of this seems to be
quite clear on the theory of natural selection, as recently expounded by
Mr. Darwin in the 'Origin of Species'" (Bates 1862, p. 512). Something
must be at work, and "this principle can be no other than natural selection, the selecting agents being insectivorous animals, which gradually
destroy those sports or varieties that are not sufficiently like [the forms
being mimicked] to deceive them" (p. 512). Furthermore, Bates ruled out
the possiblility that mimetic forms are produced in one step by sports,
because we find different degrees of mimetic accuracy. "Thus, although
we are unable to watch the process of formation of a new race as it occurs
in time, we can see it, as it were, at one glance, by tracing the changes a
species is simultaneously undergoing in different parts of the area of its
distribution" (p. 513). These changes show that the key to the formation

of a new species is small steps, not large jumps. Darwin was not unappreciative of Bates's support. That Bates's paper might not go unnoticed, he wrote a highly laudatory (and prudently anonymous) review of it in a journal that Huxley had started and was editing (Darwin 1863).

Let us turn now from the adequacy of natural selection as a mechanism to another criticism leveled against it. Even were natural selection granted a limited role in the evolutionary process, some critics thought Darwin was trying to use it in too many ways. Darwin wanted to claim that natural "selection" has nothing to do with conscious selective decisions. But there was a pervasive feeling that selection of any kind implies consciousness. At the very least, critics thought, Darwin's language was unduly anthropomorphic. Alternatively, those who were keen to find design in nature felt that through talk of "selection" even Darwin had to give God some explicit place in evolution (Young 1971).

Wallace, for one, though he did not agree with the thrust of the criticisms, felt Darwin laid himself open to attack on this score, and he urged Darwin to drop the term "natural selection" and replace it with Herbert Spencer's "survival of the fittest" (Darwin and Seward 1903, 1:269). Although Darwin did introduce Spencer's term as a synonym in later editions of the *Origin* and did admit he wished he had used the less anthropomorphic term "natural preservation," his basic reply was that such criticisms were misrepresentations. He felt he had as much right to metaphorical language as physical scientists: "It has been said that I speak of natural selection as an active power or Diety; but who objects to an author speaking of the attraction of gravity as ruling the movements of the planets? Every one knows what is meant and is implied by such metaphorical expressions; and they are almost necessary for brevity" (Darwin 1959, p. 165).

Although Wallace wanted Darwin to drop the term "natural selection," he was clearly at one with Darwin on the mechanism to which the term applied. A more serious difference between Darwin and Wallace arose over sexual selection. Wallace never doubted the validity of sexual selection through male combat, but he did argue at some length against sexual selection through female choice. Wallace thought Darwin was being too anthropomorphic in his explanation of some of the more striking aspects of sexual dimorphism. Although Wallace seems never to have denied absolutely that bright coloration in either sex might enhance sexual attractiveness, he came to feel that the explanation for bright males and drab females lies less in the brightness than in the ⟨.⟩rabness (letter from Wallace to Darwin; quoted in Vorzimmer 1970, ʃ. 200). Wallace argued that females need better camouflage than male; as they lay eggs, protect and rear the young, and so on. Thus, Wallace contended, natural selection favors inconspicuous females. On this score, perhaps with good reason, Wallace claimed to be more Darwinian than Darwin himself!

But no, &c. if still leaves the more odd thing unacc'd for—

On this question of sexual selection through female choice, Darwin again was not prepared to budge, writing (with truth) to Wallace that "we shall never convince each other" (Darwin and Seward 1903, 2:76). One suspects that this dispute reflects the different ways Darwin and Wallace arrived at their evolutionary mechanisms. For Darwin, animal and plant breeding yielded the key to evolutionary change. He saw the natural world through the lens of the domestic world, and the distinctions that came so naturally in the latter he automatically read into the former. Selection for beauty is important to the breeder—hence one expects to find some analogous process in nature. Wallace, not coming to selection in this manner, did not feel the force of the domestic world as Darwin did and lacked Darwin's emotional commitment to selection through aesthetic choice.

Finally, a German naturalist, Moritz Wagner, criticized Darwin for failing to place adequate emphasis on isolation in speciation. "The unlimited sexual intercourse of all individuals of a species must always result in uniformity" (Wagner 1868; quoted in Vorzimmer 1970, p. 179). Again Darwin was essentially unmoved, though he did pay friendly tribute to Wagner in the final edition of the *Origin* (1872), noting that Wagner had "shown that the service rendered by isolation in preventing crosses between newly formed varieties is probably greater even than I have supposed" (Darwin 1959, p. 176). As we have seen, Darwin had moved away from isolation as a necessary condition for speciation because he feared that isolated populations, frequently small, would have too little variation to fuel evolution. In a moment we shall learn how Jenkin underlined this fear. Consequently, Wagner could not make Darwin backtrack to his first and, as he came to think, inadequate thoughts on speciation.

As so often happens, we find here a certain amount of talking at cross-purposes. First, in defending isolation Wagner emphasized speciation caused by a species' splitting in two, which seems to demand isolation. In attacking isolation, Darwin emphasized one species' evolving into another, without splitting, which needs no isolation. Second, Darwin granted, as he had in the first edition of the *Origin*, that where splitting speciation occurs, some kind of separation is necessary. As before, however, he felt that ecological isolation might be adequate.

Heredity

On the question of heredity and new variation, in the first edition of the *Origin* Darwin had candidly admitted great ignorance. Two matters concern us here: Darwin's belief in "blending" inheritance and his "provisional hypothesis" of pangenesis.

Roughly speaking, at the phenomenal level there are two kinds of inheritance. First, when a black and a white animal breed, there is a blending of color in the offspring. Second, when a male and a female breed, sexual characteristics are passed on to the offspring undiluted. There is either a whole penis or none at all. One might think the former kind of inheritance is the norm and the latter the exception ("blending" inheritance). Alternatively, one might think the latter kind of inheritance the norm and the former the exception ("nonblending" or "particulate" inheritance). For the moment let us imply nothing definite about causes, though belief in the primacy of phenomenal blending or nonblending obviously will influence one's causal speculations.

Darwin, like most others, believed in a blending inheritance, and it is clear that right through the writing of the first edition of the *Origin*— though he had not then worked things out at the causal level—this belief crucially influenced his position. If one gets a new heritable characteristic, Darwin thought, other things being equal, it will not make much difference because in a generation or two its effects will be swamped by blending. To counteract this one needs many new characteristics of the same kind so that all organisms in the group will develop in the same way. It was because of this consideration that Darwin started to play down the importance of isolation. The larger the group, the more chance there is for new variations. This was also important in Darwin's preference for the smallest of variations rather than large ones. Almost every organism shows some slight difference from its fellows, whereas large changes are far less common. Hence, individual differences show more potential for the frequent change that can counteract blending than do occasional big changes. Finally, it seems that it was partially to counter the swamping effects of blending that Darwin started to advocate (through the various editions of the *Origin*) an enlarged role for the inheritance of acquired characteristics. We know he had always subscribed to this idea, but to defuse criticisms about the inadequacy of selection he started to give it a bigger part, both as an evolutionary mechanism in its own right and as providing more variation as grist for the selective mill. (Darwin emphasized other ideas too: for instance, he favored a "genetic momentum," where a variation in some direction increases the likelihood of another variation in the same direction. Nothing irreducibly teleological was intended here; only the successful would be around to keep on varying. Vorzimmer 1970 discusses these ideas in full.)

Although for many years Darwin had been relying on the most minute of variations, caused by secondary processes, he felt strongly confirmed by a line of reasoning in Jenkin's review. Arguing against the efficacy of natural selection, Jenkin claimed that no matter how efficient a new variation might be, in a generation or two it would be swamped out of existence. Using a typically Victorian example, Jenkin supposed that a white man shipwrecked among blacks would prove much superior to them. No matter how well he did, argued Jenkin, in a generation or two his white skin would be virtually swamped out of existence.

> In the first generation there will be dozens of intelligent young mulattoes, much superior in average intelligence to the negroes. We might expect the throne for some generations to be occupied by a more or less yellow king; but can any one believe that the whole island will gradually acquire a white, or even a yellow population, or that the islanders would acquire the energy, courage, ingenuity, patience, self-control, endurance, in virtue of which qualities our hero killed so many of their ancestors, and begot so many children; those qualities, in fact, which the struggle for existence would select, if it could select anything? [Jenkin 1867, p. 156]

Darwin had anticipated this objection and could handle it. Solitary white superiors are analogous to isolated large variations. But Darwin relied not on these, but on individual differences. Hence Jenkin's objection was harmless, though, as Darwin admitted to Wallace, he had never seen the point put so well. "F. Jenkin argued . . . against single variations ever being perpetuated, and has convinced me."[4] In the final editions of the *Origin* he paid due respect to Jenkin; and the inheritance of acquired characteristics, as a causal mechnism of individual differences, was geared up even further. As might be expected, Darwin's increased reliance on individual differences was in turn criticized, particularly on the grounds that such differences would be too slight to be captured by selection (Mivart 1870).

We turn next to the hypothesis of pangenesis. Although he never introduced it into the *Origin,* in the 1860s Darwin devised a causal hypothesis or theory of heredity to back up and tie together his various beliefs about new variations and how they are passed on.[5] This theory, called "pangenesis," was strongly indebted to a similar theory of Herbert Spencer, with a good dose of Richard Owen's ideas about the causes behind parthenogenesis (Geison 1969). It involved the idea that little "gemmules" were given off by the various cells of the body, circulating around and moving down to the sex organs, where they were combined into the sex cells. Through this theory Darwin thought he could explain many of the facts of heredity. First, normal individual differences are "random"—not being directed toward an organism's immediate needs. These are primarily a function of the way the gemmules are altered by external conditions that impinge on an organism's reproductive organs, and hence on the gemmules themselves. Second, one can see how acquired characteristics can be inherited. When something changes, like the strength of the blacksmith's arms, this affects the gemmules given off, and thus the acquired characteristic can be passed on to the next generation. Third, the blending of characteristics like skin color occurs because one has different sets of gemmules from each parent, which mingle and cause blended features. Fourth, Darwin thought he could explain such phenomena as atavism, where characteristics appear in the grandparent and grandchild but skip the intervening generation, and what he called "prepotency" (and we call "dominance"), where the characteristics from one parent mask the characteristics from the other parent. The gemmules are not blended out of existence in each generation but can somehow lie dormant and reappear in an undiluted form. Because Darwin thought the gemmules from different parents usually mingle rather than fuse, it is dangerous to speak without qualification of his commitment to blending inheritance. At the phenomenal level he believed blending was more common, and this is the sense in which the term has been used. But at the causal level Darwin's beliefs were in important respects nonblending or particulate. Only in hybrids did he think the gemmules themselves blend, though of course he believed that in each generation we get a mixing of units from both parents, which is a kind of general causal blending.

I am not sure that Darwin himself was greatly enthusiastic about his theory of pangenesis. He certainly never accorded it the importance or

certainty of natural selection, and it never found its way into the *Origin*.
His main motive seems to have been the need for some theory and the
want of any alternative. And this seems to have been the most he got from
his supporters, many of whom were notably cool, as were his critics.
Darwin's most vocal supporters were Huxley and Hooker in Britain, and
in America the botanist Asa Gray, all of whom had once publicly opposed
evolutionism. Conversely, the man who proved Darwin's most vitriolic
critic had once been firmly within the Darwinian camp, a protégé of
Huxley. This was the Catholic anatomist St. George Jackson Mivart
(1827–1900), who in 1870 published the most thorough and sustained
attack on Darwin's views in his *Genesis of Species* (Gruber 1960). Included
in this general attack was a treatment of pangenesis, and along with other
objections Mivart raised a criticism that has long been a favorite with
opponents of the view that the effects of conditions impinging on the body
can be transmitted through the sex cells to future generations. If this were
possible, asked Mivart, why is it that the Jews, who have been circumcis-
ing their sons for many generations, must continue to do so? Phenomena
like this show clearly that there must be something wrong with
pangenesis.[6]

Such doubts about pangensis were shared by others. Francis Galton
(1872), Darwin's cousin and certainly a far less hostile critic than Mivart,
experimented with transferring the blood of one kind of rabbit to another
kind, without appreciable effect on their offspring. Darwin's reply was
that in his discussion of pangenesis "I have not said one word about the
blood or about any fluid proper to any circulating system" (Letter to
Nature, 27 April 1871). But this hardly solves matters, for there is still
the question of how the gemmules are circulated about the body with the
most likely candidate, the blood, eliminated. Darwin's theory of
pangenesis was not an overwhelming success.

A hundred years after Darwin, this whole question of heredity and its
causes makes modern biologists feel they have progressed greatly since the
Origin. But, lest we criticize Darwin from our "superior" vantage point,
let me add some points to help us appreciate what Darwin himself did.
First, in proposing his theory of pangenesis Darwin may not have convinced many of his fellows, but he was not being stupid or reactionary by
the scientific standards of his day. Many scientists accepted the inheritance
of acquired characteristics: Owen, for example (1860). On this issue Darwin was at one with the best biologist in Britain. Moreover, believing that
gemmules somehow come from all over the body to the sex cells was not so

ridiculous, though there was doubt about the transmitting medium. If one accepts the inheritance of acquired characteristics, one is practically forced to accept some such medium. Second, as was hinted in our discussion of Spencer, this scientific belief about the inheritance of acquired characteristics and about a transmitting medium ties in nicely with some fundamental Victorian views on sexuality. The best medical opinion was that a man ought not have intercourse more often than once every ten days because an ejaculation of semen depletes the body generally, drawing on elements and fluids from throughout the frame, particularly the brain (Marcus 1966). Darwin's theory of pangenesis was therefore not some aberrant extravaganza of a brilliant mind, but was in harmony with (not to say a product of) the most respectable of beliefs—the selfsame beliefs that led to the universal conviction that the quickest route to the lunatic asylum lay in unrestrained self-abuse (Hare 1962; MacDonald 1967).

Complex Structures and Hybrids

As Darwin had anticipated, there was much criticism of his claim that something as complex as the eye could be fashioned by natural selection. Mivart (1870, chap. 2) held forth at length on this point. But Darwin had provided his defense here, and he stuck to it, though he added a subsidiary argument by Wallace pointing out that we should not think of the eye as perfect, nor should we consider every aspect of the eye absolutely essential and irreplaceable. Natural selection is opportunistic, working with what is at hand. This is the impression given by the eye, for, as Mr. Wallace had remarked, "if a lens has too short or too long a focus, it may be amended either by an alteration of curvature, or an alteration of density" (Darwin 1959, p. 342).

Although Darwin quoted Wallace to support his treatment of the eye, in hybridism we have another point where the codiscoverers of natural selection came to differ. In the *Origin* Darwin had argued that cross-sterility (members of two species unable to crossbreed at all) and the sterility of hybrids are by-products of natural selection. Two forms evolve apart, and when they come together again they cannot interbreed properly. Natural selection does not directly cause sterility. "The sterility of the hybrids could not possibly be of any advantage to them, and therefore could not have been acquired by the continued preservation of successive profitable degrees of sterility" (Darwin 1959, p. 424). But in the years immediately following the *Origin* Darwin came to doubt this conclusion for a while, at least in plants. He began to study *Primula* (primroses and cowslips) and made some fascinating discoveries: in some species of *Primula,* individuals come in two forms, some with long styles and some with short styles, and the different forms play a crucial role in ensuring that the hermaphroditic plants will be cross-fertilized (Darwin 1861; see Ghiselin 1969 and Kottler 1976). Thus, referring to figure 25, which Darwin used to illustrate his work, the left-hand form has a prominent style, with the stamens tucked away halfway down the corolla tube; the right-hand form has the stamens right up at the top and the style tucked away. Darwin discovered that in general the short-styled form is needed to fertilize the long-styled form and vice versa ("heteromorphic" unions). Short-styled crossed with short-styled, and long-styled crossed with long-styled ("homomorphic" unions) yield far less seed than a crossing of opposites (see fig. 26).

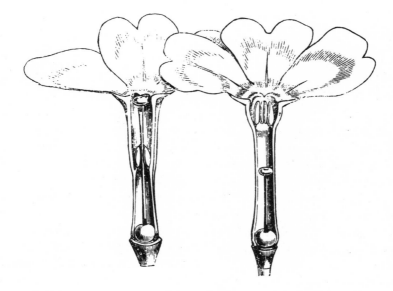

Figure 25. The two forms of *Primula: left,* long -styled; *right,* short-styled. From a paper Darwin read to the Linnean Society in 1861.

Darwin explains that there is a strong selective premium against self-fertility (because self-fertilized plants are far less vigorous than plants that have been crossed), and that this selectively fashioned self-sterility is generalized to sterility with all plants similar to the individual plant: that

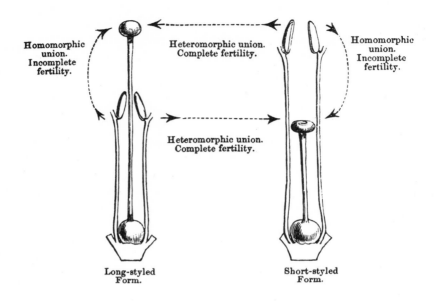

Figure 26. The possible unions of *Primula* forms and their respective fertility. From a paper Darwin read to the Linnean Society in 1861.

is, self-sterility is created by selection, but its generalization is somewhat accidental. In confirmation of this hypothesis that in at least some sense sterility could be caused by selection, Darwin suggested that the more likely self-fertilization is, the stronger should be the barriers to it. Since *Primula* are fertilized by insects that take the pollen from the stamens to the styles, the short-styled forms are more likely to be self-fertilized than the long-styled forms (as the insects probe into the plants for nectar). Hence, even among the small numbers of seeds that homomorphic unions do produce, there should be a marked difference between the long-long crosses and the short-short crosses: the short-short crosses should be far less fertile. Darwin was happy to report that this was indeed so, apparently confirming his hypothesis.

Nevertheless, Darwin soon drew back from this conclusion because he discovered that the three forms of the plant *Lythrum salicaria* do not bear out the supposition that sterility is directly related to self-fertilization (Darwin 1864). Instead of finding that degree of sterility is a straightforward function of the likelihood of self-fertilization, he found just the opposite. There was not evidence that sterility had evolved to prevent self-fertilization. Darwin therefore returned to his position that sterility was an accidental by-product of selection rather than actually fashioned by selection (see fig. 27).

Perhaps so that others would not be led astray, in the fourth edition of the *Origin* (December 1866) Darwin explained in detail why he did not ascribe cross sterility and the loss of fertility of hybrids to a selective process (Darwin 1959, pp. 443–46; Darwin and Seward 1903, 1:287–97). First, we know that fertility can be lost even when natural selection cannot have been at work. Species known to be isolated, especially during formation, have been found to be sterile when joined. Natural selection therefore cannot be a necessary condition for sterility. Second, the sterility relationship is sometimes nonsymmetrical—males of species A are sterile with females of species B, but males of species B are fertile with females of species A.

Third, and most important, Darwin thought that for selection to effect sterility would contradict the basic idea of selection. Suppose one has two groups, As and Bs, and that hybrids, ABs, are less well adapted than either parent. From the point of view of the *groups,* the sterility of As with Bs would be an advantage—no less well adapted ABs would be produced. But suppose we have a characteristic, $\emptyset$, reducing the fertility (or viability) of *individual* ABs (something of group advantage). There is no way this characteristic $\emptyset$ can be selected for, because the possessors (ABs) are at a disadvantage compared with As and Bs. Hence sterility between groups cannot be perfected by selection. (Darwin was prepared to allow for selection for sterile forms within a group, as with bees. This is easily explained if the fertile siblings or sterile forms gain an overall advantage. It is probable, however, that Darwin, unlike modern thinkers, saw a whole hive of bees as one individual, rather than seeing individual hive members as competing rivals [Wilson 1975]).

If it was indeed Darwin's intention to prevent others' being led astray, the results were quite other than he had hoped. Darwin's discussion induced Wallace to argue against him, claiming that because sterility would be advantageous to groups, it could indeed be improved and fixed by selection. "It is admitted that partial sterility between varieties does occasionally occur. It is admitted [that] the degree of this sterility varies; is it not probable that Natural Selection can accumulate these variations, and thus save the species?" (Darwin and Seward 1903, 1:294). Though each tried to see the other's point, Wallace and Darwin remained apart.

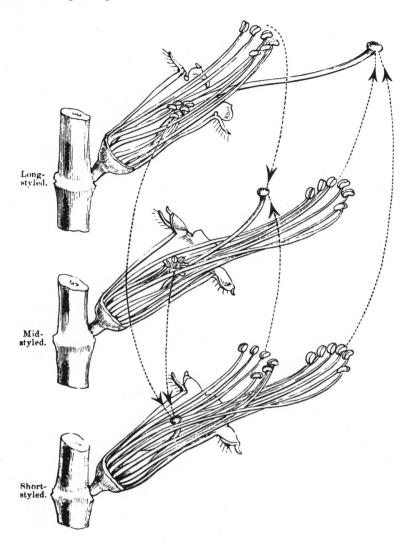

Figure 27. Flowers of the three forms of *Lythrum salicaria*, with the outer parts stripped away to show the stamens and styles of the three forms. The dotted lines show which pollen must be applied to which stigmata for fullest fertility. Had Darwin's hypothesis that sterility was fashioned to prevent self-fertilization been correct, then sterility would have been at a maximum when the inequality in length between styles and stamens was at a minimum. However, sterility was found to increase with the inequality. From a paper Darwin read to the Linnean Society in 1864.

Nevertheless, their dispute may seem more stark than it really was. To say that two groups are different species because they do not interbreed is to cover more aspects than producing infertile or inviable hybrids. Even if the hybrids are fertile, the groups may have no inclination to breed together, resulting in reproductive isolation. Moreover, this disinclination certainly can be selected for, as both Darwin and Wallace acknowledged, though Wallace wanted to link a disinclination to breed with a propensity to produce less than fully fertile hybrids (Darwin and Seward 1903, 1:290–94). Hence Darwin did not deny that some mechanisms keeping species apart can be fashioned by selection.

Geology

Of the twin subjects geology and geography, the first gave Darwin some support but also some great trouble, whereas the second gave him unalloyed comfort. Let us turn first to geology, starting with the fossil record. Chambers's critics had contended that though the fossil record may show a general progressive rise, and though it may be true that just before new orders occur, organisms of older orders look most like them, there are nevertheless insurmountable blocks to evolutionism. The oldest fossils are not really very primitive, all the basic forms are present from the beginning, and there are unbridged gaps between organisms of different species. Darwin's interpretation (or hoped-for interpretation) of the fossil record was very different from Chambers's. Darwin hypothesized branching evolution from general to specialized, with a much more subjective kind of overall progression. But, as one might expect, similar arguments were made against Darwin—for example, by John Phillips (1860), professor of geology at Oxford and president of the Geological Society (see also Rudwick 1972). Moreover, asserted Phillips, the fossil record is too well known to lend plausability to Darwin's Lyellian strategy, which refers all conflicts between paleontology and evolutionism to the inadequate preservation of past forms. Essentially, the history of the past has been spelled out, and one can no longer contend that gaps in the record are a function of imperfect preservation rather than nonexistence.

We have here all the makings of an impasse between Darwin, or any evolutionist, and the critics of evolutionism. Depending on how far one accepts the unverifiable hypothesis that the fossil record is fragmentary, one has evidence for or against evolutionary hypotheses. But shortly after the *Origin* was published one of the major paleontological objections to evolutionism started to decline. In 1861, in the limestone quarries of Solnhofen in Bavaria, was found the first specimen of *Archaeopteryx*. Though this organism was, in Owen's words (1863), "unequivocally a bird," it had unmistakable reptilian features. It was a classic example of a "missing link," as was another newly described fossil, a small bipedal dinosaur, *Compsognathus*, whose birdlike features helped close the bird-reptile gap from the other side. Huxley (1868, 1870a), who fully recognized the significance of *Archaeopteryx* and *Compsognathus*, asserted that these fossils made the "title deeds" of evolutionism as full as one could reasonably expect. Of course there still were gaps between fossil species, and so supporters of saltatory evolutionary mechanisms could turn to the fossil record for comfort. But absolute antievolutionism based on the fossil record became less tenable.

But what about Darwin's own version of evolutionism and the fossil record? Although nothing definitive appeared that made a Darwinian interpretation of the fossil record utterly compelling, enough pertinent evidence was uncovered to please Darwin and to steer Huxley toward Darwin's way of thinking.

In the early 1860s, though he was by then an evolutionist, Huxley still argued strenuously against reading progression into the fossil record, in particular progression from basic embryonic forms to specialized forms. Put bluntly, the record "negatives those doctrines" (Huxley 1862, p. 528). Part of Huxley's reason for this position was his crusade against Owen, but at the same time Huxley was denying something Darwin hoped was true. By 1870 Huxley was more positive. He still refused to give way on lower forms of life, "but when we turn to the higher *vertebrata*, the results of recent investigations, however we may sift and criticize them, seem to me to leave a clear balance in favour of the doctrine of the evolution of living forms one from another" (Huxley 1870b, p. 529). In the case of the horse, "the process by which the *Anchitherium* has been converted into *Equus* is one of specialization, or of more and more complete deviation from what might be called the average form of an ungulate mammal" (1870b, p. 535). In the 1870s Huxley's case for the horse was made even stronger, particularly by the American paleontologist O. C. Marsh, who could draw on the far richer fossil resources of the New World. Even today, the most famous of all specializing trends in the fossil record is surely the reduction in number of toes from the earliest horse (*Eohippus*) to the modern horse (see fig. 28 and Simpson 1951).

The old unilinear progression was gone.[7] In its place was branching and

Figure 28. Marsh convinced Huxley that the earliest horse was *Eohippus,* and to celebrate Huxley drew this cartoon, complete with rider!

ever-increasing specialization as organisms worked themselves into new niches. It could now be seen that the fossil record (once started) could bear such an interpretation, which in turn could be taken as evidence of evolution. There was still room for debate about causes: Huxley continued to favor saltations supplementing selection, and many late nineteenth century paleontologists, particularly in North America, were Lamarckians. But it was growing clear that not just any evolutionism would fit the fossil record, and that Darwin's theory not only fit the record but predicted and demanded the new progressionism. Someone committed to the efficacy of natural selection could now turn with more confidence to the fossil record. A poignant footnote to this story of the interpretation of the fossil record is that von Baer, whose embryology proved so crucial, not only remained opposed to evolutionism but denied any paleontological interpretation of his embryology. To the end, he insisted that ancient fossil forms are as far removed from modern embryos as are modern adult forms (von Baer 1873).

Another major antievolutionary argument based on the fossil record involves the complexity of the earliest known fossils. We have seen the remarkable way Darwin tried to circumvent this problem. In the 1860s the Darwinians—Darwin and Huxley in particular—thought that, as with progression and gaps, a newly discovered piece of evidence was decisive in the evolutionary cause. This was *Eozoön canadense*, the "dawn animal of Canada" (O'Brien 1970). This organism, identified as a foraminifer by the geologist J. William Dawson, principal of McGill University, and by W. B. Carpenter, a leading authority on *Foraminifera*, came from very old rocks in Canada, and at one stroke it pushed the fossil record back to twice what it had been. The Darwinians seized on it as evidence of the great age of life on earth, thus expanding the time available for the newly discovered earliest forms of life to evolve into the complex forms that had previously been considered the earliest. *Eozoön canadense* removed the difficulty of the sudden appearance of sophisticated life forms.

Unfortunately, the Darwinians' house was built on metamorphic sand. After an acrimonious dispute, whose unpleasantness was beginning to be the norm for scientific controversy, the inorganic origin of *Eozoön* was established; the supposed foraminifers were nothing but inorganic structures of metamorphosed rock. The Darwinians hence lost their pre-Cambrian organisms. But they were not the only ones who lost. *Eozoön*'s two strongest backers were Carpenter and Dawson. Although Carpenter was one of the first and strongest of evolutionists, Dawson was a violent antievolutionist. Appalled when the Darwinians hailed *Eozoön*, he had countered that it shows the widest gaps of all in the fossil record—the paradigm of antievolutionary evidence! So the demise of the dawn animal brought comfort to neither the evolutionists nor their opponents.

There is more to geology than paleontology, and Darwin wanted more from geology than a favorable fossil record. He needed to prove the earth old enough for the slow, gradual process of natural selection. Here, perhaps more than anywhere else, Darwin met his strongest scientific checks (Burchfield 1974, 1975). First was Darwin's absolute time calculation based on the denudation of the Weald. On the basis of a simple—his critics were to argue, dreadfully naive—calculation, Darwin had concluded that the denuding must have taken some three hundred million years. This conclusion soon came under fire. Phillips, for one, took issue with Darwin's figures and calculated that a river like the Ganges, acting more slowly than the marine action Darwin invoked, could do the job in 1.3 million years. Denouncing Darwin's calculations as an "abuse of arithmetic," Phillips (1860) shortly produced his own estimate of absolute earth time and found only 95 million years available since the beginning of the Cambrian—a period far more limited than Darwin then thought he needed.

Although Darwin had geologically qualified defenders—including J. B. Jukes, head of the Irish Geological Survey—he soon came to regret bitterly his dabbling with figures. To Lyell he wrote of "Having burnt my own fingers so consumedly with the Wealden" (Darwin and Seward 1903, 2:139), and when the third edition of the *Origin* appeared in April 1861 the calculation from the denudation of the Weald had vanished.

But this squabble over the Weald was just the start of the age-of-the-earth question. We now come to one of the most interesting of all of the disputes raised by the *Origin*, and certainly the criticism that (rightly) worried Darwin the most. Kelvin, later supported by Jenkin, claimed that physical considerations show that the world is far younger than uniformitarian geology supposes, and certainly far younger than the time span demanded by the theory of evolution through natural selection. This argument, based on the time it has taken the earth to cool to its present temperature, reduces the earth's age to a span of from 20 to 400 million years, with 98 million the most probable figure (Jenkin 1867). Moreover, the argument was based not on unknown and hypothetical geological processes, but on apparently firm calculations and data from the leader of the sciences—physics. We learn much about the various members of the Darwinian group when we look at their reactions to this objection. Here they were, supporting a highly controversial theory in the biological and geological sciences, faced with a seemingly insurmountable counterargument from physics. How would they handle it?

Huxley always strikes me as rather straightforward, inevitably preferring the simple, unsubtle solution to the complicated one. His response here was typical. Speaking as a biologist, he simply put all the responsibility on geology, then argued that biology need not worry! "Biology takes her time from Geology. The only reason we have for believing in the slow rate in the change of living forms is the fact that they persist through a series of deposits which, geology informs us, have taken a long while to make. If the geological clock is wrong, all the naturalist will have to do is to modify his notions of the rapidity of change accordingly" (Huxley 1869, p. 331). Of course, as Kelvin rather sneeringly noted, this evasion hardly makes it plausible that evolutionary change should have been caused by so leisurely a process a natural selection (Thomson 1869). But, for all his emotional identification with Darwin, Huxley put evolution first and natural selection second. And critics like Jenkin, who were using physics to cut down earth history, still allowed that evolution could happen, even given their own reduced time span.

Wallace not only had a more personal stake in natural selection, he had a penchant for ingenious arguments. His reaction too was typical, for he tried to adapt the whole geological and biological clock to the new time scale (Wallace 1870*a*). Relying on an argument about the causes of glaciation by one James Croll (1867, 1868), that glacial periods are a function of the oscillating orbit of the earth, Wallace managed to squash geology down to the physicists' time scale. He then argued that since geological phenomena are much speeded up, they might be expected to have an exaggerated effect on organisms, causing many new stresses and demands. Thus new characteristics would be acquired and inherited more quickly, quite apart from the effects of increased struggles and selective forces caused by the rapid coming and going of glaciers. Then, in an argument that has echoes of Lyell's assertion that we should not expect to see new species created because they are so rare, Wallace contended that because of the earth's orbit we are now living in a period of low glacial activity and perhaps ought not argue analogically to other periods of time. "High eccentricity would therefore lead to a rapid change of species, low eccentricity to a persistence of the same forms; and as we are now, and have been for 60,000 years, in a period of low eccentricity, *the rate of change of species during that time may be no measure of the rate that has generally obtained in the past geological epochs*" (Wallace 1870*a*, p. 454; his italics).

Darwin, like Wallace, had an emotional identification with natural selection and, of course, with the whole program of uniformitarian geology. He felt that it was all very well for nonbiologists and nongeologists like Jenkin to wipe out whole areas of theory at one swoop—they did not have to live with the biological and geological problems. On the other hand, as we well know, Darwin was extremely sensitive to and respectful of physics. If this were not enough, his son George, one of Kelvin's brightest research assistants, was ready to make matters very clear! Essentially, therefore, Darwin compromised. He went part of the way with people like Wallace, reluctantly admitting that things may have gone a bit more quickly in the past and relying a bit more heavily on evolution-speeding alternatives to selection, such as the inheritance of acquired characteristics. But then Darwin rather miserably dug in his heels and refused to defer to the physicists, hoping that someday enough time would be found for his theory (Darwin 1959, p. 728). George Darwin wrote to Kelvin in 1878, "I have no doubt however that if my father had had to write down the period he assigned at that time [of writing the *Origin*], he would have written a 1 at the beginning of the line and filled the rest up with o's.—Now I believe that he cannot quite bring himself down to [the] period assigned by you but does not pretend to say how long may be required" (unpublished letter in Kelvin Papers, D.8, Cambridge University Library; quoted in Burchfield 1974, p. 321).

Geographical Distribution

Geographical distribution was the area in which Darwin and many of his followers felt happiest and found their strongest support. We know that Darwin was converted to evolutionism essentially because of evidence from organic geographical distributions, and we know that this was also a strong consideration for Wallace. Since it was Lyell who taught Darwin and Wallace to concentrate on these facets of the world, it was appropriate and predictable that, when they started to expound their evolutionary positions, at the level of science it was geographical distributions that most strongly inclined Lyell himself to give up his lifetime opposition to evolutionism and join the Darwinian camp.

For Lyell there was far more to the organic origins question than mere science. But after much heart-searching, in the mid-1860s Lyell rather heroically decided he had to endorse evolutionism, and he also decided that, after a fashion, the chief causal mechanism must be natural selection (Lyell 1868). In these decisions geographical distribution played a key role. Wallace's paper of 1855 impressed Lyell deeply, despite Wallace's fears that it had gone generally unnoticed, for he realized at once that if Wallace's claims were true, evolution was but a short step away (Wilson 1970, p. 1). When some ten years later Lyell followed his friends over to evolutionism, distribution was again crucial. Special creation, miraculous or nonmiraculous, puts heavy emphasis on God's tendency to design being manifested in the world. Consequently, one is constrained to believe that organisms generally are where they are because they (and only they) are best suited for the local conditions. Darwin contended that this was not necessarily so. Indeed, given forms A and B, where A has evolved in a limited area with consequent limited struggles, whereas B has evolved through much larger struggles for life, were form B to come across some secluded versions of form A, in all probability form A would be wiped out. In short, when an alien type colonizes a new area, overwhelming the natives, we have evidence for Darwinism and against special creation. Lyell (1906, 2:216), knowing this commonly happened, recognized it for what it was worth and told a friend that "nothing contributed more to shake his belief in the old doctrine (which he formerly held) of the independent creation of species, than the facts of which so many have lately been recorded, relating to the rapid naturalization of certain plants in countries newly colonized by Europeans."

There were some who felt that, though geographical distribution proved evolution, it left natural selection an open question. Alfred Newton, professor of zoology and comparative anatomy at Cambridge, published a fascinating paper on extinct birds like the dodo, arguing that their relations and distributions point to a common ancestry (Newton and Newton 1869). However, although personally a Darwinian, he felt his evidence obliged him to be more reserved. "Whether this result [evolution] can have been effected by the process of "Natural Selection" must be regarded as an open question." Others, particularly those closest to Darwin, explicitly used the struggle for existence and natural selection to explain distributions and, conversely, thought their results supported their theory.

Thus Hooker (1861) discussed the distribution of Arctic plants and asserted that only by assuming evolution through natural selection could one account for various anomalies of distribution. Hooker drew special attention to the peculiarities of the flora of Greenland. Although Greenland is far closer and more similar to Arctic North America than it is to Scandinavia, its flora is relatively much closer to the flora of Scandinavia, though the flora of Greenland is sparse, with a smaller absolute number of Scandinavian forms than North America. For Hooker this demanded a Darwinian explanation. First he hypothesized that the Scandinavian flora is of great antiquity and originally covered the whole of the polar zone. Then he had a glacial period drive plants south; and finally he reduced the cold, allowing plants to return. In Greenland the plants were confined to a peninsula and could not mix with North American plants. Moreover, "many species would, as it were, be driven into the sea, that is, exterminated, and the survivors would be confined to the southern portion of the peninsula, and not being there brought into competition with other types, there could be no struggle for life amongst their progeny, and consequently no selection of better-adapted varieties" (1861, p. 254). In North America, however, plants would not be driven into the sea, and the new southern conditions—climate, aboriginal species, and so on—would cause new struggles and plenty of speciation. Hence, when the warm climate returned, many Scandinavian plants could return north along with many North American neighbors. We have, therefore, "no slight confirmation of the general truth of Mr. Darwin's hypothesis" (1861, p. 254).

Wallace was another who, in the 1860s, went on adding to the evidence for evolution and selection from geographical distributions. In a paper on Papilionidae, a family of butterflies from Malaysia, he averred that only on Darwinian principles could the various distributions be explained (Wallace 1866). Wallace noted that the more extensive the range of a particular species, the more variability there is between individuals. In the widely dispersed species, there would naturally be different conditions, hence different selective forces, and hence different forms; but full speciation is checked by the interbreeding that still goes on. In a species with a limited range, all individuals tend to face the same conditions, and hence we get more uniformity. Wallace also gave selective answers to explain the variability—for instance, Batesian mimicry, sexual selection (at this point he did not deny Darwin's selection through female choice), and his own hypothesis of camouflage for females. And he showed how we get gradations, from slight varieties to species, just as we would expect with only gradual change.

All in all, therefore, it was in the area of geographical distribution that the Darwinians felt most at home and felt that the mechanism of natural selection found fullest application and confirmation.

Morphology

For the seeker after the naturalistic explanation, the idea of evolution was a godsend to the problem of general homology. Before the *Origin*, Huxley made no attempt to hide his distaste for explanations referring to archetypes, and when the *Origin* appeared he was no less forthright. "That such verbal hocus-pocus should be received as science will one day be regarded as evidence of the low state of intelligence in the nineteenth century" (Huxley 1894). For himself, an adequate and convincing explanation of the similarities between the limbs of different mammals was that these animals had all evolved from common ancestors. Nevertheless, general homologies did not automatically convince people of the adequacy of natural selection. Such homologies seem compatible with other evolutionary mechanisms (such as those allowing large variations), and the adequacy or inadequacy of natural selection was decided on other grounds—as is still true today.

Since Owen was responsible both for strongly drawing attention to general homologies—convincing people that they must be acknowledged and explained—and for providing the explanation that so raised Huxley's ire, let us end this survey of scientific reactions to the *Origin* by a brief reference to him (MacLeod 1965). We know that Owen had previously tried to have matters two ways, praising *Vestiges* to its author and condemning it to its critics, and here we see this pattern repeated. After his brief reference at the British Association meeting of 1858, Owen let Darwin think that "at bottom, he goes an immense way with us" (F. Darwin 1887, 2:240). He then wrote a long anonymous review in the *Edinburgh Review* in which he brought up many of the standard arguments against evolution, intimated that if one must be an evolutionist one would do much better to take one's cue from *Vestiges,* then recommended to the reader the position of Professor Owen![8]

Having reaffirmed his earlier position, during the next few years Owen proceeded to backtrack and accept evolutionism after all. The exact details of Owen's new position need not concern us here, which is fortunate because no one, including Owen himself, seemed to have much idea of just what the position was. Owen certainly embraced some kind of teleological theory—a move that became very popular. But, on the principle that if you can't beat them you join them, Owen calmly suggested that the struggle for existence might have been important. Smaller forms may have evolved because "the smaller and feebler animals have bent and accommodated themselves to changes which have destroyed the larger species—they have fared better in the 'battle' of life" (Owen 1866–68). When an incredulous critic asked if this were not Darwinism, Owen pompously pointed out that he had used a similar phrase in the early 1850s. If he could not get credit for archetypes, Owen was prepared to take credit for selection. As we shall see when we discuss religion, there was more to Owen's story after the *Origin* than pure science. After the *Origin,* most scientists accepted Owen's data and, with few qualms, sloughed off his archetype theory. To some extent, even he himself did so.

A Hundred Years of Darwin

We have been considering scientific reactions to Darwin's theory item by item. Lest we miss the woods for the trees, let us stand back for a moment and try to gauge overall scientific response to Darwin—the state of things by about 1875. Two general points stand out. First, for all the scientific criticism of Darwin, a very high proportion of active scientists, particularly biologists, had followed him over to evolutionism. In Britain it is hard to think of an exception among major figures. From friends like Hooker and Huxley to foes like Owen and Mivart, British biologists were evolutionists. Moreover, people became evolutionists at a remarkable speed. In 1859 hardly anyone was an evolutionist—and those who were tended to be on the fringes of biology. Certainly, by 1875 the British biological community was in favor of evolution, but in fact one can put the date of general conversion back by some years. By 1865 most British biologists were evolutionists, and one can make a case for setting the date as early as 1862. Of course, at this date there were some exceptions, like Sedgwick (though we shall see that even he was not immutable), and if we broaden our scope to include America and Canada we find men like Agassiz and Dawson. But as a general rule, certainly in Britain, the holdouts tended to be the older men who were dropping away from active science. I am confining my remarks basically to the scientific community, particularly to those interested in organic origins; but in general, as with the emperor's new clothes, once Darwin had spoken up others followed.[9]

Second, most people were much more hesitant about Darwin's mechanism of natural selection. There were always doubts about its power to do all Darwin claimed, and by the early 1870s support for it had declined even further. This grew from concerted attacks launched against Darwin about the beginning of the decade, notably that of Mivart in his *Genesis of Species,* a powerful compendium of every significant charge that could be made. Moreover, Darwin's case was not helped by his own constant tinkering with the *Origin.* He brought out six editions between 1859 and 1872, rewriting and augmenting in response to criticism, until he had on his hands an awkward patchwork quilt of a book rather than the forceful and elegant work he had started with.

One suspects that the dual reactions to the *Origin*—acceptance of evolutionism and hesitancy about natural selection—were not unconnected. For various reasons, people did not want to accept natural selection. There was therefore a tactical advantage to accepting evolution: one could show one's "reasonableness," one could accept all of Darwin's arguments that one found attractive, one could avoid an all-out negative war, and therefore one could more easily balk at selection. Much of this book shows that people were ready for evolution itself; but it might well be that they gave in to evolution so that they might concentrate their attack on selection. Nevertheless, one suspects that even those who objected to selection found evolution made more credible by selection: a suggested mechanism, even if untenable, helped establish the plausability of evolution.

At this point we must strike a strong note of moderation. Natural selection was far from a total failure, even aside from its indirect role in promoting evolution per se. First, nearly everyone, certainly nearly every evolutionist, allowed that natural selection did exist and that it could and did cause heritable change, though many felt it was not nearly as powerful as Darwin claimed and that it must be supplemented. Second, there were some who followed Darwin in agreeing that natural selection could do almost everything he claimed. These were a minority, but they did exist. Not surprisingly, they tended to have the same research interests as Darwin, being particularly interested in geographical distributions and other areas that needed a working mechanism, not just vague platitudes about evolution. Around Darwin there was a group—Hooker, Bates, Wallace, and others—who were actually using selection in their studies and who, with Darwin, went beyond evolution to selection. They started a tradition that still exists and that still includes the strongest advocates of selection (Ford 1964; Mayr 1963; Dobzhansky 1970). Third, those who hesitated over or rejected the full adequacy of natural selection hardly formed a united front. Most thought that natural selection must be supplemented with saltations, though their reasons for and ideas of such saltations tended to differ. Others, most notably Spencer in England (1864–67), felt that in the inheritance of acquired characteristics lay the chief key to evolutionary change. But there was no question of any consensus, nor did Darwin's critics offer much that could be used to solve real problems. The people who went beyond abstract ideas of evolution to problems needing a workable mechanism were the Darwinians using selection. Drawing together these three points, it would be a mistake to think that Darwin's mechanism of natural selection was as scientifically unsuccessful as his general evolutionary thesis was successful.

In the century that followed the initial reactions to the *Origin,* a great deal happened; but in another sense many of the controversies sparked by the *Origin* still seem curiously fresh. Two unresolved items stood out during our survey of scientific reactions to Darwinism: the question of heredity and the question of the age of the earth. Answers to both these questions have been forthcoming, though they did not follow immediately on our period and though they go beyond the British community that has been our focus.

With hindsight, it is generally agreed that Charles Darwin and the others who accepted blending inheritance and the inheritance of acquired characteristics were on the wrong track, though this does not mean Darwin was foolish to accept such ideas (Churchill 1968; Provine 1971; Carlson 1966; Dunn 1965). Modern thinking on the subject dates from the work of the Moravian monk Gregor Mendel, who, unknown to virtually everyone in the 1860s, was taking an approach different from, but eventually more fruitful than, that of Darwin (Stern and Sherwood 1966; this includes translations of Mendel's key papers). Although in his published papers Mendel was guarded about how he regarded the ultimate units of heredity, essentially it seems that he accepted a nonblending phenomenal view of heredity. To explain this he supposed that the causes of heredity could be passed on unblemished and undiluted from generation to generation—thus eliminating the mixing of gemmules that we find in pangenesis. Any blending of phenomena is caused by blending of the effects of the units, rather than by blending (e.g., through mixing) of the units themselves.

Mendel's work was ignored until 1900, when it was rediscovered by three investigators separately. By that time great advances had been made in cytology, and the German biologist August Weismann had sounded the death knell of the other important element in Darwinian heredity, the inheritance of acquired characteristics: with great conviction he had argued that the sex cells have an existence independent of the rest of the body, and thus no new heritable changes can come from habit, use or disuse, and the like.

The new "Mendelian" genetics and cytology were soon brought together and developed into the "classical theory of the gene," particularly by T. H. Morgan and his co-workers in New York. Genes exist on paired chromosomes within the nuclei of the cells. These determinants of function and heredity are passed from one generation to another according to fairly simple rules. Normally they remain unchanged, but sometimes they "mutate"—that is, they change instantaneously, causing new characteristics. Genes do not change in response to needs, nor do changes in the total organism affect the genes in the sex cells. With this new gene concept many of the difficulties Darwin faced were laid to rest: no longer need a Darwinian worry about the swamping effect Jenkin had posed. Because the units of heredity stay uncontaminated through the generations, they can be preserved and promoted by selection however much in the minority they may be (Dunn 1965).

Nothing in real life is completely straightforward, and so it proved in the history of evolutionary theory. Because early twentieth-century geneticists were interested in large differences, the belief grew that the significant changes in evolution are large: Mendelian genetics was taken to prove saltationism. It solved the problems of Darwinism only to make it redundant! It was not until about 1930 that scientists realized that the important heritable variations are small, and that Mendelian genetics and Darwinian selection are complements, not rivals (Provine 1971). Looking back toward both Darwin and Mendel, and bringing all together in the modern "synthetic theory of evolution," we conclude that evolution is a function of selection working on small, heritable variations brought about by random mutation.[10]

The problems of heredity were internal to biology and thus had to be solved by biologists. The problem of the age of the earth was external to biology; it came from physics. Biologists could try to get around it, as by speeding up the process of evolution, but ultimately they had to wait for the physicists to free them from the restrictions imposed by Kelvin. The physicists accomplished this at the beginning of this century: the discovery of radioactive decay and the heat it generates implied that the earth is far older than Kelvin thought possible. Even now we have not gone back quite to the age Darwin first supposed necessary, but there is adequate time for the slowest of evolutionary processes (Burchfield 1975). If there is an afterlife for evolutionists, we might forgive Darwin for being a little smug about this.

These then were the fates of the two major problems left from our period. Today we feel that they have been answered and are no longer live issues. But some other scientific issues arising from Darwin's *Origin* still seem pertinent. There is continuing debate about the status of natural selection as a cause of evolution. No longer do scientists put forward Lamarckian or saltatory alternatives, and everyone seems to agree that major organs like the hand and the eye were fashioned by selection: but whether all of evolution is a direct function of selective forces is still a matter of controversy. Some want to argue that a goodly part of evolutionary change comes about through "drift"—through random fluctuations of characters with too little effect to be controlled by selection (Lewontin 1974).

Today's controversy about natural selection does not have the same bases as the controversy of a hundred years ago. Three other issues, however, seem very little changed. First, there was the disagreement about speciation, with Wagner insisting on geographic isolation and Darwin denying its necessity. In recent years the systematist Ernst Mayr (1963) has argued strongly for Wagner's position; but there are others who believe with Darwin that speciation does not necessarily involve spatial separation, although like him they usually invoke some kind of ecological isolation (Ford 1964; Kottler 1976). Second, there was the disagreement between Darwin and Wallace on whether selection can preserve something of value to the group rather than to the individual. The vast majority of biologists today agree with Darwin that this is not possible (Williams 1966); others are not prepared to rule out Wallace's case entirely, though they feel that selection for the group can happen only in very specialized instances (Wilson 1975). Third is the other disagreement between Darwin and Wallace—sexual selection. Undoubtedly, some biologists still feel that Darwin was unduly anthropomorphic in his notion of female choice, but in recent years there has been a fairly dramatic swing toward Darwin's position (Mayr 1972*c;* see also other articles in Campbell 1972).

One could find many other points linking the scientific reception of Darwin's theory with the present. To my mind one of the most interesting concerns the beginnings of life on earth. It is now believed that the earth is a little less than 5 billion years old, and the earliest known fossils—very primitive bacteria—go back to the oldest sedimentary rocks, more than 3 billion years ago. So here again some of Darwin's problems have vanished, though little is yet known about the first appearance of the main invertebrate groups (Maynard Smith 1975). But my aim is not to give a general history of evolutionary ideas; as I stated at the beginning, my focus is the switch to evolutionism and the extent to which this was linked to Charles Darwin. By 1875 the switch had basically been made. Darwin's last great work, the *Descent of Man,* was published in 1871, and the last edition of the *Origin* appeared in 1872. It thus seems that Darwin's personal role was over.[11] But Darwin's success, if that is an appropriate term, was limited. Most could not follow him all the way on his chief mechanism of natural selection. Though many of the reasons for this hesitancy were scientific, there were other reasons, to which we must now turn.

9 After the "Origin": Philosophy, Religion, and Politics

Philosophy   Before the *Origin,* philosophical questions were critical. After the *Origin* they continued to be so. First, quite aside from the matter of natural selection, many people found the whole idea of evolution attractive because it alone seemed to offer a *scientific* answer to the organic origins problem. Lyell put his finger on it when he wrote in his notebook: "The claim of the transmutationist to be a speculation meriting more favour than any other is that it is at present the only one which even pretends to bring the successive changes under a law or within the dominion of science" (Wilson 1970, p. 246). In the period after the *Origin,* as in earlier periods, there was a powerful metascientific drive to subsume organic origins beneath law, and Darwin greatly benefited. For example, A. C. Ramsay, professor of geology at University College, London, was converted to Darwinism almost at once, for just such philosophical reasons. As he wrote to Darwin: "The succession of small miracles required to produce certain species in a formation just a very little different from those in the preceding formation went sadly against my mental stomach" (letter from Ramsay to Darwin, 21 February 1860; quoted in Wilson 1970, p. 356).

Second, philosophical considerations played a major part in Darwin's invention of pangenesis.

One might feel that the whole excursion into the causes of heredity was rather pointless from Darwin's viewpoint. He could just posit new variations and leave matters at that. He had done this in the *Origin* and had justified his position by analogy with the domestic world. There was no need for him to delve into causes, any more than Newton needed to go beyond taking gravitation as a given. But as one sensitive to philosophy Darwin could not sit back at this point. His need to supply some such theory as pangenesis stemmed directly from the philosophy of Herschel and Whewell, who divided theories into formal or empirical and causal or physical parts and specified that the best theories include the second parts. Darwin accepted this entirely, and shortly after his discovery of natural selection we find him speculating on heredity in the philosophers' own language (de Beer et al. 1960–67, E, pp. 53–55). Darwin's invention of pangenesis nearly thirty years later was an attempt to fulfill what he felt as his philosophical obligations.

Third, in the aftermath of the *Origin* some interesting issues were raised about theory confirmation—specifically in relation to Darwin's theory (Hull 1973*a*, *b*). There was a serious, though not fully recognized, philosophical difference between Darwin and many others, including some of his closest supporters like Huxley, over the true nature of theory confirmation. By concentrating on the artificial selection analogy, Huxley always maintained an important reservation about the total efficacy of natural selection. But Darwin himself did not feel this insecurity, for he was absolutely convinced that despite all difficulties' his theory was essentially proved. Although Darwin thought the analogy from artificial selection was strong evidence for natural selection and its effects, time and again he rested the case for natural selection not on the analogy but on natural selection's wide explanatory power—on its being at the center of a consilience. "I must freely confess, the difficulties and objections are terrific; but I cannot believe that a false theory would explain, as it seems to me it does explain, so many classes of facts" (F. Darwin 1887, 1:455). On at least one occasion, when defending his appeal to all the evidence rather than to definitive direct or analogical evidence, Darwin likened his theory to the wave theory of light: something Whewell invoked when, arguing against Herschel, he claimed that the best evidence for a theory is total explanation rather than direct evidence (Darwin and Seward 1903, 2:184).

Whereas Huxley argued that the only way a theory can be definitively proved is by empiricist analogy from direct experience, Darwin argued rationalistically that a consilience is the key to confirmation. They viewed the question from rather different philosophical positions. Of course, using the terminology of the time, Darwin thought that an empiricist *vera causa* was of value: that was why he included artificial selection in the first place, and he usually referred to it even after the publication of the *Origin*. But when under fire because of the purported inadequacy of his empiricist *vera causa,* Darwin showed that he was prepared to accept (if not demand) a rationalist *vera causa:* one centered on a consilience. This was not a total change of mind, because he had always liked the rationalist *vera causa* criterion (although perhaps like Herschel he would not have called it this). But as the criterion stood the test of time Darwin liked it that little bit more! Perhaps it is fairest to say he changed his emphasis. When he converted to Lyellianism Darwin was under the influence of the empiricist criterion. Whewell showed that one had to take the rationalist criterion seriously, and so Darwin was guided in his theory by both criteria. Then as people faulted him on the empiricist criterion, he turned to the rationalist criterion. Huxley, on the other hand, insisted absolutely on an empiricist *vera causa*. Indeed, he showed this insistence in a letter to a friend, the Reverend Charles Kingsley. "[Darwin] *has* shown that selective breeding is a *vera causa* for morphological species; but he has not yet shewn it a *vera causa* for physiological species."[1]

What we know of the influences on Darwin and Huxley supports this conclusion about their philosophical differences. Darwin was influenced by Whewell and Herschel, both of whom lauded consiliences (though only Whewell based his *vera causa* concept on consiliences and though Herschel also favored empiricist *verae causae*). Huxley, on the other hand, was by his own admission close to John Stuart Mill. But Mill, in his *System of Logic* (1875, 2:19), attacked Whewell, denying that a consilience is the definitive mark of truth. Hence one would not expect Huxley to be enthusiastic about consiliences—and he was not. In this context Mill himself seemed to get the logical status of evolutionary theory mixed up—at least Darwin's position on the status of the theory. Praising Darwin, Mill added: "The rules of Induction are concerned with the conditions of Proof. Mr. Darwin has never pretended that his doctrine was proved" (Mill 1875, 2:19). But this of course was just what Mr. Darwin had pretended. Darwin, however, was relying on a method of theory confirmation that Mill repudiated. Mill did bring himself to say that Darwin's theory was not as ridiculous as it looked, but that his religious beliefs prevented a wholehearted endorsement (Mill 1874).

Fourth in the list of philosophical matters surrounding the reception of Darwinism is Platonism. Before the *Origin* some critics of evolutionism were Platonists, though the exact relationship between their particular versions of Platonism and their antievolutionism was not always straightforward. After the *Origin* some critics made it clear that they were unable to accept its doctrine because, like Plato, they saw species as having an inviolable extrasensory existence. Some of the best-known scientists who arged this way came from North America. Agassiz, for example, had reiterated such a position shortly before the *Origin* was published, in his *Essay on Classification,* intended as an introduction to a great work he was preparing on North American natural history. After the *Origin,* leading the American opposition to Darwinism, Agassiz remained implacable: "While individuals alone have a material existence, species, genera, families, orders, classes, and branches of the animal kingdom exist only as categories of thought in the Supreme Intelligence, but as such, have as truly an independent existence and are as unvarying as thought itself after it has once been expressed" (quoted by Ellegard 1958, p. 202, from a talk given in 1860). And Dawson in Canada echoed this with approbation.

A purist might justly question how Platonic some of these ideas were. One might, for example, contend that Agassiz's major debt in this respect was to Cuvier's interpretation of Aristotle, and in any case for most such idealists the essence of species was bound up with the Christian God and his creation. "The species is not merely an ideal unit, it is a unit in the work of creation" (Dawson 1860; quoted by Ellegard 1958, p. 202). The fact remains, however, that a significant factor in philosophical objections to Darwinism was an a priori commitment to species as real, immutable essences—"essentialism" (Hull 1973*b;* Mayr 1964).

There was not a great deal that Darwin could do here. There were conflicting metaphysical commitments. Darwin was clashing absolutely with essentialism, both in claiming that within any species there is variation—species do not seem to have uniquely defining sets of inviolable characters yielding their true essences—and also in claiming that the sharp border between nonspecies and species is illusory. But, metaphysical clash or not, in the *Origin* Darwin had shown that in nature the whole question of species is far from clear-cut. One not only has variation within species, one also has different pairs of groups that range all the way from complete interfertility to complete intersterility: just what one would expect given evolution. Of course the determined critic could still draw an a priori dividing line somewhere on this spectrum; but many of those not prepared to go all the way with Darwin still conceded that he had done all that could reasonably be expected (Ellegard 1958, p. 209).

So far we have seen Darwin more or less in control of the philosophical situation. But, finally, let us note that—Platonism apart—Darwin was also attacked philosophically. Many of these criticisms meant little, and Darwin recognized this. One of the most popular objections was that he was not sufficiently "Baconian," or that he had violated the true path of induction. Sedgwick wrote of Darwin's having *"deserted . . . the true method of induction"* (F. Darwin 1887, 2:248). Such criticisms, Sedgwick's particularly, usually meant merely that Darwin had done what the critics did not want done. If challenged to explicate the correct methodology, one suspects that most critics would simply say the correct path would lead to their own position.

But one philosophical critic was of much tougher fiber, and Darwin recognized it. This was William Hopkins—physicist, geologist, and incredibly successful Cambridge mathematics coach. Hopkins knew what the right scientific methodology was. It was precisely that adopted by Darwin himself, aimed at producing a model of science based on Newtonian astronomy as seen through Whewell's eyes. Hopkins (1860) endorsed the hypothetico-deductive model, distinguishing the formal parts of theories (the "geometrical laws of the phenomena") from the physical causes. In the case of astronomy, Kepler supplied the formal part and Newton the physical causes. Now, claimed Hopkins, the greatness of Newton's theory lies in the exactness with which claims deduced from his premises about physical causes correspond with facts in the world. But when we come to Darwin's theory, the situation is quite opposite. The phenomenal facts, for example, the fossil record, cannot be deduced from Darwin's claims about natural selection. At best Darwin shows that the phenomenal facts *could* have been caused by natural selection—that there is nothing inconsistent in this—rather than proving that they actually have been so caused. But, Hopkins contended, this "may be" philosophy will not do. Because it failed to satisfy his Newtonian ideal, Hopkins thus felt justified in rejecting Darwin's theory.

Darwin's reply to Hopkins was more a grumble than a refutation. "I believe that Hopkins is so much opposed because his course of study has never led him to reflect much on such subjects as geographical distribution, classification, homologies, and c., so that he does not feel it a relief to have some kind of explanation" (F. Darwin 1887, 2:327). One might well sympathize with Darwin's sentiment, but this hardly denies Hopkin's point about the lack of rigor in the *Origin*. Indeed, the point could hardly be denied, for Darwin never actually proves anything with full deductive rigor, and he is often reduced to the most sketchy of treatments. Of course one might argue that Newtonian astronomy is not the right model for evolutionary theories—C. S. Peirce asserted this, as have many philosophers in this century.[2] One may not want to draw the same drastic conclusions as Hopkins (and no one not already against Darwin did seem to want to), but one feels that Darwin was hoist by his own petard, for he shared Hopkins's philosophy of science. He tried to make his theory conform to this philosophy and took pride in the nature of his theory judged by this standard. In short, the philosophical factors in the reception of Darwinism cover something of a spectrum. In some respects, as with Platonism, Darwin more than held his own. In some respects, as with his differences with Huxley, rival positions were based on different philosophies. And in some respects, perhaps as with Hopkins, Darwin was on the defensive.

Religion

What really took the Darwinian controversy beyond the realm of science was religion: the way Darwin's ideas were considered to impinge upon and contradict religious truths. Nevertheless, in the 1860s the row over Darwin's ideas was only one part of a huge religious controversy (Benn 1906). Moreover, it was if anything a minor part, for at the beginning of the decade German higher criticism finally flooded into Britain. First came *Essays and Reviews,* published just after the *Origin* in February 1860. This work was a compilation from the pens of seven liberal Anglicans (only one of whom was a layman), showing the influence of the German critical approach to the Bible and advocating a less dogmatic and less conservative Christianity. It included an essay by Baden Powell objecting to the attempt to make religious belief dependent upon the literal acceptance of miracles, one by Jowett on the need for reason in understanding Scripture, and another by one Mark Pattison showing how religious thought had developed in the early part of the eighteenth century, clearly implying that such development might be expected to continue in the second half of the nineteenth century. Although the volume was written mainly by clergymen, or rather because of this, it caused a tremendous outcry, with conservatives of all kinds within the church rushing to condemn it. Bishop Samuel Wilberforce of Oxford fulminated against it in the *Quarterly Review,* the archbishop of Canterbury denounced it, and prosecutions for heresy were initiated, though they were unsuccessful.

If all this were not enough, there was the case of Bishop Colenso of Natal—a colonial bishop, but a bishop nonetheless. Famous as an author of arithmetic textbooks, to answer queries from his flock Colenso turned his skills on the Old Testament, discovering to his surprise that many of its claims just could not be true. He calculated that one man's reading aloud supposedly was audible to two million people (even with crying babies), that six men had 2,748 sons, and that every priest had to eat eighty-eight pigeons a day. After performing this kind of reductio, Colenso, like the authors of *Essays and Reviews,* called for a more liberal interpretation of the Bible. And like them he became the center of a storm of theological controversy.

One might wonder why these works should cause so much trouble. People like Sedgwick and Whewell had taken a liberal attitude toward the Bible for years, and though there had been some objectors they were not from the central body of the church. The answer is that these liberal broad churchmen were threatening the very compromise the Sedgwicks and the Whewells had erected: Give science the period before man and we will concede the literal truth of the Bible for the period after. This may have called for a judicious reinterpretation of the universality of the Flood, but it had worked fairly well. Now Colenso with his calculations was threatening the whole period dealing with man—wherever would the rot end?

One might think this controversy over higher criticism could not have come at a worse time for Darwin and his theory. Surely this would make churchmen even more opposed to new and threatening ideas, whether they came from Germany or biology. But, though it may be true that some reactions to Darwin were made more shrill by the general calamity, it is quite possible that Darwin and his ideas (at least as they extended to evolution) gained from the fights over *Essays and Reviews* and Colenso. Conservative churchmen—who were not going to have anything to do with Darwin's ideas anyway—could not give their full attention to the fight against the *Origin*. A correspondent wrote from Oxford in 1861: "The book [the *Origin*] I have no doubt would be the subject still of a great row, if there were not a much greater row going on about *Essays and Reviews*" (Abbot and Campbell 1897, 1:291). It may be that the members of the scientific community were freer to get on with the things they wanted to do, such as accepting evolutionism, because the group that would harass them most—conservative churchmen—had their energies directed elsewhere. The other aspect of this story favors Darwin and his ideas even more. The controversy shows that by the 1860s an ever-growing number of Victorians could no longer accept dogmatic religion centered on a literal reading of the Bible. When a leading essayist and poet like Matthew Arnold (1873, p. 23) could write of religion as being only "ethics heightened, enkindled, lit up by feeling," an evolutionist could take heart.

Among people concerned about science, we find that those who took religion at all seriously reacted to Darwin's ideas in ways ranging from enthusiastic acceptance to outright rejection. At one end, Baden Powell slipped into his contribution to *Essays and Reviews* (1860, p. 139) a sympathetic endorsement of the *Origin*. "Mr. Darwin's masterly volume . . . now substantiates on undeniable grounds the very principle so long denounced by the first naturalists,—*the origination of species by natural causes:* a work which must soon bring about an entire revolution of opinion in favour of the grand principle of the self-evolving powers of nature." At the other extreme we find Sedgwick (1860), who declared himself unswayed by Darwin's arguments and continued to support miracles—not just any kind of miracles, but miracles of the kind Whewell had worked out in the 1840s in response to *Vestiges*. He admitted that some kind of phenomenal law governs the introduction of new species: "But here, by *law,* I mean order of succession, and not a law like that of gravitation, out of which the actual movements of our system follow by mechanical succession." There are therefore no natural causes bringing about new species—we must appeal to miraculous interventions, although this involves no breaking of natural law. "The hypothesis does not suspend or interrupt an established law of Nature. It does suppose the introduction of a new phenomenon unaccounted for by the operation of any *known* law of Nature; and it appeals to a power above established laws, and yet acting in harmony and conformity with them" (1860; unlike Whewell, Sedgwick was always a successionist).

So much for general reactions. What about the details? Even if we ignore people's scientific responses, we might expect two things. First, religious people trying to grapple with Darwin's ideas would not be worried by such things as the great age of the earth that Darwin's theory demanded. We have seen that Sedgwick and Whewell had never objected to things of this nature, though many religious people certainly seized upon Kelvin's calculations to belabor Darwinism. What would stick in the craws of the traditionally religious would be man and final causes. Second, some people, particularly people like Herschel and Lyell, would try to find some kind of compromise between the extremes of Sedgwick and Baden Powell. Both these expectations are fulfilled.

Man: The Facts

In the *Origin* Darwin had said virtually nothing about man except for that almost-final comment: "Light will be thrown on the origin of man and his history" (1859, p. 488). But no one was fooled by the brevity of this reference, and almost every attack on Darwin brought in the "monkey question." Yet, though it was religion that made the debate over man so intense, there was a factual level—as close to pure science as anything. Here we see the themes of science, philosophy, and religion intertwine.

The question at issue was whether man is in any essential way different from other animals, particularly from the great apes. That he is was the position of Richard Owen (1858*b*), who, with all his great authority, had asserted that man's brain "presents an ascensive step in development" because, among other things, it alone has "the 'hippocampus minor' which characterise[s] the hind lobe of each hemisphere." This claim probably led to the most famous of the clashes between the Darwinians and their opponents, that between Huxley and Bishop Wilberforce at the British Association meetings at Oxford in 1860 (L. Huxley 1900, 1:192–204). At a midweek meeting Owen restated his claim about the differences between man and other animals, and Huxley contradicted him flatly, promising to make good his denial in print shortly thereafter. Infuriated, Owen primed Wilberforce, who returned to the attack on Darwinism at a Saturday meeting. Apparently somewhat charmed by his own rhetoric, Wilberforce asked Huxley (also on the program) whether it was through his grandfather or his grandmother that he claimed descent from monkeys. Trading comments of this sort with Huxley was not safe, as the bishop soon found out. To the delight of the Darwinians the story got around that Huxley had retorted that he would rather be descended from a monkey than from a bishop of the Church of England. More probably, but hardly less scathingly, he replied that if faced with the question, "'would I rather have a miserable ape for a grandfather, or a man highly endowed by nature and possessed of great means and influence, and yet who employs these faculties and that influence for the mere purpose of introducing ridicule into a grave scientific discussion'—I unhesitatingly affirm my preference for the ape."[3] After this came a long pro-Darwinian discourse by Hooker, and the opponents were left crushed on the field.

Huxley did not forget his promise to Owen. Early in 1861, in the *Natural History Review,* he argued in some detail against Owen, specifically showing that "the *hippocampus minor* is neither peculiar to, nor characteristic of, man, as it is found in certain of the higher quadrumana." Although Owen went on asserting his position, Huxley's reply seems to have been virtually the end of the matter. The great hippopotamus debate ended in victory for that noted expert in necrobioneopalaeonthydrochthonanthropopithekology, Professor Ptthmllnsprts! (This was Kingsley's version of events in his parody in *Water Babies.*) No essential differences could be found between the brain of man and that of the ape. This is not to say that there were no relative differences—Huxley admitted there were and conceded that the differences were greater between man and the next primate down than between any succeeding two types of apes. So there was still room for debate about anatomical differences. But Huxley himself (1863) downplayed the disparity because the differences between man and the highest apes were less than between upper and lower apes.

Parallel with this debate about human anatomy were some exciting developments in the understanding of man's fossil history (Oakley 1964; Gruber 1965). Independent of our evolutionary story, man's existence was being pushed back drastically further than the 6,000 years hallowed by tradition. Greatest credit for this must go to Jacques Boucher de Crèvecoeur de Perthes, controller of customs at Abbeville in France, who, despite ridicule and disregard by virtually all in the scientific community (including Darwin), was able to establish through the remains of tools that man coexisted with mammals now extinct. In 1859 British scientific opinion was steered dramatically toward acceptance of this claim through the conversion and subsequent championing of the British geologist Joseph Prestwich, who was able to support the validity of Boucher de Perthes's position with similar evidence of ancient men from Brixham in Devon. By 1868 the evidence of man's great antiquity was so convincing that even Sedgwick was prepared to admit it (Clark and Hughes 1890, 2:440). Sedgwick remained a violent antievolutionist until his death. But since man's recent appearance was a central element in his compromise between revealed religion and scientific creationism, this was a major concession with significance far beyond the actual number of years involved. Sedgwick himself may never have become an evolutionist, but his credibility as an authoritative alternative was hardly maintained. Searching for reasons why Darwin's ideas succeeded, we ought not neglect the extent to which the position of his most conservative scientific opponents was coming apart, even when he had no part in the causes.

Aside from the antiquity of man, the possibility of actual fossil links between man and nonhumans was more open to contention. In 1830 in Grotte d'Engis in Belgium some skulls of possible links were found, and in 1856 in Neanderthal in Germany other remains were discovered. These were on the borderline with modern man, and the authorities were divided on their interpretation. Huxley (1863) decided that Neanderthal man was not an intermediary form between man and the anthropoids. On the other hand, William King, anatomist at Queen's College, Galway, was prepared to make a distinct species, *Homo neanderthalensis* (Ellegard 1958, p. 110). It was not until 1891 that the young Dutch army doctor Eugène Dubois made the great breakthrough with his discovery of Java man, a much more convincing link between man and apelike ancestors. Hence, throughout this period critics of evolutionism (Darwinian or otherwise) could parade the case of the "missing link" as an objection to man's evolutionary ancestry. On the other hand, the evolutionists were not destitute of facts on which to build hopes (Ellegard 1958).

Man: The Interpretation

Whatever the fascination of the various facts about man, they were not really of most importance. Depending on one's a priori convictions, one could draw completely different conclusions from the same facts. Thus Huxley (1863, p. 125), having argued that man and the apes are on a par anatomically, concluded: "But if Man be separated by no greater structural barrier from the brutes than they are from one another—then it seems to follow that if any process of physical causation can be discovered by which the genera and families of ordinary animals have been produced, that process of causation is amply sufficient to account for the origin of man." In contrast, the duke of Argyll calmly conceded Huxley his anatomical facts and contended that behavioral and mental phenomena are crucial, and that there is thus a barrier between the apes and man that natural selection could never cross. "Whatever may be the anatomical difference between Man and the Gorilla, that difference is the equivalent, in physical organization, of the whole mental difference between a Gorilla and a Man" (Argyll 1869, p. 51).

For Huxley, who was relatively unfettered by orthodox religious belief, the question of man was fairly easy to answer. Forget religion and let the "facts" speak for themselves. For someone on the other side of the barrier, for whom religion was paramount and the Bible was the authority, the question was again fairly easy to answer. God created man miraculously in his own image. The person desperately troubled by the question was the man in the middle—the one who wanted to roll with the advances of science and who saw great virtues in evolutionism (and perhaps in natural selection) but who was also keen to see man set apart, the favored of God.

Such a man was Charles Lyell, the perfect example of the scientific/religious tensions caused by Darwin's theory. We know that Lyell was desperate to find a scientific solution to the organic origins questions, and that he was gradually forced to realize that only evolutionism would provide it. But still there was the question of man, long of concern to Lyell. As soon as Darwin let him in on the secret of natural selection, in the spring of 1856 (Wilson 1970, p. 54), Lyell again began to worry about man, and he went on worrying. He was explicit about the religious nature of his concern: "It is small comfort or consolation to me, who feels that Lamarck or Darwin have lessened the dignity of their ancestry, making them out to be with[t] souls, to be told, "Never mind, you will be succeeded in unbroken lineal descent by angels who, like Superior Beings spoken of by Pope, 'Will show a Newton as we show an ape'" (Wilson 1970, p. 382).

In the *Antiquity of Man,* published in 1863, Lyell allowed natural selection a secondary position but seemed to favor as the main cause of new species a kind of guided saltatory mechanism. I am not sure whether we can say at this point that Lyell was an evolutionist, and probably it is a matter of definition anyway. He was certainly closer to letting new kinds of organisms appear through laws from forms not too far different, but he still supposed teleological impulses. But the facts—particularly the facts of geographical distribution—were too strong. Though Lyell never became a full-blooded evolutionist, he did finally become an anemic one. In 1868, in the tenth edition of his *Principles* (2:492), Lyell seemed at first to have accepted both evolution and selection: "Was Lamarck right, assuming progressive development to be true, in supposing that the changes of the organic world may have been effected by the gradual and insensible modification of older pre-existing forms? Mr. Darwin, without absolutely proving this, has made it appear in the highest degree probable." Moreover, Lyell had the courage to admit that man's intellect could be improved by natural selection; more generally he suggested that "if progressive development, spontaneous variation, and natural selection have for millions of years directed the changes of the rest of the organic world, we cannot expect to find that the human race has been exempted from the same continuous process of evolution" (1868, 2:492–93). But Lyell then showed that his conversion to evolutionism was hedged with restrictions, for having said such supportive things of Darwin and his mechanism, he immediately took most of them back. Darwin explains, we are told, not the creation of species of organisms—rather, he shows that species were created by law, not miracle. Reverting almost to his 1830s position, Lyell reemphasized the way God shows design, even though he may work through law. Hence, though it is probably true that Lyell finally staggered across the threshold of evolutionism, to the end the tensions between his science and his religion, particularly his fears for man, made him constantly ambivalent on the subject of new species.

Lyell was not alone in worrying about the evolution of man, particularly through selection. Whatever he may have said about selection working on small changes, Lyell probably always favored special, saltatory pushing, at least for man. This position found favor, from Darwin's critics like Owen and Mivart right through to—of all people—the codiscoverer of natural selection, Wallace (Smith 1972; Kottler 1974). In his early years Wallace had had little or no faith, but on his return to England from the East he became enamored with quasi-religious speculations such as spiritualism. Although for the rest of the organic world Wallace never once wavered from his commitment to natural selection, for man he became more and more inclined toward a teleological saltatory position. Having listed characteristics of man that he felt could not initially have been the product of natural selection—man's brain or mind ("the large brain of the savage man is much beyond his actual requirements in the savage state"), man's hairlessness, man's feet, hands, and voice, and so on—Wallace inferred that "a superior intelligence has guided the development of man in a definite direction, and for a special purpose, just as man guides the development of many animal and vegetable forms."[4] A second factor probably helped change Wallace's original views on man, a factor stemming from another "disreputable" cause he espoused: phrenology. This belief that units of the brain correspond to units of the mind and that the sizes of the brain units strongly influence the mental characters associated with them, holds it essential that these characters are irreducible, gross units of analysis; they can exist only in their entirety. With a psychology like this it is easy to see why Wallace would have been troubled by a theory of natural selection working on small variations, for such a theory would preclude the large steps needed to create the units of the brain. For this reason also Wallace was tempted to a teleological, saltatory mechanism for man.

How did Darwin react to all this compromising with religion? As might be expected, not favorably. Increasingly he lost patience with Lyell, showing little sympathy for the older man's turmoil, and he was downright appalled at Wallace (F. Darwin 1887, 3:116). In 1871 Darwin himself spoke out on man, in his *Descent of Man.* This is a strange book, for most of it is given over to a very general discussion of sexual selection, ostensibly to show that many of the differences within the human species, between races and between sexes, are caused by this mechanism. But the strange balance of the book in no wise changes the fact that here Darwin applied his evolutionary speculations to man entirely, as though he had no thought of religious objections. We find Darwin asserting that man shows variability like other animals, that man has a geometric potential to increase, and that hence we should expect natural selection. Then Darwin went on to explore in detail just how and why one might expect man's various characteristics, physical and intellectual, to have evolved. Thus Darwin wrote that "in the rudest state of society, the individuals who were the most sagacious, who invented and used the best weapons or traps, and who were best able to defend themselves, would rear the greatest number of offspring" (Darwin 1871, 1:196). Thus, through selection, man's intelligence would have evolved.

The only point where Darwin might be said to have weakened was over the question of morals. We have seen that in the nonhuman world Darwin was adamantly in favor of individual selection rather than group selection. Even when he toyed with the idea that selection might have caused some sterility, the starting point was the advantage to the individual primula in avoiding self-fertilization. Nevertheless, when it came to human morals, Darwin sometimes thought a kind of group selection would be needed. "It is extremely doubtful whether the offspring of the more sympathetic and benevolent parents . . . would be reared in greater numbers than the children of selfish and treacherous parents of the same tribe" (Darwin 1871, 1:163; see Kottler 1976). At other times, however, Darwin thought that a kind of enlightened self-interest, what modern thinkers call "reciprocal altruism," could cause morality (Trivers 1971). With human evolution, "each man would soon learn that if he aided his fellow-men, he would commonly receive aid in return" (Darwin 1871, 1:163). Ability to behave this way obviously could be caused by individual selection.

Whether we should conclude that Darwin was a group selectionist or an individual selectionist with respect to morality is probably unanswerable. He would not have known himself. The important point is that, either way, for one of the discoverers of natural selection, there was to be no compromise with religion: man, no less than any other organism, must be explained in purely natural terms.

The Argument from Design

When our story began, about 1830, we saw the argument from design centered on adaptation, the "utilitarian" argument, in full sway. There were a few minor difficulties like male nipples, but most things were taken to show direct purpose. In the thirty years following, for various reasons we have examined, that argument came under fire and was somewhat displaced by other natural theological arguments that, because they did not center on adaptation, posed less of a threat to evolutionary hypotheses, if they did not positively invite them. The weakening of the utilitarian argument was surely one reason for the incredible speed with which the biological community converted to evolutionism. On the other hand, in Britain in the 1860s the utilitarian argument still had considerable force and was no doubt a factor in many people's reluctance to accept natural selection, other than to a very limited extent. Selection, working through blind law, claims to explain adaptation, and there remained a strong feeling that unguided law could not lead to intricately complex organic adaptation. Therefore, people felt that scientifically Darwin's theory must be wrong, and religiously they often found it offensive, because by denying or downplaying adaptation one cuts away at the foundations of natural theology.

For some of the older men like Sedgwick and Whewell, this was the end of the matter. Darwin's book is *"utterly false"* because "it repudiates all reasoning from final causes; and seems to shut the door upon any view (however feeble) of the God of Nature as manifested in His works."[5] Others, though they felt strongly about adaptation and final causes, tried to go some way with Darwin. Almost invariably these were the same people who had thought Darwin's theory needed an extra push for man; so they simply introduced more pushes through the course of evolution— law-bound, yet directed, saltations introducing adaptations. This seems to have been Herschel's position. When he received his copy of Darwin's book (Herschel, Whewell, and Sedgwick all received copies from the author, as did those who had stayed closer to Darwin, like Henslow and Lyell), Herschel was reported as referring to its incorporating the law of "higgledy-piggledy" (a response that upset Darwin; Darwin and Seward

1903, 1:191). But Herschel's studied, public reaction was more sympathetic. He argued that, because of final causes, Darwin's theory just would not do. For the organic world to come into being, "an intelligence, guided by a purpose, must be continually in action to bias the directions of the steps of change—to regulate their amount—to limit their divergence—and to continue them in a definite course" (Herschel 1861, p. 12 n). But Herschel seems to have thought that, so long as one supplemented Darwin's theory with some specially guided law of variation and made special provision for man, there might well be something to be said for the *Origin*. Granting such a law, "we are far from disposed to repudiate the view taken of this mysterious subject in Mr. Darwin's work" (1861, p. 12 n). Writing to Lyell two years later, Herschel made it clear that he himself favored some kind of saltatory evolutionary theory, with occasional law-bound jumps from one species to another, where adaptation-causing design could take effect, presumably aided by the backup action of natural selection. He spoke of the "idea of *Jumps* . . . as if for instance a wolf should at some epoch of lapine history take to occasionally littering a dog or a fox among her cubs" (unpublished letter from Herschel to Lyell, 14 April 1863; Herschel Papers, Royal Society). He added that such a process would introduce *"mind, plan, design,* and to the . . . obvious exclusion of the haphazard view of the subject and the casual concourse of atoms."[6]

We have here another point at which science and religion came together and ought not be considered as separate. We know there were scientific reasons why many favored saltationism. Now we see religious reasons, and though for many they may have been more basic, those wanting designed saltations could legitimately turn to science to strengthen their case. Indeed, philosophy is not absent at this point. Lyell, for one (1863, p. 505), realizing that saltations come dangerously close to violating the empiricist concept of *vera causa,* dwelt carefully on the phenomenon of geniuses born of ordinary parents—an analogy with the evolution of man from an orangutan! Of course, one does not absolutely have to be a saltationist to be a teleological evolutionist. Asa Gray (1876), botanist, professor at Harvard, and North American champion of Darwin, wanted both teleology and evolution through small changes—God directing the minute variations. However, whether the direction is supposed to come through large or small variations, Darwin responded no more enthusiastically to compromises between science and religion centered on adaptation than he had to compromises centered on man (F. Darwin 1887, 2:373,

377–78). He justifiably thought that such compromises ignored or denied the whole function of his mechanism of natural selection—to explain organic adaptation through normal laws, thus making direct interventions or special guidance quite unnecessary. In the *Origin* itself, Darwin argued against the claim that one must suppose some special design for organic characteristics, on the grounds that consistency then demands inventing ad hoc purposes for such things as male nipples—which would not be acceptable in the physical sciences and therefore ought not be acceptable in the biological sciences (Darwin 1859, p. 453). Against Herschel, Darwin repeated a version of this argument: "astronomers do not state that God directs the course of each comet and planet. The view that each variation has been providentially arranged seems to me to make Natural Selection entirely superfluous, and indeed takes the whole case of the appearance of new species out of the range of science" (Darwin and Seward 1903, 1:191). Darwin had learned from Herschel that the model for science is Newtonian astronomy. As many of the brightest students are wont to do, Darwin turned his teacher's words back on him.

One final point: Darwin disagreed with those who wanted to explain adaptation through guided laws. But he agreed that adaptation is important and must be explained. Therefore we have one more reason why Huxley was not totally committed to natural selection. For Huxley the utilitarian argument never had great charm, and so he downplayed the importance of adaptation. He went so far as to deny that the color of birds and butterflies or the color and shape of flowers could have any adaptive value (Huxley 1854–58, p. 311). Given his insensitivity to the very things on which other Darwinians were centering their case for natural selection, Huxley automatically felt free to argue for saltations. He had no great urge to explain such adaptation. Thus we have the seeming paradox that whereas many supplemented Darwinism with saltations as a way to bring in design effects, Huxley supplemented Darwinism with saltations because he felt no need to bring in design effects! Of course the paradox is really no paradox, because the teleologists thought the saltations were guided by God, whereas this was the furthest thing from Huxley's mind.

The Darwinians and Their Society

So far we have been dealing with the reception of Darwinism at the intellectual and emotional level—at the level of thought, loosely construed. But there was a more tangible level involved—a world of human relations and of social and political factors. For all the reservations we must draw, it is clear that Darwin's ideas scored some striking successes—especially if we consider matters from the viewpoint of the scientific community and remember the reception of *Vestiges.* If we are to understand fully why Darwin achieved such success, we must go beyond pure ideas and look at the scientists in their human settings.

At the most general level, the period we are considering—the fifteen years from 1860 to 1875—was a time of stability and prosperity in Britain, at least until the agricultural depressions started in 1873 (Tholfsen 1971). This was true despite great political developments, including the Reform Bill of 1867. When Darwin published the *Descent of Man* (1871), one reviewer chided him for "revealing his zoological conclusions to the general public at a moment when the sky of Paris was red with the incendiary flames of the Commune" (quoted in Houghton 1957, p. 59); but in general one does not get this reaction. Certainly one does not sense the tension evident in the reviews of *Vestiges.* Undoubtedly this is partly because the 1860s were a very different time from the 1840s. In the 1840s, many people justifiably thought that revolution was just around the corner. In the 1860s, many started to conclude that God was a middle-class Englishman after all. Hence evolutionism was less threatening than it had been, apart from the fact that ideas one has been living with for years are less worrisome than fresh heretical ideas.

The "respectability" that evolutionism gained after the *Origin* was in large part due to the respectability of the Darwinians themselves. After *Vestiges,* people claimed that evolutionism leads to atheism, which is immoral and threatening to the stability of society. But the Darwinians gave the lie to this. To a man, they were exemplars of the most boring Victorian respectability—hardworking to the point of neurosis (and in Huxley's case beyond), earnest, and good family men of impeccable sexual propriety. They were not outsiders. Conversely, they shared many of the values and norms of their contemporaries. Huxley, for example, for all his agnosticism, when a member of the London school board in the early 1870s strongly endorsed the moral value of compulsory Bible study (L. Huxley 1900, 1:363–64).

Their attitude toward women illustrates the conventionality of the Darwinians. Huxley, to his great credit, was much concerned with upgrading women's education, but he was still convinced that women are by nature men's intellectual inferiors (1900, 1:449). He blocked their attendance at the Geological Society, and though in theory he had no objection to a man's marrying his dead wife's sister (a much-debated Victorian topic, involved with the threat an unmarried sister might pose to the unity of the home), when his own daughter married her dead sister's husband, he did not approve (1900, 2:231). Darwin was even more a child of his time. In the *Descent of Man,* with all the emphasis on sexual differences as determined by sexual selection through male combat and female choice, one can well imagine the status of women. "Man is more courageous, pugnacious, and energetic than woman, and has a more inventive genius" (Darwin 1871, 2:316). In compensation, woman has "greater tenderness and less selfishness" (1871, 2:326). Shades of Sedgwick and Brewster here! The Darwinians may have been rebels in some respects —in other respects, they were anything but (Greene 1977).

Moreover, the Darwinians were not merely respectable members of Victorian society; in important ways they were positive and valued forces in that society. For instance, this was a time of great concern about education at all levels, from the elementary level, where the children of the working classes were getting virtually none, to the university level, where the British seemed so anachronistic compared with the Germans. Huxley was incredibly active and useful here and was recognized as rendering invaluable service (Bibby 1959). Hooker, too, was a man of great value. At Kew, botanists were classifying plants sent from all over the empire and experimenting to see what could be grown where: Could new commercially valuable crops be raised in different parts of the world? Thanks to their work, in 1861 cinchona (quinine) was introduced from South America to India and Ceylon, and in the 1870s rubber trees were transplanted from South America to the East Indies. Although much of this work was done under the rubric of humanitarian benefits to mankind in general, it was critical to British imperialism, which was of increasing economic importance as the growing industrialization of Europe and America threatened Britain's preeminence. Quinine permitted white men to function in tropical areas despite malaria, and rubber transformed the Malay Peninsula.

Hence, quite apart from the fact that these latter projects depended on principles important to Darwinism, men like Huxley and Hooker carried weight in Victorian society. Indeed, a row between Hooker and his government superior in the early 1870s probably contributed to the defeat of the Gladstonian government (MacLeod 1974). One might not like the Darwinians' ideas; but their example went a long way to counter the argument that those concepts were antisocial or otherwise corrupting. One could hardly be hostile toward Hooker, who through his Himalayan travels was a key figure in introducing the rhododendron, a shrub that soon graced the gardens of aristocrats and gentry all over Britain.

The Darwinians and the Scientific Community

So far we have been considering Victorian society as a whole. Let us now restrict our gaze to the scientific community. First, one suspects that a major factor in Darwin's success was his own status as a scientist. He was not some unknown speculator, but a man of proved scientific stature both in geology and in zoological systematics. As early as 1851, Huxley, who liked doing this sort of thing, had rated Darwin high on the list of British biologists[7]—long before he became Darwin's supporter. Darwin had therefore earned the right to be listened to with respect, and undoubtedly he was. Obviously it is easier to be rude about someone when both you and he are anonymous; but there is a striking difference in attitude between the reviews of *Vestiges* and of the *Origin*. We know that the most unpleasant things were said about *Vestiges* and about its author; but, with Darwin and the *Origin*, even where the opposition was strongest, the tone was one of regret that so good a scientist had been seduced by so vile a hypothesis. The position Darwin had earned for himself clearly allowed his theory a much fairer hearing.[8]

Second, Darwin himself had aided his theory's reception. It did not arrive on the scene friendless, even though Darwin had to publish before he wanted to. The inner group of Darwinians—in England Hooker, Huxley, and Lyell (let us include him at least on the basis of personal friendship) and in America Asa Gray—had been introduced to the theory in previous years. Although no one seemed to understand its full force until the *Origin*, some of the most active and articulate scientists in Britain were ready to do battle for it at once (or, as in Lyell's case, to see it had a fair hearing). This paid dividends. When Wallace's paper arrived, Darwin's good friends Hooker and Lyell saw that the ideas were published at once through one of the most prestigious scientific bodies, the Linnaean Society. Hooker started using (and thereby endorsing) natural selection even before the *Origin* appeared, in his *Flora of Tasmania* (1860). Huxley wrote an enthusiastic review of the *Origin* at Christmas 1859 for that most widely read of organs, the *Times* (reprinted in Huxley 1894). Gray took up cudgels against Agassiz at Harvard (Gray 1876; Dupree 1959). And through the 1860s Darwin's friends promoted the *Origin*—in reviews, by applying its ideas, at the British Association, at the various societies to which they belonged, and in other ways that we shall learn. No theory was introduced to the world with a better set of friends.

Darwin's friends and supporters did not just descend like manna from heaven. They were there through Darwin's own efforts. Of all the work published between the time Darwin completed his *Essay* and the time Wallace forced his hand, the thing that ought to have brought him most comfort was Owen's 1851 critique of Lyell. Here a foremost authority in the field was arguing for just such a view of the fossil record as Darwin desperately needed. This was an argument by a man Darwin respected— even to the extent of considering him as a possible editor of the *Essay* should he himself die (Darwin and Wallace 1958, p. 36)—and by his own admission a man he liked (Darwin and Seward 1903, 1:75). What was Darwin's response when Huxley (1854), with all the authority gained by four years of cutting up jellyfish, savagely criticized and ridiculed Owen? Darwin wrote a most flattering letter to Huxley, saying that "the way you handle a great Professor [Owen] is really exquisite and inimitable" (Darwin and Seward 1903, 1:75). One can well imagine what an ego boost that letter must have been for the young and rather insecure Huxley! And one cannot help but feel that Huxley's emotional commitment to Darwin in the 1860s, despite his lack of enthusiasm for selection, was in part due to the way Darwin had cultivated him. In reading their correspondence one is struck by the deep feeling between them (and Hooker also). Darwin needed Huxley, but Darwin also fulfilled a need for Huxley—he supplied Huxley with a kind of crippled older brother to defend. I am not criticizing Darwin for cultivating Huxley; obviously he could not be friendly with both Huxley and Owen. I wish to show how Darwin personally campaigned for Huxley's support. Charles Darwin was the father of his theory both intellectually and socially.

A third point about the scientific community is that Darwin's ideas were greatly aided by significant changes in the scientific world between the 1830s and the 1860s (Cardwell 1972). Science had become much more professional (by the criteria introduced early in this book), and the links between science and organized religion had been much weakened. The men concerned with the organic origins problem in the 1830s, within whose group Darwin came to scientific maturity, were generally associated with the older universities and thus were professionally and emotionally linked with Christianity. Henslow, Sedgwick, and Whewell had to become clergymen to keep their jobs, though this does not mean they were hypocrites. Their science and their religion were inevitably intertwined. But by the 1860s this group was old, dying out, and no longer actively involved in science. Buckland was already dead; Henslow and Baden Powell died almost as soon as the decade opened; Sedgwick fulminated against evolutionism, but it is clear that no one really took him seriously any more—he had become one of the ancient monuments at Cambridge.

Whewell was bound up in university administration and in any case died in 1866; Herschel's great scientific heyday was over. As Darwin rather cruelly remarked to Herschel, in reply to his inability to accept all of Darwin's ideas, the tempo had passed from the old group to the younger men.[9] The old Oxford and Cambridge network was no longer in a position to stop evolutionism.

One must be fair. The old group helped prepare the way for Darwin's ideas, particularly in their desire to give science a place along with religion. And even in the 1860s the members indirectly helped the Darwinian cause. In the mid-1860s a declaration, ostensibly allowing science and religion to harmoniously coexist, but really suggesting that science defer to religion, was circulated among scientists. Of the old group only Sedgwick signed, and Herschel brushed it off roughly and publicly (Brock and MacLeod 1976). They had fought that battle already! But it was these older people who belonged to the group that would provide the strongest explicit opposition to Darwin's ideas, and they were dying out. Moreover, old or young, those who most wanted to oppose Darwin personally seem to have been the least suited to do it in a coordinated way. Owen managed to alienate just about everyone of consequence in the scientific community. And Mivart too was socially inept. He impugned the moral integrity of one of Darwin's sons and thus got himself labeled "not quite a gentleman" (Gruber 1960). Even in the 1880s, Hooker was prepared to blackball Mivart were he proposed for that favorite club of the Victorian intellectual aristocracy, the Athenaeum.[10]

More positively, we find that in the thirty years or so since our story started there had been phenomenal growth in secular higher education and in professional science, particularly in London. And so in the 1860s as opposed to the 1830s we have a body of scientists with no debts or ties to the established church. This was obviously central to the building of a Darwinian party and to the success of evolutionism, for almost every man who was positively involved belonged to this London group. Excepting Lyell, Darwin was one of the few Oxford or Cambridge men, and he had long been separated from his alma mater. Thus we have the rise of University College, London, from its very young state when we last saw it. W. B. Carpenter was educated at University College, was for some years a professor there and professor of physiology at the Royal Institution, and in 1856 became registrar of the University of London. As soon as the *Origin* appeared, he tacked a proevolutionary passage onto a paper on Foraminifera that he was writing for the *Philosophical Transactions* (Carpenter 1860; Carpenter believed in a directed evolution). Similarly, Michael

Foster, from a Baptist family and a very close student of Huxley, was educated at University College and joined its staff in the 1860s. We also have the Government School of Mines, founded in 1851—one Darwinian, Ramsay, was professor of geology there and another, Huxley, became professor of natural history in 1854. And in another part of the city at the Royal Institution, John Tyndall, who fought against the church alongside Huxley, was professor of natural philosophy.

These were the teachers. In addition we have Hooker (Scottish-educated) at Kew. Wallace and Bates, neither with a university background, made their livings from collections and writings (and in Bates's case, as assistant secretary to the Royal Geographical Society). Herbert Spencer, not so much a Darwinian but certainly a strong evolutionist, self-educated (or, depending on one's viewpoint, uneducated), also made his living by writing. J. B. Jukes, although Cambridge-educated, worked for the Geological Survey. And so the list grows. We have men who were involved in science, science education, and science application. We have men who had no professional ties with the church. We have men who were active in the spread and support of evolutionism and (in many cases) of full-blown Darwinian natural selection.

Hence, in the 1860s we see a body of men positively involved in promoting elements of Darwinism or evolutionism in one way or another, and their very existence was possible because in the 1860s in Britain, as opposed to the 1830s, there were far greater opportunities for secular scientists, particularly in education and government posts. In the 1830s such a party could not have formed and held together. Given the way higher education was so church-oriented, such a party would not have been likely even to start. In short, Darwin's ideas benefited from the evolution of the structure of British science. (One suspects this was a two-way affair. Huxley was happy to promote Darwinism, or at least his version of it; but, equally, Darwinism gave him good material to support his case for professional science as he saw it.)

One must not exaggerate. There was no sharp break between the old and the new. Lyell (and Darwin!) straddled the two groups, and though, particularly with respect to the relation of science and religion, the old group had extremists like Sedgwick at one end of the spectrum and the new group had extremists like Hooker at the other end, there were some in both groups more or less in the middle: Lyell, Herschel, Carpenter, and even Wallace after a fashion. And this is not to mention Owen and Mivart, who, though bitterly against the Darwinians personally, took

middle positions intellectually. One must not think the various compromises sketched in earlier sections belonged exclusively to scientists of
one generation. But there was an undoubted overall shift in attitude
toward the organic origins problem, and in part this was a function of
changes in the organization of science.[11]

The Scientific Societies

One can show that the new group (and like-minded friends) not only set
out to promote Darwin, but in certain respects were very successful at it. I
am speaking less about ideas than about scientific politics. Take the
learned societies, through which one would expect scientific ideas to be
discussed and promulgated. Strictly speaking, the British Association was
not a learned society, but we find that at its meetings Darwinism, particularly evolutionism, got thorough and generally sympathetic treatment. Toward the end of the decade (in line with a general trend
spearheaded by Mivart) criticism of the adequacy of natural selection itself
somewhat increased, but after 1862 only nonprofessionals criticized evolution (Ellegard 1958). An informal tradition grew up of letting Darwinians
and non-Darwinians alternate as president, aside from support in the various sections. But the learned societies proper stayed curiously on the side-
lines of the Darwinian debate (Burkhardt 1974). The societies' tradition, or
at least their practice, was to avoid controversy, at least in their published
proceedings and transactions. Hence they concentrated on presenting substantial, descriptive material, avoiding too much speculation—or, more
precisely, too much openly provocative speculation. But this was by no
means a universal rule, and we find that the Darwinians certainly did not
stand still, even though they preferred to fight in the more public arena of
books, magazines, lectures, the British Association, and the like. A paper
by Huxley on *Archaeopteryx* exemplifies this (1867–68). In a version read
before the Royal Society on 30 January 1868 there is no direct reference to
evolutionism. In a version presented to the Royal Institution on 7 February 1868, *Archaeopteryx* is used as a key piece of evidence in a proevolutionary argument.

If we take the pertinent societies in roughly descending order of prestige, starting with the Royal Society, then (coming together) the
Geological and Linnaean societies, the Zoological Society, and the Entomological Society, we tend to find that the less prestigious the society,
the more open the speculative discussion became and the more flagrant the

Darwinians were in putting across their message. Certainly, the Entomological Society went in most for open discussions of evolution; and by the end of the 1860s, with first one Darwinian, John Lubbock, and then another firmly ensconced as president, giving pro-Darwinian discourses, Darwinism virtually constituted orthodoxy. Of course the study of insects was perhaps the area of science most amenable to studies not only of evolution by also of natural selection. Hence, quite apart from the society's being less formal and cautious than other societies, Darwinism had a special claim to its members' interests and sympathies.

The Darwinians had less of a direct ascendency in the Zoological and Geological societies, at least in the early part of the 1860s. Nevertheless, they published a number of papers that were proevolutionary, even supporting natural selection. In the Geological Society, at both ends of the decade we get presidential addresses from Huxley that are pro-Darwinian—at least Huxley's version of pro-Darwinian—and in the center we get more of the same from Ramsay (Burkhardt 1974). In the Linnaean Society the Darwinians had, if not a firm grip, at least a very strong entry. The *Proceedings*, by tradition, tended to be fairly free of theory, though Darwin's papers on plant forms and sexuality were published there, and they were confidently open in their evolutionism. On the other hand, the president through the 1860s, George Bentham, who worked with Hooker at Kew, treated the members to a yearly review of the evolutionary debate—with a slant that was definitely sympathetic to Darwin. And the more prestigious publication of the society, the *Transactions* (open to both botany and zoology), seems to have been regarded by the Darwinians as a house organ. At least it carried papers by Hooker, Wallace, and Bates sympathetic both to evolution and to natural selection (also Trimen 1868). Scientifically, these were the best of all the pro-Darwinian writings, for they were not mere propaganda and special pleading but actually used natural selection to solve problems.[12]

The Royal Society of the 1860s was a very different organization from that of the 1830s. Major reforms had been effected, and rigid standards of scientific excellence were demanded for admission. In the 1870s the Darwinians took over the Royal Society entirely—Hooker became president and Huxley was secretary (and later president). In the 1860s their power was not so great, and General Edward Sabine, the president, was opposed to Darwin's ideas. But in ways subtle and not-so-subtle, the Darwinians fought back. Darwin's friends, particularly Huxley, saw to it that in 1864 he was awarded the society's highest honor, the Copley medal, and when

Sabine tried to emphasize that the honor was in no way awarded because of the *Origin*, they made him modify his language. Wallace got a Royal Medal in 1868, and his paper on natural selection was expressly included in the award's citation.

Although the publications of the society tended by custom to be relatively free of theory, here too—particularly in the *Transactions*—we see the Darwinians positively and effectively at work. When W. H. Flower submitted a paper in 1862 that contained strong support for Huxley's position on man against Owen, the referee—who just happened to be Huxley himself—was very enthusiastic and felt that, both in manner and in matter, the "memoir appears to me to be eminently worthy of a place in the Philosophical Transactions" (referee's report RR.4.97, Royal Society; oddly, Huxley misidentifies the author). Similarly, a paper submitted by W. K. Parker (1866) describing the skull of the ostrich, after denying any intention of dabbling with theory, carefully noted that there was no way the ostrich skull could be part of the backbone. In reviewing this, Huxley wrote: "I have no hesitation in very strongly recommending the publication of the chief part of this Memoir" (referee's report RR.5.173, Royal Society). Since Flower, the other referee, was also enthusiastic, the paper appeared.

In contrast, Principal Dawson of McGill, the promoter of *Eozoön canadense*, deservedly considered a worldwide authority on the geology of the carboniferous system and of fossil plants, was invited in 1870 to give the Bakerian lecture at the Royal Society. This was the second most senior lecture of the society and until that date had been included almost automatically in the *Transactions*. Dawson, who like most Canadian Presbyterians took his religion seriously, was injudicious enough to finish a long technical paper by coming out against evolution. "The botanist and zoologist expects sooner or later to arrive at elementary specific types, which, if to be accounted for at all, must be explained on some principle distinct from that of derivation."[13]

Of the two referees, P. M. Duncan, geologist and fellow-traveler with the Darwinians, thought that something might be salvaged were the inappropriate comments on the origins of species dropped. "I advise its being printed as it stands with the exception of the chapter on the origin of species & C. This is quite out of place and the author does not appear to be aware of the British opinions upon persistent species & C" (referee's report RR.7.24, Royal Society). The other referee, Hooker, felt that Dawson's "fruitless and even groundless speculation" made the whole paper unsuitable (referee's report RR.7.25, Royal Society). And so Dawson had to be content with a two-page abstract in the *Proceedings of the Royal Society* (Dawson 1869–70). It is reputed that when the *Origin* appeared, Whewell refused to allow a copy in the Trinity library—apparently two could play that game.

It is heartwarming that Duncan's sterling position did not go unrecognized. In 1872 Huxley recommended acceptance of an evolutionary paper by Duncan "as the facts he brings forward are of great paleontological interest" (referee's report RR.7.91, Royal Society). Nor did the Darwinians have a blanket objection to the Bakerian lecturer. In 1873, when Parker gave the lecture—in which he praised Huxley for weaning him from Owen and making him an evolutionist—the referee, one T. H. Huxley, reported that "the paper is a very valuable one" (referee's report RR.7.192, Royal Society), and with such an endorsement it duly appeared in the *Transactions* (Parker 1873).

Finally, it is worth bringing up Alfred Newton's paper on the dodo (Newton and Newton 1869). Published in the *Philosophical Transactions* in 1869, the paper was strongly proevolutionary but rather hesitant about natural selection. The two referees were Huxley and Owen. Huxley wrote: "The memoir is . . .well worthy of publication in the Transactions of the Royal Society" (referee's report RR.6.200, Royal Society). In almost identical terms, Owen wrote: "I . . . find in it every characteristic which makes it deserving and desirable to be published in the Transactions of the Royal Society" (referee's report RR.6.201, Royal Society). These reactions show that by the end of the 1860s in England one had to be an evolutionist to stay abreast of the scientific stream and to publish in the leading scientific journals. Darwin's defenders and critics came together on this, and the swing to evolutionism probably came much earlier in the decade. Also, Owen's reaction makes us question how great and prolonged the scientific opposition to a goodly proportion of Darwin's ideas really was. It is clear that many scientists, ranging from those close to Darwin, like Lyell, to those who opposed him, like Owen, felt the need for a supplement to natural selection. But, given the warmth of Owen's report (on a paper by a man known to be Darwinian, though he did praise some of Owen's ideas), one wonders if Owen and others like him ever put up quite the opposition that traditional reports would have offered. As the years went by and the Darwinians had clearly "won," they may have exaggerated their victory—the dragons they killed grew that much bigger. Much of Owen's reputed opposition undoubtedly sprang from personal dislike of Huxley and Hooker rather than from scientific conviction. Even in his British Association address in 1858, in mentioning Darwin's and Wallace's papers Owen used his own earlier speculations on the struggle for existence to lead up to their "ingenious" evolutionary ideas (Owen 1858*a*). Had Darwin larded the *Origin* with flattering references to Professor Owen's "penetrating insights"—and there was plenty of chance to do so—one wonders what the subsequent course of the evolutionary debate might have been.

In addition to the societies and their journals, Huxley started his own journal, the *Natural History Review*, which ran for a short time in the 1860s. This was a sympathetic home for Darwinian ideas—Huxley's own arguments against Owen on man appeared there. Although this journal failed, Darwinians were also involved in founding *Nature* in 1870 (MacLeod 1969). The early issues remind one of Mr. Micawber's newspaper from New South Wales. One finds an address by Huxley, a review by Huxley, an introduction by Huxley. No doubt, somewhere in its pages there is report of a song by Huxley Junior.

The Older Universities

We find that the London network sympathetic to Darwinism also aided the spread of Darwin's ideas in other ways. We know that members of this group were concerned to see that key scientific posts in Britain were filled by men who shared their philosophy of science, understanding this term in its broadest sense as including science education, freedom from religious pressure, and so on. In the late 1860s a small group, including Hooker, Huxley, and Spencer, formed the "X-Club." Although primarily an opportunity for busy friends to socialize, the club rapidly started to function as a powerful group concerned with decisions significant to the course of science—grants, honors, posts, and the like (MacLeod 1970). This influence naturally favored those sympathetic to Darwin's ideas.

We have positive evidence of the way the influence exerted by the Darwinians, Huxley in particular, brought direct dividends for the Darwinian cause in regard to posts at the older universities that were filled at the beginning and end of the decade after the *Origin*. In 1860 George Rolleston was appointed Linacre Professor of anatomy and physiology at Oxford. Apparently Huxley had a major share in the election of Rolleston over a rival candidate strongly supported by Owen.[14] After the election, Rolleston wrote to Huxley: "I suppose you will have heard the result to which you yourself have largely contributed. I shall set to work so as never to give you cause to regret the share you have had in my promotion" (letter from Rolleston to Huxley, 1860, Huxley Papers, Imperial College, 25.148). Apart from Rolleston's proving to be a very able teacher of the new biology, within a year Huxley's efforts were paying direct dividends. In particular, Rolleston became an active and articulate supporter of Huxley in the battle with Owen over the nature of man (Rolleston 1884). And as the years went by Benjamin Jowett made repeated though unsuccessful efforts to get Huxley himself to take a post at Oxford.

At Cambridge at the end of the decade it was Michael Foster, close friend and assistant of Huxley, whose good offices were being promoted:

April 2 [1870]
Trinity Coll. Cambridge

My dear Huxley:

I read your letter to the Seniority yesterday. Your suggestion as to the Praelectorship of Physiology and the man to fill it [Foster], was most favorably received. . . .

If Dr. Carpenter, whom we have also consulted, gives the same advice, I have no doubt it will be followed. . . .

If he is appointed, we will set to work about establishing the Physiological Laboratory.

Very truly yours,
W. G. Clark[15]

Again Huxley's (and Carpenter's) work paid dividends, for Foster proved to be one of Cambridge's most brilliant and influential teachers. But he was not too proud to turn to his old master for examination questions.[16]

The Darwinians were active at Cambridge in other ways too. Through the 1860s one sees, for example, a distinct change in the examination papers. In 1863 Hopkins was one of the examiners and one of the questions asked students to give "evidences of design" in the biological sciences.[17] By the end of the decade, Hooker and Flower were examiners and students were being asked to analyze the concept of the struggle for existence.[18] Since Charles Darwin's son Francis got a first on this exam, being a Darwinian apparently was no hindrance!

Keeping matters in perspective, as early as the 1830s moves had been afoot to tone down the church influence at Oxford and Cambridge and to improve the place of the sciences (Campbell 1901). Sedgwick had been much involved in upgrading science, and about 1850 Cambridge had started science exams (although at first they were to be taken only after one had completed one's degree). Also, laws were enacted to allow noncon-formists to attend and graduate from the universities, and it is clear that the number of clerical fellows was on the decline from the 1850s on. No claims can be made that the men close to Darwin were responsible for all or even most of these changes—Sedgwick, who at one point played an important role, was the archetypal anti-Darwinian. In a sense, therefore, Darwin and his friends were lucky. The tide was flowing in their favor, with scientific posts available at the ancient universities as never before and a fresh receptivity to new ideas and methods. But it is clear that the Darwinians did their part in keeping up the momentum of change at Oxford and Cambridge, and they seized or created opportunity whenever possible.

Social Darwinism

The major omission from this discussion of the reception of Darwin's ideas has been the reaction to the Darwinian controversy by the general public, people not particularly interested in or knowledgeable about science (for-tunately, Ellegard 1958 considers this problem). My concern has been the scientific community, for here we have problems worth trying to explain. First, by and large scientists rejected *Vestiges* and other pre-*Origin* argu-ments for evolution; but after Darwin's *Origin* they rapidly switched to evolutionism. Why should this have been so? Second, even though scien-tists became evolutionists, many shrank from full acceptance of Darwin's mechanism of natural selection. Again, Why?

As we go beyond the scientific community into general Victorian middle-class society, the opposition to evolutionism, let alone natural selection, rises. Hence these questions start to lose their relevance. Moreover, there is no great mystery about why Darwin's ideas were found so objectionable. These ideas were taken to conflict with fundamental religious truths; therefore they were ridiculed and rejected. But one can-not properly make a rigid separation between scientists and their society, and so as we approach the end of our analysis it is appropriate to make some comments about beliefs outside scientific circles. Despite opposition to various aspects of Darwinism, even in the wider Victorian context we find elements that helped Darwin's ideas to spread—both outside and within the scientific community. This is just what our study of the period before the *Origin* would lead us to expect: one thinks of the way many nonscientists responded to the ideas in *Vestiges,* particularly as presented by Tennyson.

One suspects that some of the extrascientific elements favoring Darwinism did not have a great effect. It is unlikely that there were many followers of the Presbyterian James M'Cosh, who championed natural selection because it confirmed his belief that God picks out an elect (Passmore 1968, p. 535). But one thing particularly commended evolution, even elements of selection, to the man in the street. This was the way evolutionary views could be taken to justify various, often contradictory, theses about the nature and development of society: theses that collectively go under the title "social Darwinism" (Himmelfarb 1968). Thus some welcomed evolution because they found in it support for a general progressive trend they wanted to read into the development of human society. Some favored natural selection, based on the struggle between members of the same species, because they could then justify extreme laissez-faire economics, claiming that the crudest cutthroat business practices have biological sanction. Some favored natural selection, based on the struggle between different species, because they could then argue for state controls as one pushed militaristically and imperialistically against other nations!

These various doctrines owed as much to Herbert Spencer as to Charles Darwin, if not more, and particularly in America Spencer was praised equally by academics and by barons of industry like Andrew Carnegie (Hofstadter 1959). Because such social views are out of favor today, there is a tendency to deny Darwin any hand in their paternity. But this is not quite accurate. On occasion Darwin did explicitly disavow social Darwinian views (he was indignant at the assertion that he had proved "'might is right' and therefore that Napoleon is right, and every cheating tradesman is also right"; F. Darwin 1887,2:262). But in the *Descent of Man* Darwin worried that modern medical techniques such as vaccination were preserving the unfit, adding that "no one who has attended to the breeding of domestic animals will doubt that this must be highly injurious to the race of man" (Darwin 1871, 1:168). And toward the end of his life he wrote of natural selection as helping the progress of civilization. "The more civilized so-called Caucasian races have beaten the Turkish hollow in the struggle for existence" (F. Darwin 1887, 1:316). Consequently, the relationship between biological Darwinism and social Darwinism is far from clear-cut (see Himmelfarb 1968; Greene 1977). The same holds of the relationship between Spencer's views and all the social doctrines they were used to support. Spencer claimed, for instance, that the struggle for existence among men ultimately causes its own demise.

But the actual relationship between the evolutionists with their biological doctrines and the various social doctrines supposedly supported by biology was not really important. What was important was that people, including some biologists, attempted to make biology support various social doctrines they fancied. And the result was that even at the general level the battle over evolution (even over selection) was not one-sided. Among the nonscientific public there was greater opposition to evolutionism than among scientists, and even greater opposition to selection. But for a number of reasons, not always based on exact truth, for certain members of the public evolutionary doctrines struck responsive chords. We may be reaching the point where Darwin's personal credit or blame for the ideas involved diminishes; the important fact is that in Britain in the 1860s and 1870s evolutionary ideas generally started to make some progress (Burrow 1966). Some of the influence was from biology; conversely, some scientists no doubt were made more secure in their scientific beliefs because these beliefs fitted in with social and political convictions they shared with various segments of society. One should not conclude that what happened in the scientific community either did not affect or was not affected by the Victorian world at large.

Conclusion

The full development of social Darwinism carries us beyond 1875, the end of the period we are considering. Judging from the situation in that year, in many ways Darwin and his ideas were obvious successes. Politically, the Darwinians were virtually in control. Hooker was president of the Royal Society and Huxley was secretary. A similar state obtained in other societies, and at the universities key posts were held by men close to Darwin and his circle. In their beliefs the scientists were evolutionists, and though this was not all Darwin's work, it would be churlish to deny that he was primarily responsible. Had Darwin (and Wallace) never existed, scientists would still have become evolutionists—perhaps before 1900—but it was Darwin who precipitated and expedited the change.

In other respects the successes were more limited. Many people were not convinced that natural selection was nearly as powerful as Darwin claimed; to most it was clear that the problem of heredity remained unsolved; and the physicists' restrictions on the age of the earth cast a pall over biological theorizing. In other realms the same tale can be told. More than a few people were unconvinced by the attempt to explain man's arrival purely in natural terms, and the lack of conviction extended to naturalist explanations of adaptation.

We have been following an evolving scientific community through some half a century.[19] When we joined them the members were just focusing on organic origins as a major scientific problem: the "mystery of mysteries." It does not necessarily follow that because one is a scientist one will find any particular scientific problem significant or interesting, but we have reviewed a number of reasons why the British community at that time was drawn to the organic origins problem. The question was highlighted by the members' scientific interests: geology, geographical distributions, and the like. It impinged directly upon philosophical systems being developed at that time. And it had immediate ramifications in religion, both revealed and natural, that were magnified by the peculiar organization of British science, where so many scientists necessarily had formal links with the established church.

This community espoused the organic origins problem, applying to it skills and abilities that stemmed from those very factors that made the problem significant and interesting. Driven by their own internal forces and prodded by external influences such as Continental science and speculators on the fringe of science, like Chambers, the community produced Darwin's *Origin* and accepted it to the degree we have seen. Moreover, the very things that carried the British that far were in great part the things that stopped them from going on. Darwin was led to selection because he saw adaptation as a significant facet of the organic world. Others could not accept selection as fully adequate for exactly the same reason.

By about 1875 the resources of the British were virtually depleted. The resources that had taken them this far could take them no further and in some respects were inhibiting further growth. Scientifically, they needed two basic things. (Two things, that is, that biologists could supply. They also needed a revised age for the earth.) First, they needed new ideas, approaches, and techniques to solve the problems of heredity, the introduction of new characters, and so on. But the British were not well fitted to fill this need. For example, they were not skilled in one of the most obvious technical approaches to heredity, microscopy. Hence it was not surprising that others took over and solved the problems of heredity. The second need was for detailed, long-term studies of Darwin's putative mechanism for evolutionary change, natural selection. For this the British were much better fitted and, through the work of Hooker, Bates, and Wallace, had made a good start. But showing the full strength and ubiquity of natural selection was not to be done in a day or two, or even a year or two. Our scientific community could set up a research program, but it took time and effort to reap the dividends: dividends that have been abundantly realized in recent years.

We have seen a community take up a problem and wrestle with it. By about 1875 that community had done what it could, and so the problem was passed on to others of different places and different times.[20] But that is someone else's story.

10 *Overview and Analysis*

I have told the story from 1830 to 1875.[1] The bare bones of the tale can be quickly reviewed. In Britain in 1830 there was division of opinion on the organic origins question, but virtually no one was an evolutionist. Some favored an unknown but essentially law-bound originating mechanism; others supported miracles. In 1844, when Chambers published his *Vestiges,* almost no member of the professional scientific community accepted its central message; but in 1859, when Darwin published the *Origin,* scientists concerned with the origins of organisms were rapidly converted to evolutionism. Yet even Darwin had only limited success, for many refused to accept fully his mechanism for evolutionary change, natural selection. All the threads around which our narrative has been structured throw light on this course of events.

First, matters could be considered at the purely *scientific* level. Many facts that were inexplicable if not downright anomalous from a nonevolutionary viewpoint fitted into place for an evolutionist. And knowledge of such facts grew during our half-century. Lyell brought systematic order to these problems, but he did so despite himself, to highlight the difficulties for a nonevolutionist. Perhaps the most dramatic new knowledge was the distribution of finches and tortoises on the Galápagos

Islands, as revealed by Darwin himself. For any nonevolutionist this caused problems. One could certainly suppose that God put different finches on different islands, but this seems pointless at best, if not a direct violation of God's good sense. And Lyell, supposing laws but not evolution, was in no better position. If the finches came from elsewhere, how and why were they distributed as they were? If the finches started life on the various islands, how could this happen except by a process suspiciously like evolution?

Perhaps no facts were quite as dramatic as those of geographical distribution, but other areas did not comfort nonevolutionists. In 1830 people read the fossil record more or less as they pleased. But as the years went by, evidence built up for a kind of sequential development, ruling out a Lyellian steady state, and eventually the record negated the transcendental progressionism favored by those like Agassiz. Hence evolutionism came to appear more reasonable, especially an evolution like Darwin's, with no inevitable progression. The gaps in the record started to close, though the record was always sufficiently incomplete to permit causal interpretations other than selection. Much the same is true of morphology. It was clear that homologies needed explanation, and though Owen's archetype theory sufficed for a while, it had serious scientific flaws quite aside from the more philosophical objections opposed to the whole notion of archetypes.

Science also shows why people rejected *Vestiges,* though reactions to it certainly were not exclusively scientific. *Vestiges* was saddled with all kinds of untenable assumptions, like spontaneous generation; and the positive evidence was not as strong for Chambers as for Darwin. Chambers's main scientific argument came from the fossil record, and in 1844 there was much less reason to see it as evolutionary than there was in the 1860s. Chambers, and Lamarck for that matter, made no real attempt to solve one of the major puzzles—the origins of species as opposed to the origins of organisms. For Lamarck species were an embarrassment; for Chambers species were an accidental by-product of evolution. But for Darwin species were a natural consequence of basic principles.[2]

Finally, science throws light on reactions to Darwin's doctrines aside from his general evolutionism. There were good scientific reasons to doubt that selection working on the smallest of variations could be effective enough to cause evolution. Darwin's speculations on the nature and causes of heredity and variation left much to be desired; and physics apparently showed that the time available for evolution was far too short for such a leisurely mechanism as selection. Science also shows the rationale of those who went furthest with Darwin: people like Wallace, Hooker, and Bates were interested in the same sorts of problems as Darwin, and they needed a working mechanism as opposed to a general background of evolutionism.

Philosophy yields much the same tale as science (Ruse 1979). Throughout our period scientists felt a metascientific urge to explain through law, bound up with a general belief that science ought, as much as possible, to imitate Newtonian physics (particularly the Newtonian astronomy of the 1830s). For all their subtlety, the nonevolutionists fell short of their own ideals, in others' eyes if not in their own. Whewell and Sedgwick divorced the organic origins question from science—to less conservative scientists a cure somewhat worse than the disease—and people like Lyell and Herschel knew well that their nonevolutionism came dangerously close to violating their empiricist *verae causae* principles. A major motive behind Chambers's work clearly was his desire to explain through law, and the same holds for Darwin and for those who responded favorably to his writings. Good Newtonian science requires law, and in the end only evolution could provide it.

Here in the philosophical realm, perhaps more than anywhere, we can see the advantage Darwin had over Chambers and understand why Darwin succeeded and Chambers did not. For all his talk of being Newtonian, Chambers really made no attempt to provide a *vera causa* for evolution, and his critics quickly noted this. Darwin certainly did not satisfy everyone, friend or foe, with his mechanism of natural selection; but in the *Origin* he made a systematic attempt to conform to criteria of scientific excellence, and the powerful effect of the *Origin* in large part sprang from this care. However, for empiricists like Huxley, employing an analogical *vera causa* concept, Darwin's failure to demonstrate that selection did lead to species change left the mechanism unproved.

Platonism, perhaps the epitome of idealistic philosophies, raises different questions from Newtonianism. Despite internal differences, nobody on either side in the Darwinian Revolution questioned the need to be Newtonian. The issue was how one could best accomplish this. But with Platonism we see the decline and expulsion of a particular metaphysics. Owen, Whewell, Agassiz, and others were Platonists, and this manifested itself in their science in various ways. Huxley, on the other hand, was not a Platonist: as an empiricist he criticized his opponents (specifically Owen) for introducing their idealistic metaphysics into science. The Darwinian Revolution went as it did, and Darwin pushed out Owen, in part because the nineteenth century brought a decline of such idealism and a rise of empiricism. We can justly give Darwin credit for his successful New-tonianism. But though he did bring some strong arguments against the notion of absolutely distinct and essentially uniform classes— a funda-mental tenet of Platonism and other forms of idealism like Aristotelianism—he can hardly claim the same credit for the overall de-cline of idealism. This decline was due to many factors, such as changes in the educational system (no longer did every schoolboy have an uninter-rupted diet of the classics) and a weakening in the influence of religion, which was intimately connected with idealism. But, whatever the causes, Darwin's science, though he himself incorporated rationalist elements with his empiricism, was far more in tune with the anti-idealistic philosophy of science of men like Huxley than were Owen's explicitly idealistic scientific systems. Even discounting what he took to be Darwin's shortcomings in satisfying the empiricist *vera causa* criterion, Huxley found ancestors more to his philosophical taste than archetypes, and Dar-winism benefited accordingly. (But remember Huxley's debt to Carlyle, and Carlyle's debt to Plato!)

Concerning *religion,* the major question might be not how religion helped the cause of evolutionism, but why it failed to suppress it (Ruse 1975*b*). Even at the beginning of our period people had become accus-tomed to modifying their religion to accommodate the advances of science; so, while evolutionism may have been a serious problem, the threat was not really new. In our half-century we see religion continuing a retreat that had started long before. Yet religion was still a major force behind the nonevolutionary positions formulated in the 1830s. Religious people were terrified by the threat evolution posed to the special status of man, and both those for law and those for miracle cared desperately that full place be left for God's designing powers. Religion was also a major element in the reaction against Chambers. With good reason, he was considered to undermine the dignity of man and to leave no room for God's design. And much the same points came up again after the *Origin.* For various reasons these points were becoming less powerful—man's special status was coming under fire for reasons outside Darwinism, and Darwin had made some attempt to come to grips with adaptation—but many still felt the need to soften Darwin's speculations with elements of religion. However, the Darwinian Revolution was not exclusively a war of science against religion, even if the two are considered distinct. In certain ways, unintentional and intentional, religion aided the coming of evolutionism—even Darwin's version. Unintentionally, evolutionism was undoubtedly helped by those like Agassiz and Miller, who made so much of the progressive nature of the fossil record, and the same holds true of the morphological version of the argument from design. By stressing homologies, such natural theologians were preparing the way for an evolutionary interpretation. Even Whewell, though he was never an evolutionist, helped spread the idea of law when he adopted Owen's archetypes and emphasized the morphological version of the design argument. And the utilitarian design argument, focusing upon and highlighting the importance of adaptation, contributed to Darwin's discovery of natural selection. Intentionally, evolutionism was helped by those who saw the highest mark of God's power as his ability to work through unbroken law—who saw God as a super leader of industry. This approach to religion spurred Chambers, Baden Powell, and many others. And somewhere on the borderline between intentionality and unintentionality we must find a place for Lyellian geology. Lyell found his approach to geology attractive because it alone satisfied his deistic concept of theology. But Lyellian geology, molded as it was by religion, was probably the major influence in bringing about Darwinian evolutionism, though the parent was not altogether happy with the child. Thus it is clear that evolutionism, and even selection, came because of religion as well as despite it.

Finally, we must consider *social and political* factors. Between 1830 and 1875 we see an evolution of British society, and in particular British scientific society. In the 1830s science as a profession (particularly biology, geology, and the like) was just beginning, and because of the peculiarities of British higher education there were strong links between science and organized religion. The next forty to fifty years saw those links weaken and break, until it was possible for someone without independent means both to become a professional biologist or geologist and to owe nothing whatever to the church. Evolutionism, particularly Darwinian evolutionism, gained by this, though the opportunity had to be used. Darwin succeeded where Chambers failed because Darwin had earned respect as a scientist and because he had built up a group prepared to battle for his ideas. But given that the Church of England was an integral part of the establishment, the Darwinian Revolution also reflected the way power and authority were being loosed from the hold of an exclusive minority and shared by a wider middle class. In this most important change in Victorian Britain, the Darwinian Revolution was part cause and part effect.

I have taken up the various threads separately. They have crossed, intertwined, and run together, and often when advance in one area seemed blocked another came into play. When Darwin was faulted on religious grounds because he had paid too little attention to God as designer, he replied on philosophical grounds that such an explanation is neither required nor allowed in the physical sciences and therefore ought not be required or allowed in the biological sciences. Similarly, the science/religion compromise formulated by Sedgwick and Whewell, depending centrally on man's recent origin, was badly jolted by scientific discoveries about man's long history.

But separately or together, the various elements of our story point to one important conclusion, hinted at the beginning of this book: The Darwinian Revolution cannot be considered a single thing. It had different sides, different causes, and different effects. Often it is portrayed as a triumph of science over religion; but, though there is some truth to this idea, as a total assessment of the Darwinian Revolution it is far from adequate. The supposed triumph of science over religion was questionable, more was involved than science and religion, and in some respects religion helped the cause of science. It probably is a mistake to say that in the coming of evolutionism certain things were essential, whereas others were not. It makes little sense to compare the relative merits of, say, Bates's work on mimicry with Huxley's writing of supportive referee's reports for all who favored evolutionism. I would feel very uncomfortable with an analysis of the Darwinian Revolution that belittled these points to concentrate solely on the "real" issues, such as man's place in the natural scheme of things. In its way the Darwinian Revolution was one of the most significant movements in man's history. That it had many sides, intellectual and otherwise, should be no surprise.[3] Indeed, we should have expected this.

Notes

Chapter 1

1 Excellent background material on the idea of evolution can be found in Greene 1959; Rudwick 1972; and Goudge 1973.

2 Burkhardt 1977 discusses in some detail Lamarck's position in the scientific community and speculates on why he did not gain a band of followers. For my treatment of Cuvier I am much indebted to Russell 1916; Coleman 1964; and Bowler 1976*a*.

Chapter 2

1 In a notebook comment early in 1839 Darwin wrote, "being myself a geologist" (de Beer et al. 1960–67, E, p. 156). Because de Beer's transcription of Darwin's notebooks was published over several years, with some pages out of sequence, I shall refer throughout to Darwin's pagination.

Chapter 3

1 We can infer that this is a direct reply because Whewell added it to the back of his manuscript just as he was completing the *History* early in 1837, the very time Lyell made his criticism (the manuscript is in the Whewell papers, Trinity College, Cambridge).

2 Darwin's three major geological works are *Coral Reefs* (1842); *Volcanic Islands* (1844); and *South America* (1846): collectively they constitute the *Geology of the Voyage of the Beagle*. Ideas from these works were published in papers in the late 1830s. I refer to the combined collection, reprinted as *Coral Reefs* (1910).

3   I discuss Lyell here. But tribute must also be paid to the
    influence of Alexander von Humboldt, particularly his
    Personal Narrative (1814–29), which speculates about
    the world's geological balance (with specific references to
    South America). Darwin spoke of this work in the high-
    est terms, and a copy was on the *Beagle*. But Lyell too
    probably learned from Humboldt, and so one should not
    think of Lyell and Humboldt as contrary influences.
    And, as a geologist, Darwin clearly thought of himself
    as essentially Lyellian. While it is obvious that Hum-
    boldt influenced Darwin toward travel and science, and
    while Darwin adopted much of Humboldt's quasi-
    scientific style in his travel book (*Journal of Researches*), I
    do not see Humboldt as a great force shaping Darwin's
    evolutionism. Though Humboldt makes some mention
    of population data, which was to be important to Dar-
    win, the *Origin*'s immediate sources do not seem very
    Humboldtian. But see Egerton 1970.

4   Credit must be given to Mayr 1964; Ghiselin 1969; and
    Hull 1973*a, b* for showing that philosophy was a key
    element in the Darwinian Revolution.

5   In British politics one had the liberals (Whigs), the party
    of the aristocrats and the middle class, and the conserva-
    tives (Tories), the party of the crown (before Victoria)
    and the gentry, including most of the established
    clergy. My use of the terms "liberal" and "conservative"
    is not intended to be political, though there are obvious
    connections. The Tories resisted change, using their
    Christian theism to bolster their belief in the right or-
    dering of things. The Whigs, behind the reform move-
    ment, saw the propriety of gradual, continual change.
    Lyell was a Whig; Whewell was a Tory. But, perhaps
    showing that we must be wary of assuming too strong a
    link between religion, science, and politics, we find that
    the conservative catastrophist Sedgwick was strongly in
    the Whig camp.

Chapter 4

1   Figure 19, in chapter 6, shows that the discovery of the
    Stonesfield mammals moved the first appearance of
    mammals back to the oolite in the middle of the Secon-
    dary, rather than the Eocene, at the beginning of the
    Tertiary—in modern terms, back to the Jurassic in the
    middle of the Mesozoic, rather than the Eocene-
    Paleocene at the beginning of the Cenozoic.

Chapter 5

1   With the notable exception of Robert Grant, professor of comparative anatomy and zoology at University College, London. I will mention him later.

2   "Popular information on science" (1838). I ascribe these unsigned articles to Miller primarily by elimination. They are the only popular articles on geology in the pertinent issue of the journal, and we know that in that year the Chambers brothers published such work by Miller (Chambers 1872, p. 262). Also, a year later Miller admitted his enthusiasm for Buckland, who gets prominent treatment in the articles (Miller 1839, p. 138).

3   See *Vestiges* 1845*a*, p. 493; Geology versus development 1850, p. 359; Sedgwick 1845, p. 19; Brewster 1844, p. 477; Bowen 1845, p. 441. Whewell continued to feel a fondness for the hypothesis. In 1864 William Huggins, relying on Kirchhoff's method of using Fraunhofer lines to determine chemical composition, showed that some nebulae are irresolvably gaseous; but since the nebulae contain fewer elements than the stars, he felt unable to accept the nebular hypothesis.

4   Goethe seems to have been the first man with this idea; but Owen acknowledged his own debt as being to Oken (Owen 1848, p. 73). See Russell (1916) for a detailed discussion of this theory.

Chapter 6

1   Lyell 1851, p. lxiii. For details, look ahead to figure 19, Owen's summary of the state of the fossil record in 1860. The progressionists of the 1840s put the invertebrates in the Silurian, the bony fish in the Devonian, the reptiles as post-Carboniferous, and birds and mammals in the Tertiary (Cenozoic) period. Lyell's claim therefore was that the Stonesfield fossils implied birds and mammals, which were significantly pre-Tertiary.

2   Again, refer to Owen's diagram (fig. 19). Lyell wanted to locate Cetacea in the Cretaceous sea. Progressionists thought Cetacea first appeared in the Eocene.

3   Huxley supported George Rolleston, the successful candidate. "I daresay you will have seen Dr. Wallers Testimonials and among them the very strong one from Owen. It has been hinted to me that some of this strength is due to a feeling of antagonism . . . to you"

(Rolleston to Huxley, 20 May 1860. Huxley Papers, Imperial College, 25.147).

4   Huxley, by invitation, wrote an essay on Owen's achievements, which was included in *The Life of Richard Owen* (Owen 1894).

5   On this matter see Benn 1906 and Willey 1949.

6   Tennyson's commitment to evolutionism is controversial. I myself think he was an evolutionist, but what really counts here is that he can certainly be taken for one. See Killham 1958 and Millhauser 1971.

7   Ross 1973, pp. 34–35, from sections 55 and 56. This passage was written in direct reaction to a reading of Lyell's *Principles* in 1837 (Ross 1973, pp. 120–26).

8   Ross 1973, pp. 89–90, from the Epilogue, probably written early in 1845. In *Vestiges* Chambers had written: "Is our race but the initial of the grand crowning type? . . . There is in this nothing improbable on other grounds . . . . There may then be occasion for a nobler type of humanity." This passage was reprinted in a review in the *Examiner* that Tennyson read. See Killham 1958, p. 85, for details. Notice how the poet echoes "race," "crowning," "nobler type"—apart from the identity of sentiments, including a shared commitment to a recapitulatory theory.

9   This article appeared in the *Westminster Review*. Presumably for tactical reasons, in this essay Spencer was not quite as categorical as usual.

10 Forbes 1854. In the 1840s, Forbes had done important and influential work on British organic distributions. See Forbes 1846.

Chapter 7

1   Darwin 1969, p. 49. Huxley later sneered at Grant, but this is not the impression Darwin gives. See F. Darwin (1887, 2:188).

2   This is from the notebook Darwin marked on the cover "R. N." It is quoted in Herbert 1974, p. 2 7 n.

3   There are four notebooks directly related to e species question, marked by Darwin B, C, D, and E These are transcribed and published in de Beer et al. 960–67. There are also two notebooks, M and N, that al essentially with man. These have been transcribed nd published in Gruber and Barrett 1974. For con ience, I refer to Darwin's labels and pagination.

4   De Beer et al. 1960–67, E, p. 60. With his commitment to branching, we know that at this time Darwin could not have accepted orthodox progression. His actual views will be discussed later.

5   Without denying that reading Brewster's review of Comte must have given Darwin a strong psychological boost toward the rightness of Newtonianism, I cannot follow some recent commentators (e.g., Schweber 1977) in arguing that the review was *the* key philosophical influence on Darwin. Apart from anything else, Comte denied that science is a search for causes: Darwin followed Herschel and Whewell in considering this the ultimate aim of science. Theologically speaking, however, Comte's open atheism may have influenced Darwin.

6   One can work out dates from comments in the species notebooks (see Ruse 1975*a*). Darwin's copies of these books are in the Darwin collection, University Library, Cambridge.

7   Youatt 1831, read in December 1839; Youatt 1834, read in March 1840; and Youatt 1837, read in October 1840.

8   See correspondence with Thomas Poole (in British Museum). Wedgwood, at one point, had 2,000 sheep, including 300 Merinos he was trying to introduce into England.

9   He was president of the Etruria branch. Note that the SDUK published Youatt's books.

10   For example, "Daines Barrington says cock birds attract females by song do they by beauty" (de Beer et al. 1960–67, C, p. 178).

11   As we know, the notion of a consilience, though not tied directly to a *vera causa,* also occurred in Herschel's *Discourse.* Whewell first wrote explicitly of a consilience, tying it to a *vera causa,* in the *Philosophy* (1840). I am not sure Darwin read this at this time, though he did respond enthusiastically to a detailed review by Herschel, explicitly discussing consilience and *verae causae:* "from Herschel's Review Quart. June 41 I see I MUST STUDY Whewell on Philosophy of Science" (Darwin, "Books to be Read"; Herschel's review was in the *Quarterly Review*). But, as pointed out, Darwin knew Whewell's philosophy of science from the *History* and from personal contact. See also Thagard 1977, who shows that at some point Darwin may have read the *Philosophy.*

12 Darwin does mention man's geometric bre∈   ∖g rate in
    the *Origin* (p. 64).

13 Like nearly everyone else, I find myself rele   ng Wal-
    lace to the notes. This is unfair, since Walla   eally did
    discover natural selection as an evolutionary   :hanism,
    but not totally unfair. Wallace's creative   rk came
    twenty years after Darwin's, he did not wr: ∖ut a full
    theory, and he did not form a party of sup   ers.

Chapter 8

1   Wilson 1970, p. 57. The Philosophical   b was a
    ginger group within the Royal Society tryi   :o better
    the state of science. George Busk was a n   ∙list and
    surgeon.

2   Others, most notably Spencer, wanted to r   ∶ the in-
    heritance of acquired characteristics the chi   olution-
    ary mechanism.

3   Letter from Harvey to Darwin, 24 August 1   Quoted
    in Vorzimmer 1970, p. 61.

4   F. Darwin 1887, 3:107. For more on Darw: views on
    variation, see Vorzimmer 1970 and Bowler 1   . Bowler
    suggests that Jenkin's criticism shifted Dar∖   from be-
    lieving that selection could work on distin∈   riations,
    large or very small, to believing that the cru   variation
    exists in a range in a population and *by d∈   *on* will
    always be favorable. There may be truth in   Darwin
    knew of and accepted Quetelet's finding tl   riation
    falls in ranges about means (Darwin 1977, 2   and in
    his letter to Wallace he admitted that even b   •ading
    Jenkin he had a longing for single variati   ∖ough
    this could be read as *large* variations. How∈   en in
    the sixth edition of the *Origin* Darwin often   aria-
    tions as discontinuous and not necessarily gu   ∖in a
    population (e.g., Darwin 1959, pp. 326–   haps
    it is best to conclude that Jenkin pushed D∈   ∶her
    down a road he had already embarked c   hat
    Darwin turned even more to variation seer   lly
    discontinuous in a group, with more chanc∈   ∖le
    variation.

5   The hypothesis was introduced in *Variation*   ∫-
    ti∙cation.

6   Mivart 1871, pp. 241–42. Against thi   ∖,
    Lamarckians argued that the doctrine of i∤   of
    acquired characteristics applied only to fe∂   ∶d
    in response to stress—for example, thick-sl   s.

Circumcision is an imposed mutilation and thus irrelevant. But the objection certainly counts against Darwin, who allowed the inheritance of all bodily changes, externally or internally caused.

7   For more details, see Bowler 1976*a*. We know that Chambers, eclectic as always, had added stirps to his progression, but this hardly denied its essential unilinear nature.

8   Owen 1860. All the reviews in the *Edinburgh Review* were anonymous, so there was nothing discreditable in Owen's being likewise.

9   Impressions like these are difficult, if not impossible, to confirm; but insofar as this is possible, Ellegard 1958 does so. In the next chapter I shall cite scientific literature to add to the evidence.

10   Loci classici of the synthetic theory include Dobzhansky 1937; Mayr 1942; Simpson 1944, 1953; and Stebbins 1950. A first-class modern account is Dobzhansky et al. 1977.

11   It may seem that I am committing the post hoc fallacy, assuming that because the switch to evolution followed the *Origin,* Darwin alone caused it. Given the time of the switch, the fact that so many acknowledged a debt to Darwin, and other factors covered here, this possibility of a causal connection obviously has a strong foundation. But I shall return to this in the next chapter and willingly allow that Darwin was not the sole cause.

Chapter 9

1   Huxley Papers, 19.212. Quoted in Bibby 1972, pp. 43–44. Calling the *verae causae* concepts "empiricist" and "rationalist" is my own terminology.

2   I discuss this fully in Ruse 1973*b*. See also Goudge 1961 and Hull 1974.

3   Letter to F. Dyster, 9 September 1860. Huxley Papers, Imperial College, London, 15.115. Quoted in Bibby 1959, p. 69.

4   Wallace 1870*b*, p. 359. This was an expansion of similar remarks made the year before in a review of Lyell's works (Wallace 1869).

5   Clark and Hughes 1890, 2:359–66. The preface to the 7th edition of Whewell's *Bridgewater Treatise* (1863) contained similar sentiments.

6   Unpublished letter from Herschel to Lyell, 14 April

1863; Herschel Papers, Royal Society. His
avoid constant qualification, I am assuming
properly speak of Herschel and Lyell as "evc
though under a strict interpretation of the
introduced "evolution" (and "law"), one som
they are at best on the borderline. The imp
is that the Herschels and the Lyells though
almost evolutionists, and they were indeed
they had been. Certainly they were now adr
organisms came from other organisms, clos

7   L. Huxley 1900, 1:102. Ironically, but
    Huxley had put Owen at the top of the lisi

8   The point being made has been studied and s
    recent work in the sociology of science. Se
    Cole 1973.

9   Letter from Darwin to Herschel, 1860; Dar
    University Library, Cambridge.

10 Letter from Hooker to Huxley, 1888; Hux
    Imperial College, London, 3.335.

11 I am not denying that Owen's and Mivart
    contained more overtly anti-Darwinian eler
    say, Carpenter's; yet Lyell seems not to
    Mivart's *Genesis of Species* completely unpala

12 Significantly, the Entomological Society w
    less run by fellows of the Linnean Society.

13 "On the Pre-carboniferous Floras of Nc
    America, with Especial Reference to That (
    (Devonian) Period." The complete paper is (
    the Royal Society.

14 Owen had more success in 1862, when hi
    rather than Bates got a British Museum pos

15 Huxley Papers, Imperial College, London, .

16 "My dear Foster, I send you seven teasers for
    ship Examination . . ." Letter from Husley tc
    September 1874. Huxley Papers, Imperi
    London, 4.92.

17 *Cambridge University Almanack and Regis*
    Natural Science Tripos Papers for 1863, qu

18 *Cambridge University Almanack and Regis*
    Natural Science Tripos Papers for 1870, qu

19 I use the term "community" to cover both th
    the younger networks. In their primes bot
    torchbearers for British biology, and there
    and continuity.

20 Darwin is sometimes portrayed as retiring in
    crushed by criticism (see Vorzimmer 1970

truer to suppose that he realized the frontal attack of the *Origin* had served its purpose and that the time had come for selection to prove its worth as a tool for working scientists.

Chapter 10

1   Because I shall be recapitulating, I shall not duplicate earlier references.
2   Admittedly, Darwin too saw reproductive isolation as in part a by-product of selection.
3   At this point my thoughts run parallel to those of Hodge 1974.

Bibliography

Abbott, E., and Campbell, L. 1897. *The life and letters of Benjamin Jowett, M.A.* London: Murray.

Adamson, J. W. 1930. *English education, 1789–1902.* Cambridge: Cambridge University Press.

Agassiz, L. 1840. On the development of the fish in the egg. In *Report of the tenth meeting of the British Association for the Advancement of Science,* p. 129. London: Murray.

————. 1842. On the succession and development of organized beings at the surface of the terrestrial globe, being a discourse delivered at the inauguration of the Academy of Neuchatel. *Edin. New Phil. J.* 23:388–99.

————. 1849. *Twelve lectures on comparative embryology.* Boston: Redding.

————. 1859. *Essay on classification.* London: Longman, Brown, Green, Longmans, and Roberts and Trübner.

Altick, R. D. 1957. *The English common reader.* Chicago: University of Chicago Press.

Argyll, G. J. 1869. *Primeval man.* London: Macmillan.

Arnold, M. 1873. *Literature and dogma.* London: Smith Elder.

Babbage, C. 1830. *Reflections on the decline of science in England and on some of its causes.* London: Fellowes.

————. 1838. *Ninth Bridgewater treatise.* 2d ed.

285

London: Murray.

Baer, K. E. von. 1828. *Über Entwickelungsgeschichte der Thiere, Beobachtung und Reflexion.* Königsberg: Bornträger.

————. 1873. The controversy over Darwinism. *Augsburger Allgemeine Zeitung* 130:1968–88 (translated and reprinted in Hull 1973*b*, pp. 416–25).

Barlow, N. 1967. *Darwin and Henslow: The growth of an idea. Letters, 1831–1860.* Berkeley: University of California Press.

Barry, M. 1836–37. On the unity of structure in the animal kingdom. *Edin. New Phil. J.* 22:116–41, 345–64.

Bartholomew, M. 1973. Lyell and evolution: An account of Lyell's response to the prospect of an evolutionary ancestry for man. *Brit. J. Hist. Sci.* 6:261–303.

Bates, H. W. 1862. Contributions to an insect fauna of the Amazon Valley. *Trans. Linn. Soc. Lond.* 23:495–566.

Bayne, P. 1871. *Life and letters of Hugh Miller.* London: Strahan.

Beales, D. 1969. *From Castlereagh to Gladstone, 1815–1885.* London: Nelson.

Becker, B. H. 1874. *Scientific London.* London: King.

Ben-David, J. 1971. *The scientist's role in society.* Englewood Cliffs, N. J.: Prentice-Hall.

Benn, A. W. 1906. *The history of British rationalism in the nineteenth century.* New York: Russell and Russell; reprinted 1969.

Best, G. 1971. *Mid-victorian Britain: 1851–1875.* London: Weidenfeld and Nicolson.

Bibby, C. 1959. *T. H. Huxley: Scientist, humanist, and educator.* London: Watts.

————. 1972. *Scientist extraordinary: The life and scientific work of Thomas Henry Huxley, 1825–1895.* Oxford: Pergamon.

Blyth, E. 1837. On the psychological distinctions between man and all animals and the consequent diversity of human influence over the inferior ranks of creation, from any mutual and reciprocal influence exercised among the latter. *Mag. Nat. Hist.* 1:1–9, 77–85, 131–41.

Bowen, F. 1845. *Vestiges. . . . North Amer. Rev.* 60:426–78.

Bowler, P. J. 1974. Darwin's concepts of variation. *J. Hist. Med. Allied Sci.* 29:196–212.

———. 1975. The changing meaning of "evolution." *J. Hist. Ideas* 36:95–114.

———. 1976a. *Fossils and progress.* New York: Science History Publications.

———. 1976b. Malthus, Darwin, and the concept of struggle. *J. Hist. Ideas* 37:631–50.

———. 1976c. Alfred Russel Wallace's concepts of variation. *J. Hist. Med. Allied Sci.* 31:17–29.

———. 1977. Darwinism and the argument from design: Suggestion for a reevaluation. *J. Hist. Biol.* 10:29–43.

Brewster, D. 1837a. *The history of the inductive sciences* by Whewell. *Edin. Rev.* 66:110–51.

———. 1837b. Review of Comte's *Cours de philosophie positive. Edin. Rev.* 67:271–308.

———. 1844. *Vestiges. . . . North Brit. Rev.* 3:470–515.

———. 1854. *More worlds than one: The creed of the philosopher and the hope of the Christian.* London: Murray.

Briggs, A. 1973. *Victorian people: A reassessment of persons and themes.* Rev. ed. Chicago: University of Chicago Press.

Brock, W. H., and MacLeod, R. 1976. The "scientists' declaration": Reflexions on science and belief in the wake of *Essays and Reviews,* 1864–5. *Brit. J. Hist. Sci.* 9:39–66.

Brongniart, A. 1821. Sur les charactères zoologiques des formations. *Ann. Mines* 6:537–72.

———. 1823. *Mémoire sur les terrains de sédiment supérieurs Calcavo-Trappéens du Vicentin.* Paris.

Buckland, W. 1820. *Vindiciae geologicae; or, The connexion of geology with religion explained.* Oxford: Oxford University Press.

———. 1823. *Reliquiae diluvianae.* London: Murray.

————. 1836. *Geology and mineralogy*. Bridgewater Treatise no. 6. London: Pickering.

Buckle, H. T. 1890. *History of civilization in England*. New York: Appleton.

Burchfield, J. D. 1974. Darwin and the dilemma of geological time. *Isis* 65:300–321.

————. 1975. *Lord Kelvin and the age of the earth*. New York: Science History Publications.

Burkhardt, F. 1974. England and Scotland: The learned societies. In *The comparative reception of Darwinism*, ed. T. F. Glick. Austin: University of Texas Press.

Burkhardt, R. W. 1970. Lamarck, evolution and the politics of science. *J. Hist. Biol.* 3:275–98.

————. 1972. The inspiration of Lamarck's belief in evolution. *J. Hist. Biol.* 5:413–38.

————. 1977. *The spirit of system: Lamarck and evolutionary biology*. Cambridge: Harvard University Press.

Burn, W. L. 1964. *The age of equipoise*. London: Allen and Unwin.

Burrow, J. W. 1966. *Evolution and society: A study in Victorian social theory*. Cambridge: Cambridge University Press.

Burstyn, H. L. 1975. If Darwin wasn't the *Beagle's* naturalist, why was he on board? *Brit. J. Hist. Sci.* 8:62–69.

Butts, R. 1965. Necessary truth in Whewell's theory of science. *Amer. Phil. Quart.* 2:1–21.

————. 1968. *William Whewell's theory of scientific method*. Pittsburgh: University of Pittsburgh Press.

Cambridge University. 1851. *Cambridge University Callender*.

————. 1865. *Cambridge University almanac and register*. Cambridge: University of Cambridge Press.

————. 1872. *Cambridge University almanac and register*. Cambridge: University of Cambridge Press.

————. 1875. *Cambridge University almanac and reg-*

ister. Cambridge: University of Cambridge Press.

Campbell, B. 1972. *Sexual selection and the descent of man*. Chicago: Aldine.

Campbell, L. 1901. *The nationalization of the old English universities*. London: Chapman and Hall.

Cannon, W. F. 1961*a*. The impact of uniformitarianism: Two letters from John Herschel to Charles Lyell, 1836–1837. *Proc. Amer. Phil. Soc.* 105:301–14.

————. 1961*b*. John F. W. Herschel and the idea of science. *J. Hist. Ideas* 22:215–39.

————. 1964*a*. William Whewell, F.R.S. (1794–1866): Contributions to science and learning. *Notes Roy. Soc. Lond.* 19:176–91.

————. 1964*b*. Scientists and broad churchmen: An early Victorian intellectual network. *J. Brit. Studies* 4:65–88.

Cardwell, D. S. L. 1972. *The organization of science in England*. 2d ed. London: Heinemann.

Carlson, E. A. 1966. *The gene: A critical history*. Philadelphia: Saunders.

Carlyle, T. 1872. *Chartism*. London: Chapman and Hall.

————. 1896–1901. Shooting Niagara: And after? In *Works of Thomas Carlyle*. Vol. 5. *Miscellaneous essays*, ed. H. D. Traill, pp. 29–30.

————. 1937. *Sartor resartus*, ed. C. F. Harrold. New York: Odyssey.

Carpenter, W. B. 1839. *Principles of general and comparative physiology*. London: Churchill.

————. 1860. Researches on Foraminifera. *Phil. Trans. Roy. Soc. Lond.*, 1860:150.

Chambers, R. 1844. *Vestiges of the natural history of Creation*. 1st ed. London: Churchill.

————. 1845. *Explanations: A sequel to the "Vestiges of the natural history of Creation."* London: Churchill.

————. 1853. *Vestiges of the natural history of Creation*. 10th ed. London: Churchill.

————. 1884. *Vestiges of the natural history of Creation*. 12th ed. Edinburgh: Chambers.

Chambers, William. 1872. *Memoir of Robert Chambers*. Edinburgh: Chambers.

Checkland, S. G. 1951. The advent of academic economics in England. *Manchester School of Economic and Social Studies* 19:43–70.

Churchill, F. B. 1968. August Weismann and a break from tradition. *J. Hist. Biol.* 1:92–112.

Clark, G. K. 1963. *The making of Victorian England.* Oxford: Oxford University Press.

Clark, J. W., and Hughes, T. M. 1890. *The life and letters of the Reverend Adam Sedgwick.* Cambridge: Cambridge University Press.

Cochrane, J. L. 1970. The first mathematical Ricardian model. *Hist. Pol. Econ.* 2:419–31.

Cole, J. R., and Cole, S. 1973. *Social stratification in science.* Chicago: University of Chicago Press.

Coleman, W. 1964. *Georges Cuvier, zoologist: A study in the history of evolution theory.* Cambridge: Harvard University Press.

Coleridge, S. T. 1895. *Letters.* London.

Colp, R. 1977. *To be an invalid.* Chicago: University of Chicago Press.

Cordier, L. 1827. Essai sur la température de l'intérieur de la terre. *Mem. Acad. Roy. Sci.* 7:473–555.

Crane, D. 1972. *Invisible colleges: Diffusion of knowledge in scientific communities.* Chicago: University of Chicago Press.

Croll, J. 1867. On the eccentricity of the earth's orbit, and its physical relations to the Glacial Epoch. *Phil. Mag.* 33:119–31.

———. 1868. On geological time, and the probable date of the Glacial and Upper Miocene Period. *Phil. Mag.* 35:363–84, 36:141–54, 362–86.

Cuvier, G. 1822. *Essay on the theory of the earth.* 4th ed. Edinburgh: Blackwood.

Darwin, C. 1838. On certain areas of elevation and subsidence in the Pacific and Indian oceans, as deduced from the study of coral formations. *Geol. Soc. Proc.* 2:552–54.

———. 1839*a*. Observations on the parallel roads

of Glen Roy, and of other parts of Lochaber in Scotland, with an attempt to prove that they are of marine origin. *Phil. Trans. Roy. Soc. Lond.* 1839:39–82.

———. 1839*b*. Note on a rock seen on an iceberg in 16° south latitude. *Geogr. Soc. J.* 9:528–29.

———. 1839*c. Journal of researches.* London: Colburn.

———. 1840. On the connexion of certain volcanic phenomena in South America: And on the formation of mountain chains and volcanoes, as the effect of the same power by which continents are elevated. *Trans. Geol. Soc. Lond.* 5:601–31.

———. 1842. *The structure and distribution of coral reefs.* London: Smith Elder.

———. 1844. *Geological observations on the volcanic islands.* London: Smith Elder.

———. 1846. *Geological observations on South America.* London: Smith Elder.

———. 1851*a. A monograph of the sub-class Cirripedia, with figures of all the species. The Lepadidae; or, Pedunculated cirripedes.* London: Ray Society.

———. 1851*b. A monograph of the fossil Lepadidae; or, Pedunculated cirripedes.* London: Palaeontographical Society.

———. 1854*a. A monograph of the sub-class Cirripedia, with figures of all the species. The Balanidae (or sessile cirripedes); the Verrucidae, etc., etc., etc.* London: Ray Society.

———. 1854*b. A monograph of the fossil Balanidae and Verrucidae of Great Britain.* London: Palaeontographical Society.

———. 1859. *On the origin of species by means of natural selection.* London: Murray.

———. 1861. On the two forms, or dimorphic condition, in the species of *Primula,* and on their remarkable sexual relations. *J. Proc. Linn. Soc. (Bot.)* 6:77–96.

———. 1863. Review of Bates' "Contributions to an insect fauna of the Amazon Valley." *Nat. Hist. Rev.* 3:219–24.

———. 1864. On the sexual relations of the three

forms of *Lythrum salicaria. J. Proc. Linn. Soc. (Bot.)* 8:169–96.

———. 1868. *The variation of animals and plants under domestication.* London: Murray.

———. 1871. *Descent of man.* London: Murray.

———. 1910. *Coral reefs; volcanic islands; South American geology.* London: Ward Lock.

———. 1959. *The origin of species by means of natural selection.* Variorum text, ed. M. Peckham. Philadelphia: University of Pennsylvania Press.

———. 1969. *Autobiography,* ed. N. Barlow. New York: Norton.

———. 1977. *The collected papers of Charles Darwin.* Ed. P. H. Barrett. Chicago: University of Chicago Press.

Darwin, C., and Wallace, A. R. 1958. *Evolution by natural selection.* Cambridge: Cambridge University Press.

Darwin, E. 1794–96. *Zoönomia.* London.

Darwin, F. 1887. *The life and letters of Charles Darwin, including an autobiographical chapter.* London: Murray.

Darwin, F., and Seward, A. C. 1903. *More letters of Charles Darwin.* London: Murray.

Dawson, J. W. 1860. *Archaia.* Montreal: Dawson.

———. 1869–70. On the pre-Carboniferous floras of the north-eastern America, with especial reference to that of the Erian (Devonian) period. *Proc. Roy. Soc. Lond.* 18:333–35.

Deas, H. D. 1959. Crystallography and crystallographers in England in the early nineteenth century: A preliminary survey. *Centaurus* 6:129–48.

de Beer, G. 1963. *Charles Darwin: Evolution by natural selection.* London: Nelson.

de Beer, G., et al. 1960–67. Darwin's notebooks on transmutation of species. *Bull. Brit. Mus. (Nat. Hist.),* hist. ser., 2:27–200; 3:129–76.

Disraeli, B. 1847. *Tancred; or, The new crusade.* London: Colburn.

Dobzhansky, T. 1937. *Genetics and the origin of*

species (3d ed., 1951). New York: Columbia University Press.

———. 1970. *Genetics of the evolutionary process.* New York: Columbia University Press.

Dobzhansky, T., et al. 1977. *Evolution.* San Francisco: W. H. Freeman.

Duncan, D. 1908. *Life and letters of Herbert Spencer.* London: Williams and Norgate.

Duncan, P. M. 1872. On the structure and affinities of *Guynia annulata,* Cunc., with remarks upon the persistence of Palaeozoic types of Madreporaria. *Phil. Trans. Roy. Soc. Lond.* 162:29–40.

Dunn, L. C. 1965. *A short history of genetics.* New York: McGraw-Hill.

Dupree, A. H. 1959. *Asa Gray: 1810–1888.* Cambridge: Harvard University Press.

Egerton, F. N. 1968. Studies of animal populations from Lamarck to Darwin. *J. Hist. Biol.* 1:225–59.

———. 1970. Humboldt, Darwin, and population. *J. Hist. Biol.* 3:325–60.

Eiseley, L. 1958. *Darwin's century.* New York: Doubleday.

Ellegard, A. 1958. *Darwin and the general reader.* Göteborg: Göteborgs Universitets Arsskrift.

Explanations. . . . 1846. *Brit. Quart. Rev.* 3:178–90.

———. 1848. *Westminster For. Quart. Rev.* 48:130–60.

Flower, W. H. 1862. On the posterior lobes of the cerebrum of the Quadrumana. *Phil. Trans. Roy. Soc. Lond.* 152:185–201.

Forbes, E. 1846. On the connection between the distribution of the existing fauna and flora of the British Isles and the geological changes which have affected their area, especially during the epoch of the Northern Drift. *Mem. Geol. Survey Gt. Brit.* 1:336–432.

———. 1854. On the manifestations of polarity in the distribution of organized beings in time. *Proc. Roy. Inst.* 1:428–33.

Ford, E. B. 1964. *Ecological genetics.* London: Methuen.

Foster, M., and Lankester, E. R., eds. 1901. *The scientific memoirs of Thomas Henry Huxley.* London: Macmillan.

Fourier, J. B. J. 1827. Mémoire sur les températures du globe terrestre et des espaces planetaires. *Mémoires de l'Academie Royale des Sciences de l'Institut de France* 7:569–604.

Gale, B. 1972. Darwin and the concept of a struggle for existence: A study in the extrascientific origins of scientific ideas. *Isis* 63:321–44.

Galton, F. 1872. On blood relationship. *Proc. Roy. Soc. Lond.* 20:394–402.

Geison, G. L. 1969. Darwin and heredity: The evolution of his hypothesis of pangenesis. *Bull. Hist. Med.* 24:375–411.

Geology versus development. 1850. *Fraser's Mag.* 42:355–72.

George, W. 1964. *Biologist philosopher: A study of the life and writings of Alfred Russel Wallace.* London: Abelard-Schuman.

Ghiselin, M. 1969. *The triumph of the Darwinian method.* Berkeley: University of California Press.

Gillespie, C. C. 1951. *Genesis and geology.* Cambridge: Harvard University Press.

Gordon, A. 1894. *William Buckland.* London: Murray.

Gordon, M. M. 1870. *The home life of Sir David Brewster.* Edinburgh: Edmonston and Douglas.

Goudge, T. 1961. *The ascent of life.* Toronto: University of Toronto Press.

————. 1973. Evolutionism. In *Dictionary of the history of ideas.* New York: Scribners.

Gould, S. J. 1977. *Ontogeny and phylogeny.* Cambridge: Harvard University Press.

Gray, A. 1846. Explanations of the *Vestiges. North Amer. Rev.,* vol. 62.

————. 1876. *Darwiniana.* New York: Appleton.

Greene, J. C. 1959. *The death of Adam.* Ames: Iowa State University Press.

————. 1962. Biology and social theory in the nineteenth century: Auguste Comte and Herbert Spencer. In *Critical problems in the history of science,* ed. M. Claggett. Madison: University of Wisconsin Press.

—————. 1977. Darwin as a social evolutionist. *J. Hist. Biol.* 10:1–27.

Gridgeman, N. T. 1970. Charles Babbage. In *Dictionary of scientific biography,* 1:354–56, New York: Scribner's.

Grinnell, G. 1974. The rise and fall of Darwin's first theory of transmutation. *J. Hist. Biol.* 7:259–73.

Gruber, H. E., and Barrett, P. H. 1974. *Darwin on man.* New York: Dutton.

Gruber, J. W. 1960. *A conscience in conflict: The life of St. George Jackson Mivart.* Philadelphia: Temple University Press.

—————. 1965. Brixham Cave and the antiquity of man. In *Context and meaning in cultural anthropology,* ed. M. E. Spiro, pp. 373–402. New York: Free Press.

—————. 1968. Who was the Beagle's naturalist? *Brit. J. Hist. Sci.* 4:266–82.

Haeckel, E. 1883. *History of creation.* 3d ed. London: Kegan Paul, Trench.

Hare, E. H. 1962. Masturbatory insanity: The history of an idea. *J. Ment. Sci.* 108:1–25.

Harrison, J. 1971. Erasmus Darwin's view of evolution. *J. Hist. Ideas* 32:247–64.

Harrison, J. F. C. 1971. *The early Victorians, 1823–51.* London: Weidenfeld and Nicolson.

Herbert, S. 1968. *The logic of Darwin's discovery.* Ph.D. diss., Brandeis University.

—————. 1971. Darwin, Malthus, and selection. *J. Hist. Biol.* 4:209–17.

—————. 1974. The place of man in the development of Darwin's theory of transmutation. Part 1. To July 1837. *J. Hist. Biol.* 7:217–58.

—————. 1977. The place of man in the development of Darwin's theory of transmutation. Part 2. *J. Hist. Biol.* 10:243–73.

Herschel, J. F. W. 1827. Light. In *Encyclopaedia metropolitana,* ed. E. Smedley et al. London, 1845.

————. 1831. *Preliminary discourse on the study of natural philosophy*. London: Longman, Rees, Orme, Brown, and Green.

————. 1832. On the astronomical causes which may influence geological phenomena. *Trans. Geol. Soc. Lond.* 3:293–99.

————. 1833*a*. *Treatise on astronomy*. London: Longman.

————. 1833*b*. On the absorption of light by coloured media, viewed in connexion with the undulatory theory. *Phil. Mag.*, vol. 3.

————. 1841. History . . . and philosophy of the inductive sciences . . . by William Whewell. . . . *Quart. Rev.* 135:177–238.

————. 1845. Presidential address to the British Association for the Advancement of Science, Cambridge, June 19, 1845. Reprinted in *Essays from the Edinburgh and Quarterly Reviews, with addresses and other pieces*, pp. 634–82. London: Longman, Brown, Green, Longmans, and Roberts, 1857.

————. 1850. Quetelet on probabilities. *Edin. Rev.* 185:1–30.

————. 1861. *Physical geography*. Edinburgh: Black.

Himmelfarb, G. 1962. *Darwin and the Darwinian Revolution*. New York: Anchor.

————. 1968. Varieties of social Darwinism. In *Victorian minds*. London: Weidenfeld and Nicolson.

Hodge, C. 1872. *Systematic theology*. London: Nelson.

Hodge, M. J. S. 1971. Lamarck's science of living bodies. *Brit. J. Hist. Sci.* 5:323–52.

————. 1972. The universal gestation of nature: Chambers' *Vestiges* and *Explanations*. *J. Hist. Biol.* 5:127–51.

————. 1974. England. In *The comparative reception of Darwinism*, ed. T. F. Glick, pp. 3–31. Austin: University of Texas Press.

————. 1977. The structure and strategy of Darwin's "long argument." *Brit. J. Hist. Sci.* 10:237–46.

Hoff, K. E. A. von. 1822–24. *Geschichte der durch Überlieferung nachgewiesen natürlichen Veränderungen der Erdoberflache.* Gotha.

Hofstadter, R. 1959. *Social Darwinism in American thought.* New York: Braziller.

Hooker, J. D. 1853. Introductory essay. In *Flora Novae-Zelandia.* Part 2 of *The botany of the Antarctic voyage of "Erebus" and "Terror."* London: Lovell, Reeve.

———. 1856. *Geographie botanique raisonnée* par M. Alph de Candolle: A review. *Hooker's New J. Bot.* 8:54–64, 82–88, 112–21, 151–57, 181–91, 214–19, 248–56.

———. 1860. Introductory essay to *Flora tasmaniae;* Part 3 of *The botany of the Antarctic voyage of "Erebus" and "Terror."* London: Lovell, Reeve.

———. 1861. Outlines of the distribution of Arctic plants. *Trans. Linn. Soc. Lond.* 23:251–348.

Hooykaas, R. 1959. *The principle of uniformity.* Leiden: Brill.

Hopkins, W. 1860. Physical theories of the phenomenon of life. *Fraser's Mag.* 61:739–52; 62:74–90.

Houghton, W. E. 1957. *The Victorian frame of mind.* New Haven: Yale.

Hughes, T. 1861. *Tom Brown at Oxford.* Philadelphia: Porter and Coates.

Hull, D. L. 1965. The effect of essentialism on taxonomy: Two thousand years of stasis. *Brit. J. Phil. Sci.* 15:314–26; 16:1–18.

———. 1967. The metaphysics of evolution. *Brit. J. Hist. Sci.* 3:309–37.

———. 1973*a.* Charles Darwin and nineteenth-century philosophies of science. In *Foundations of scientific method: The nineteenth century,* ed. R. S. Westfall and R. Giere. Indiana: Indiana University Press.

———. 1973*b. Darwin and his critics.* Cambridge: Harvard University Press.

———. 1974. *Philosophy of biological science.* Englewood Cliffs, N. J.: Prentice-Hall.

Humboldt, Alexander von. 1814–29. *Personal narrative of travels to the equinoctial regions of the new continent, during the years 1799–1804* (Engl. trans.). London: Longman, Hurst, Rees, Orme, and Brown.

Hume, D. 1779. *Dialogues concerning natural religion.* London.

Hutton, J. 1795. *Theory of the earth.* Edinburgh.

Huxley, L. 1900. *Life and letters of Thomas Henry Huxley.* London: Macmillan.

————. 1918. *Life and letters of Sir Joseph Dalton Hooker.* London: Murray.

Huxley, T. H. 1851–54. On the common plan of animal forms. *Proc. Roy. Inst.* 1:444–46.

————. 1854. Vestiges. . . . *Brit. For. Med.-Chirurg. Rev.* 13:425–39 (*Scientific memoirs,* supp., pp. 1–19).

————. 1854–58. On natural history, as knowledge, discipline, and power. *Proc. Royal. Inst.* 2:187–95. (*Scientific memoirs,* 1:305–14).

————. 1856. Lectures on general natural history. *Med. Times Gaz.* 12:481–84.

————. 1857–59. On the theory of the vertebrate skull. *Proc. Royal Soc. Lond.* 9:381–457 (*Scientific memoirs,* 1:538–606).

————. 1858. On the agamic reproduction and morphology of aphis. *Trans. Linn. Soc. Lond.* 22:193–236 (*Scientific memoirs,* 2:26–80).

————. 1861. On the zoological relations of man with the lower animals. *Nat. Hist. Rev.,* pp. 67–84 (*Scientific memoirs,* 2:471–92).

————. 1862. Anniversary address of the president. *Quart. J. Geol. Soc. Lond.* 18:xl–liv (*Scientific memoirs,* 2:512–29).

————. 1863. *Evidence as to man's place in nature.* London: Williams and Norgate.

————. 1867–68. Remarks upon the *Archaeopteryx lithographica. Proc. Roy. Soc. Lond.* 16:243–48 (*Scientific memoirs,* 3:340–45).

————. 1868. On the animals which are most nearly intermediate between birds and reptiles. *Geol. Mag.* 5:357–65 (*Scientific memoirs,* 3:303–13).

————. 1869. Anniversary address of the president. *Quart. J. Geol. Soc. Lond.* 25:xxviii–

liii (*Scientific memoirs*, 3:397–426; *Essays*, 8:308–42).

———. 1870*a*. Further evidence of the affinity between the dinosaurian reptiles and birds. *Quart. J. Geol. Soc. Lond.* 26:12–31 (*Scientific memoirs*, 3:465–86).

———. 1870*b*. Anniversary address of the president. *Quart. J. Geol. Soc. Lond.* 26:xxiv–lxiv (*Scientific memoirs*, 3:510–50).

———. 1894. *Darwiniana: Essays*, vol. 2. London: Macmillan.

———. 1898–1903. *Scientific memoirs*. London: Macmillan.

Inglis, B. 1971. *Poverty and the industrial revolution*. London: Hodder and Stoughton.

Irvine, W. 1955. *Apes, angels, and Victorians*. New York: McGraw-Hill.

Jenkin, F. 1867. The origin of species. *North Brit. Rev.* 42:149–71.

Jenyns, L. 1863. *Memoir of the Rev. John Stevens Henslow . . . late rector of Hitcham and professor of botany in the University of Cambridge*. London: Van Voorst.

Kavalowski, V. 1974. The "vera causa" principle: An historico-philosophical study of a metatheoretical concept from Newton through Darwin. Ph.D. diss. University of Chicago.

Killham, J. 1958. *Tennyson and "The Princess": Reflections of an age*. London: Athlone Press.

Kingsley, C. 1863. *Water babies*. London: Macmillan.

Kosmos and Vestiges of the Natural History of Creation. 1845. *Westminster For. Quart. Rev.* 44:152–203.

Kottler, M. J. 1974. Alfred Russel Wallace, the origin of man, and spiritualism. *Isis* 65:145–92.

———. 1976. Isolation and speciation. Ph.D. diss. Yale University.

Kuhn, T. 1970. *The structure of scientific revolutions*. 2d ed. Chicago: University of Chicago Press.

Lack, David. 1947. *Darwin's finches*. Cambridge: Cambridge University Press.

Lamarck, J. B. de. 1809. *Philosophie zoologique*.

Paris. Trans. as *Zoological philosophy* by H. Elliot. London: Macmillan, 1914.

Laudan, L. 1971. William Whewell on the consilience of inductions. *Monist* 55:368–91.

Lewontin, R. C. 1974. *The genetic basis of evolutionary change.* New York: Columbia University Press.

Limoges, C. 1970. *La sélection naturelle.* Paris: Presses Universitaires de France.

Litchfield, H. 1915. *Emma Darwin: A century of family letters.* London: Murray.

Lovejoy, A. O. 1936. *The great chain of being.* Cambridge: Harvard University Press.

Lurie, E. 1960. *Louis Agassiz: A life in science.* Chicago: University of Chicago Press.

Lyell, C. 1830–33. *The principles of geology.* 1st ed. London: John Murray.

———. 1835. On the proofs of a gradual rising of the land in certain parts of Sweden. *Phil. Trans. Roy. Soc. Lond.* 1835:1–38.

———. 1851. Anniversary address of the president. *Quart. J. Geol. Soc. Lond.* 7:xxv–lxxvi.

———. 1863. *Antiquity of man.* London: Murray.

———. 1868. *Principles of geology.* 10th ed. London: Murray.

Lyell, K. 1881. *Life, letters and journals of Sir Charles Lyell, Bart.* London: Murray.

Lyell, Mrs. H. 1906. *Life of Sir Charles J. F. Bunbury, Bart.* London: Murray.

MacDonald, R. H. 1967. The frightful consequences of onanism—notes on the history of a delusion. *J. Hist. Ideas* 28:423–31.

McKinney, H. L. 1972. *Wallace and natural selection.* New Haven: Yale University Press.

Macleay, W. 1819–21. *Horae entomologicae.* London.

MacLeod, R. 1965. Evolutionism and Richard Owen. *Isis* 56:259–80.

———. 1969. The genesis of *Nature. Nature* 224:423–40.

———. 1970. The X-Club: A scientific network in late Victorian England. *Notes Rec. Roy. Soc.* 24:305–22.

————. 1974. The Ayrton incident: A commentary on the relations of science and government in England, 1870–73. In *Science and values,* ed. A. Thackray and E. Mendelsohn, pp. 243–78. New York: Humanities Press.

Malthus, T. R. 1826. *An essay on the principle of population.* 6th ed. London.

Mandelbaum, M. 1958. Darwin's religious views. *J. Hist. Ideas* 19:363–78.

Marchant, J. 1916. *Alfred Russel Wallace: Life and reminiscences.* New York: Harper.

Marcus, S. 1966. *The other Victorians: A study of sexuality and pornography in mid-nineteenth century England.* London: Weidenfeld and Nicolson.

Matthew, P. 1831. *On naval timber and arboriculture.* London.

Maynard Smith, J. 1975. *The theory of evolution.* 3d ed. Harmondsworth: Penguin.

Mayr, E. 1942. *Systematics and the origin of species.* New York: Columbia University Press.

————. 1963. *Animal species and evolution.* Cambridge, Mass.: Belknap.

————. 1964. Introduction to C. Darwin, *On the origin of species: A facsimile.* Cambridge: Harvard University Press.

————. 1972a. Lamarck revisited. *J. Hist. Biol.* 5:55–94.

————. 1972b. The nature of the Darwinian Revolution. *Science* 176:981–89.

————. 1972c. Sexual selection and natural selection. In *Sexual selection and the descent of man,* ed. B. Campbell. Chicago: Aldine.

————. 1977. Darwin and natural selection. *Amer. Sci.* 65:321–27.

Meteyard, E. 1871. *A group of Englishmen 1795–1815.* London: Longmans, Green.

Mill, J. S. 1872. *System of logic.* 8th ed. London: Longmans, Green, Reader, and Dyer.

————. 1873. *Autobiography.* London: Longmans.

————. 1874. Theism. In *Three essays on religion.* New York: Holt.

————. 1875. *System of logic.* 9th ed. London: Longmans.

Miller, H. 1839. Gropings of a working man in geology. *Chambers's Edin. J.* 7:109–10, 137–39.

————. 1841. *The old red sandstone.* In *Collected works.* Edinburgh: Nimmo, 1869.

————. 1847. *Footprints of the Creator; or, the asterolepis of Stromness.* Edinburgh: Constable.

————. 1854. *My schools and schoolmasters.* Edinburgh: Black.

————. 1856. *The testimony of the rocks; or, Geology in its bearings on the two theologies, natural and revealed.* Edinburgh: Constable.

Millhauser, M. 1954. The scriptural geologists: An episode in the history of opinion. *Osiris* 11:65–86.

————. 1959. *Just before Darwin: Robert Chambers and "Vestiges."* Middletown, Conn.: Wesleyan University Press.

————. 1971. *Fire and ice: The influence of science on Tennyson's poetry.* Lincoln: Tennyson Society.

Mivart, S. J. 1870. *Genesis of species.* London: Macmillan.

————. 1871. *Genesis of species.* 2d ed. London: Macmillan.

Müller, J. 1838–42. *Elements of physiology.* London: Taylor and Walton.

Murchison, R. I. 1854. *Siluria: The history of the oldest known rocks containing organic remains, with a brief sketch of the distribution of gold over the earth.* London.

Newman, F. W. 1845a. *Vestiges .... Prosp. Rev.* 1:49–82.

————. 1845b. *Explanations,* a sequel to the *Vestiges. Prosp. Rev.* 2:33.

Newman, J. H. 1830. Rev. H. H. Milman's *History of the Jews. Brit. Critic,* vol. 16.

Newton, A., and Newton, E. 1869. On the osteology of the solitaire or didine bird of the island of Rodriguez, Pezophaps solitaria (Gmel.). *Phil. Trans. Roy. Soc. Lond.* 159:327–60.

Oakley, K. P. 1964. The problem of man's antiquity. *Bull. Brit. Mus. (Nat. Hist.), Geo. Ser.,* vol. 9, no. 5.

O'Brien, C. F. 1970. *Eozoön canadense:* "The dawn
animal of Canada." *Isis* 61:206–23.

Ogilvie, M. B. 1975. Robert Chambers and the
nebular hypothesis. *Brit. J. Hist. Sci.* 8:214–32.

Ospovat, D. 1976. The influence of Karl Ernst von
Baer's embryology, 1828–1859: A reappraisal
in light of Richard Owen's and William B. Carpenter's "Palaeontological application of von
Baer's law." *J. Hist. Biol.* 9:1–28.

———. 1977. Lyell's theory of climate. *J. Hist.
Biol.* 10:317–39.

Owen, R. 1834. On the generation of the marsupial animals, with a description of the impreg-
nated uterus of the kangaroo. *Phil. Trans. Roy.
Soc. Lond.* 1834:333–64.

———. 1846. Report on the archetype and
homologies of the vertebrate skeleton. *Report of
the sixteenth meeting of the British Association for the
Advancement of Science,* pp. 169–340. London:
Murray.

———. 1848. *On the archetype and homologies of the
vertebrate skeleton.* London: Voorst.

———. 1849. *On the nature of limbs.* London:
Voorst.

———. 1851. *Principles of geology* by Sir Charles
Lyell. . . . *Quart. Rev.* 89:412–51.

———. 1855. *Lectures on the comparative anatomy
and physiology of the invertebrate animals.* 2d ed.
London: Longman, Brown, Green, and
Longmans.

———. 1858*a.* Presidential address. *Report of the
twenty-eighth meeting of the British Association for
the Advancement of Science,* pp. xlix–cx. London:
Murray.

———. 1858*b.* On the characters, principles of
division, and primary groups of the class Mammalia. *J. Linn. Soc. (Zool.)* 2:1–37.

———. 1860. Darwin on the origin of species.
Edin. Rev. 111:487–532.

———. 1861. *Paleontology.* 2d ed. Edinburgh:
Black.

———. 1863. On the *Archaeopteryx* of von Meyer,
with a description of the fossil remains of a

long-tailed species, from the lithographic stone of Solenhofen. *Phil. Trans. Roy. Soc. Lond.* 153:33–47.

————. 1866–68. *On the anatomy of vertebrates.* London: Longmans, Green.

Owen, Rev. R. 1894. *The life of Richard Owen.* London: Murray.

Paley, W. 1819*a*. *Evidences of Christianity.* In *Collected works.* London: Rivington.

————. 1819*b*. *Natural theology.* In *Collected works.* London: Rivington.

Parker, W. K. 1866. On the structure and development of the skull in the ostrich tribe. *Phil. Trans. Roy. Soc. Lond.* 156:113–83.

————. 1873. On the structure and development of the skull in the salmon (salmo salav. L.). *Phil. Trans. Roy. Soc. Lond.* 163:95–145.

Passmore, J. 1968. *A hundred years of philosophy.* 2d ed. Harmondsworth: Penguin.

Peel, J. D. Y. 1971. *Herbert Spencer: The evolution of a sociologist.* London: Heinemann.

Phillips, J. 1860. *Life on the earth: Its origin and succession.* London: Macmillan.

Playfair, J. 1802. *Illustrations of the Huttonian theory of the earth.* Edinburgh: Creech.

Popular information on science. 1838. *Chambers's Edin. J.* 6:114–15, 122–23, 139–40, 186–87, 202–3, 226–27, 298–99, 314–15, 379–80.

Porter, R. 1976. Charles Lyell and the principles of the history of geology. *Brit. J. Hist. Sci.* 9:91–103.

Powell, B. 1833. *Revelation and science.* Oxford: Parker.

————. 1834. *History of natural philosophy.* London: Cabinet Cyclopaedia.

————. 1838. *The connexion of natural and divine truth.* London: Parker.

————. 1855. *Essays on the spirit of the inductive philosophy.* London: Longman, Brown, Green, and Longmans.

————. 1860. On the study of the evidences of Christianity. In *Essays and Reviews.* London:

Longman, Green, Longman, and Roberts.

Provine, W. B. 1971. *The origins of theoretical population genetics.* Chicago: Chicago University Press.

Quetelet, M. A. 1842. *Treatise on man.* Edinburgh: Chambers.

Rolleston, G. 1884. *Scientific papers and addresses.* Oxford: Clarendon Press.

Ross, R. H. 1973. *Alfred, Lord Tennyson "In Memoriam": An authoritative text, backgrounds and sources of criticism.* New York: Norton.

Rudwick, M. J. S. 1963. The foundation of the Geological Society of London: Its scheme for co-operative research and its struggle for independence. *Brit. J. Hist. Sci.* 1:325–55.

———. 1969. The strategy of Lyell's *Principles of Geology. Isis* 61:5–33.

———. 1972. *The meaning of fossils.* London: Macdonald.

———. 1974. Darwin and Glen Roy: A "great failure" in scientific method? *Stud. Hist. Phil. Sci.* 5:97–185.

Ruse, M. 1971. Natural selection in *The Origin of Species. Stud. Hist. Phil. Sci.* 1:311–51.

———. 1973*a.* The nature of scientific models: Formal *v.* material analogy. *Phil. Soc. Sci.* 3:63–80.

———. 1973*b. The philosophy of biology.* London: Hutchinson.

———. 1975*a.* Darwin's debt to philosophy: An examination of the influence of the philosophical ideas of John F. W. Herschel and William Whewell on the development of Charles Darwin's theory of evolution. *Stud. Hist. Phil. Sci.* 6:159–81.

———. 1975*b.* The relationship between science and religion in Britain, 1830–1870. *Church History* 44:505–22.

———. 1975*c.* Charles Darwin and artificial selection. *J. Hist. Ideas* 36:339–50.

———. 1975*d.* Charles Darwin's theory of evolution: An analysis. *J. Hist. Biol.* 8:219–41.

————. 1976. The scientific methodology of William Whewell. *Centaurus* 20:227–57.

————. 1977. William Whewell and the argument from design. *Monist* 60:244–68.

————. 1981. Philosophical factors in the Darwinian Revolution. In *Pragmatism and purpose,* ed. F. Wilson. Toronto: University of Toronto Press.

Russell, E. S. 1916. *Form and function.* London: Murray.

Schweber, S. S. 1977. The origin of the *Origin* revisited. *J. Hist. Biol.* 10:229–316.

Scrope, P. 1830. *Principles of geology . . .* by Charles Lyell . . . vol. 1. *Quart. Rev.* 43:411–69.

Sebright, J. 1809. The art of improving the breeds of domestic animals, in a letter addressed to the Right Hon. Sir Joseph Banks, K.B. London.

Sedgwick, A. 1831. Presidential address to the Geological Society. *Proc. Geol. Soc. Lond.* 1:281–316.

————. 1833. *Discourse on the studies of the university.* Cambridge: Cambridge University Press.

————. 1845. *Vestiges. . . . Edinburgh Rev.* 82:1–85.

————. 1850. *Discourse on the studies of the University of Cambridge.* 5th ed. Cambridge: Cambridge University Press.

————. 1860. Objections to Mr. Darwin's theory of the origin of species. *Spectator,* 24 March 1860.

Simpson, G. G. 1944. *Tempo and mode in evolution.* New York: Columbia University Press.

————. 1951. *Horses.* Oxford: Oxford University Press.

————. 1953. *The major features of evolution.* New York: Columbia University Press.

Smith, R. 1972. Alfred Russel Wallace: Philosophy of nature and man. *Brit. J. Hist. Sci.* 6:177–99.

Spencer, H. 1850. *Social statics.* London: Chapman.

————. 1852a. The development hypothesis.

Leader. Reprinted in *Essays*, 1:377–83.

———. 1852*b*. A theory of population, deduced from the general law of animal fertility. *Westminster Rev.*, n.s., 1:468–501.

———. 1855. *Principles of psychology*. London: Longman.

———. 1857. Progress: Its law and cause. *Westminster Rev.* Reprinted in *Essays*, 1:1–60.

———. 1864–67. *Principles of biology*. London: Williams and Norgate.

———. 1868. *Essays: Scientific, political, and speculative*. London: Williams and Norgate.

———. 1904. *Autobiography*. London: Williams and Norgate.

Stauffer, R. 1975. *Charles Darwin's natural selection*. Cambridge: Cambridge University Press.

Stebbins. G. L. 1950. *Variation and evolution in plants*. New York: Columbia University Press.

Stern, C., and Sherwood, E. R. 1966. *The origin of genetics*. San Francisco: Freeman.

Stevenson, R. L. 1887. *Memoir of Fleeming Jenkin*. Edinburgh: Longmans.

Swainson, W. 1835. *A treatise on the geography and classification of animals*. London: Longman, Rees, Orme, Brown, Green, and Longman.

Tennyson, Alfred Lord. 1847. *The Princess*. London. Reprinted in *The works of Tennyson*, ed. H. Tennyson, pp. 165–217. London: Macmillan, 1913.

———. 1851. *In Memoriam*. London. Reprinted in *The works of Tennyson*. ed. H. Tennyson, pp. 247–86. London: Macmillan, 1913.

Thagard, P. 1977. Darwin and Whewell. *Stud. Hist. Phil. Sci.* 8:353–56.

Tholfsen, T. R. 1971. The intellectual origins of mid-Victorian stability. *Pol. Sci. Quart.* 86:57–91.

Thomson, W. (Lord Kelvin). 1869. Of geological dynamics. *Pop. Lect.* 2:73–131.

Todhunter, I. 1876. *William Whewell D.D.: An account of his writings with selections from his liter-*

ary and scientific correspondence. London: Macmillan.

Trimen, R. 1868. On some remarkable mimetic analogies among African butterflies. *Trans. Linn. Soc.* 26:497–522.

Trivers, R. 1971. The evolution of reciprocal altruism. *Quart. Rev. Biol.* 46:35–57.

Tuckwell, W. B. 1909. *Pre-tractarian Oxford.* London: Smith Elder.

Vestiges. . . . 1845*a. Brit. Quart. Rev.* 1:490–513.

———. 1845*b. Brit. For. Med. Rev.* 19:155–81.

Vorzimmer, P. 1963. Charles Darwin and blending inheritance. *Isis* 54:371–90.

———. 1965. Darwin's ecology and its influence upon his theory. *Isis* 56:148–55.

———. 1969. Darwin, Malthus, and the theory of natural selection. *J. Hist. Ideas* 30:527–42.

———. 1970. *Charles Darwin: The years of controversy.* Philadelphia: Temple University Press.

Wagner, M. 1868. *Die Darwin'sche Theorie und das Migrationsgesetz der Organismen.* Munich: Straub.

Wallace, A. R. 1855. On the law which has regulated the introduction of new species. *Ann. Mag. Nat. Hist.* 16:184–96. Reprinted in *Lamarck to Darwin,* ed. H. L. McKinney, Lawrence, Kans.: Coronado Press, 1971.

———. 1858. On the tendency of varieties to depart indefinitely from the original type. *J. Proc. Linn. Soc. (Zool.)* 3:53–62.

———. 1866. On the phenomena of variation and geographical distribution as illustrated by the Papilionidae of the Malayan region. *Trans. Linn. Soc. Lond.* 25:1–72.

———. 1869. Sir Charles Lyell on geological climates and the origin of species. *Quart. Rev.* 126:359–94.

———. 1870*a.* The measurement of geological time. *Nature* 1:499–401, 452–55.

———. 1870*b.* The limits of natural selection as applied to man. In *contributions to the theory of natural selection.* London: Macmillan.

————. 1898. *The wonderful century: Its successes and failures.* London: Sonnenschein.

————. 1905. *My life: A record of events and opinions.* London: Chapman and Hall.

Wells, W. C. 1818. An account of a female of the white race of mankind, part of whose skin resembles that of a negro; with some observations on the cause of the differences in colour and form between the white and negro races of man. In *Appendix to two essays: One upon single vision with two eyes, the other on dew.* London: Archibald Constable.

Whately, E. W. 1889. *Personal and family glimpses of remarkable people.* London: Hodder and Stoughton.

Whewell, W. 1824. A general method of calculating the angles made by any planes of crystals, and the laws according to which they are formed. *Phil. Trans. Roy. Soc. Lond.* 1824:87–130.

————. 1829. Mathematical exposition of some doctrines of political economy. *Cambr. Phil. Trans.* 3:191–230.

————. 1831*a.* Mathematical exposition of some of the leading doctrines in Mr. Ricardo's "Principles of Political Economy and Taxation." *Cambr. Phil. Trans.* 4:155–98.

————. 1831*b. Preliminary discourse . . .* by J. F. W. Herschel. . . . *Quart. Rev.* 45:374–407.

————. 1831*c. Principles of geology,* vol. 1, by Charles Lyell. *Brit. Crit.* 17:180–206.

————. 1832. *Principles of geology . . .* by Charles Lyell . . . , vol. II. *Quart. Rev.* 47:103–32.

————. 1833. *Astronomy and general physics.* Bridgewater Treatise no. 3. London: Pickering.

————. 1837. *History of the inductive sciences.* London: Parker.

————. 1839. Presidential address to the Geological Society. *Proc. Geol. Soc. Lond.* 3:61–98.

————. 1840. *Philosophy of the inductive sciences.* London: Parker.

————. 1845. *Indications of the Creator.* London: Parker.

————. 1846. *Indications of the Creator.* 2d ed. London: Parker.

————. 1850. Mathematical exposition of some doctrines of political economy. *Cambr. Phil. Trans.* 9:128–49.

————. 1853. *On the plurality of worlds.* London: Parker.

————. 1857. *History of the inductive sciences.* 3d ed. London: Parker.

————. 1860. *Philosophy of discovery* (3d part of 3d ed. of *Philosophy of the inductive sciences*). London: Parker.

————. 1863. *Bridgewater Treatise.* 7th ed. London: Pickering.

Wilberforce, S. 1860. *On the Origin of Species . . . .* *Quart. Rev.* 108:225–64.

Wilkinson, J. 1820. Remarks on the improvement of cattle, etc., in a letter to Sir John Saunders Sebright, Bart. M.P. Nottingham.

Willey, B. 1949. *Nineteenth-century studies.* London: Chatto and Windus.

Williams, G. C. 1966. *Adaptation and natural selection.* Princeton: Princeton University Press.

Wilson, D. 1974. Herschel and Whewell's version of Newtonianism. *J. Hist. Ideas* 35:79–97.

Wilson, E. O. 1975. *Sociobiology: The new synthesis.* Cambridge: Harvard University Press.

Wilson, L. 1970. *Sir Charles Lyell's scientific journals on the species question.* New Haven: Yale University Press.

————. 1971. Sir Charles Lyell and the species question. *Amer. Sci.* 59:43–55.

————. 1972. *Charles Lyell, the years to 1841: The revolution in geology.* New Haven: Yale University Press.

Winsor, M. P. 1969. Barnacle larvae in the nineteenth century. *J. Hist. Med. All. Sci.* 24:394–409.

————. 1976. *Starfish, jellyfish, and the order of life.* New Haven: Yale University Press.

Woodward, H. B. 1907. *The history of the Geological Society of London.* London: Geological Society.

Youatt, W. 1831. *The horse, with a treatise on draught.* London: Library of Useful Knowledge.

————. 1834. *Cattle, their breeds, management, and diseases.* London: Library of Useful Knowledge.

————. 1837. *Sheep, their breeds, management, and diseases.* London: Library of Useful Knowledge.

Young, R. M. 1971. Darwin's metaphor: Does nature select? *Monist* 55:442–503.

Afterword: Two Decades Later

The Darwinian Revolution appeared originally in 1979, some twenty years after the hundredth anniversary of the publication of the *Origin of Species*. Naturally, this centennial was a time of great celebration, coinciding as it did with the one hundred and fiftieth anniversary of Darwin's birth. The centennial was also the point at which serious study of the history of evolutionary biology began in earnest, partly because this was the date around which history of science generally was taking off as a fully professional subject. It was in part also because the Darwin celebrations flagged serious scholars to an important time in the history of Western thought. Although several books dedicated to Darwin and his achievements appeared at that time, it was realized that much more work was needed to bring full understanding and appreciation. That score of years leading up to *The Darwinian Revolution* was therefore one of great activity, as documents were unearthed and edited and published, as the main figures in the whole episode were researched and described and explained and put in place, and as much effort was made to locate Darwinism not only in the history of biological science but also in the Victorian context.

I would like to thank Helena Cronin, David L. Hull, and Robert J. Richards for comments on an earlier version of this *Afterword*.

In writing my book, I was able to take advantage of such activity. Indeed, it was precisely because this activity had yielded such rich dividends that I was inspired to write *The Darwinian Revolution.* Although my book certainly contained its own share of original research and interpretation— especially with respect to philosophical matters, a function of my own background and interests—I determined to write an overview. I wanted to give a picture of Darwin and his revolution that was emerging from the new primary sources and research based on them, not to mention on many other related factors, within and without biology. This is the essence of my book and it is for this reason that I must share credit. The book's ongoing popularity shows that there was a good story to tell, that this was a story hitherto untold, and that there was (and is) an audience who wanted to hear, learn, and understand.

Because *The Darwinian Revolution* continues to sell, you might think that the story has been told, that it needs no new telling, and that one should leave well enough alone. In a sense, this is true. Those scholars who went before me did ferret out the essential facts in the Darwinian revolution, and they did give enough guidance that a synthesizer could then provide a full and adequate interpretation, one that needs no basic revision. For this reason, if for no other, I do not now offer you a completely new version of *The Darwinian Revolution.* The book as it stands in its original form is correct in its basic facts, true in its interpretation, and sufficiently self-contained as not to need massive extension and a change or redirection of focus.

Yet time does move on and there has been much new work in the second twenty years, from the publication in 1979 of *The Darwinian Revolution* down to the present day. In this *Afterword,* therefore, I would like to bring you up to date. I am not going to cover everything: there are already straightforward reviews of the Darwin literature, including one by me (Ruse 1996a). But I will give you some flavor of the intervening debates and discoveries, and then leave it to you to decide what everything ultimately means. To begin, I will simply go over the Darwinian revolution, following the chronology given in the main text of this book. I will relate Darwin's achievements to the history of evolutionism as a whole, and will try to assess his lasting contributions to science. My discussion will be framed against my own subsequent work on the role of cultural factors in science and how these connect to the urges of professional scientists to produce mature work that will win respect and support.

Before Darwin

My story began at the beginning of the nineteenth century, with the clash in France between the evolutionist Jean Baptiste de Lamarck and his great critic, the father of comparative anatomy, Georges Cuvier. This was surely the right place to begin, although now we have a greater appreciation that evolution was not simply a one-person show: before and after Lamarck there were many people who were edging toward (if not slipping right into) one form or another of developmental belief. In France itself, place must be found for Lamarck's supporters and fellow travellers. Although he gets brief mention in my book, more space should be given to Etienne Geoffroy Saint-Hilaire, who raised the ire of Cuvier even more than did Lamarck (Appel 1987). In England, we realize how full and persistent an evolutionist was Charles Darwin's grandfather, Erasmus Darwin, and we sense that he probably had influence much beyond his own century (McNeill 1987). And in Germany, we know that developmentalism of one sort or another was virtually *de rigeur* in the biological community, although it was hardly pre-(Charles)-Darwinian, being much more centred on transcendental morphology and homologies (Richards 1992). Some of these developmentalists were full-blooded evolutionists, but many Germanic transformists—*Naturphilosophen,* to use their own name—were more interested in the idea of upward development as shown through bodily forms *(Baupläne)* than in the actual physical evolution from one organism to another. Notably and influentially, this was the position of the philosopher Hegel (1817).

At the same time as one recognizes the early evolutionists, one has now a greater understanding of the thinking and situation of the critics of evolutionism. The main text discusses the objections of Cuvier and is sympathetic to his reasoning, showing that his opposition to transformism was deep and profound. His was not simply the knee-jerk reaction of a bigot, one whose motivation was extra-scientific (primarily religious) rather than based on facts and theory. Cuvier was a great scientist, and he brought his science to bear appropriately and critically on evolution. Today, however, we recognize with greater sensitivity the pertinence of the social standing of the critics, especially Cuvier, who was as much a bureaucrat as a scientist. He rightly saw evolution as being a radical and subversive doctrine (more on this point later) and hence out of the interest of his masters, as well as from personal conviction, Cuvier opposed it with every ounce of his being. Evolution was not just wrong—it was bad and corrupting. As one who had lived through the French Revolution, Cuvier thought (with reason) that ideas can be too dangerous by half (Outram 1984).

Some of the most exciting recent research on the history of evolutionism focuses in on the 1830s. My text concentrates on the Oxbridge circle: Henslow, Whewell, Sedgwick, and others. They were crucial players and my emphasis is quite properly on them. I do just barely refer to others, hinting that (despite the Huxley dismissal) there may be a little more to Robert Grant, for many years a medical anatomist first in Scotland and then in London, an evolutionist and a man whom Darwin met during his two years in Edinburgh. We now know, thanks particularly to the seminal studies of the historian Adrian Desmond (1989), that Grant represented a whole different dimension to the evolutionary debate: a dimension, influenced by both French and German thought, at a lower social level than the men I discuss. His transformism represented a radical threat to the happy harmony between science and religion that was being forged by my respectable Anglican clergymen scientists at the ancient universities. I am not sure of the lasting influence and significance of Grant's work. Perhaps, in the next decade, such evolutionism as his was reflected and continued by the *Vestiges of the Natural History of Creation,* although Robert Chambers may simply have been working in parallel, as it were, responding to similar stimulae. The point is that, when once started in the nineteenth century, evolution was a plant—a rank weed, its critics would have charged—that never wilted or disappeared or succumbed to its harsh environment. More research may show that such was also the case in France and Germany (Laurent 1987).

Moving to the 1840s, I feel little reason to add to my original portraits of Chambers and Richard Owen. Indeed, here I feel almost smug. The Darwinians, Thomas Henry Huxley particularly, had painted Owen in the blackest of colors—so much so, in fact, that I began to smell a rat. While not even the strongest revisionist could turn Owen into a warm and cuddly human being, I felt (and my researches supported the belief) that Owen had to be seen as an important scientist in his own right. Although Owen may have been the outsider in the Darwinian revolution, it surely had to be the case that his science played its part in that revolution. The work he did in anatomy and palaeontology, among other disciplines, had to be significant. My suspicions have been confirmed, and today no student of the period would fail to give Owen a full and important place in the coming of evolution, even in the coming of Darwinian evolution. He himself might not have been pleased at being given such a place, but it was his and fairly and honestly earned. The scientific base on which Darwin erected his great unification—palaeontology, morphology, systematics, embryology, and more—owed much to the labors of Richard Owen (Owen 1992; Rupke 1994).

In *The Darwinian Revolution* it would have been historically false (and from a literary viewpoint disastrous) had I let the scientists of the 1850s— T. H. Huxley and Herbert Spencer, particularly—swallow up the book, leaving only a bit part for Charles Darwin. Yet, even as I was writing, I felt that I was skating quickly over depths that should be explored. Since 1979, some of the most important research has occurred on and around these men. I will return to the significance of all of this later in this *Afterword*, but here let me say that I was unfair to the character of Huxley. He will never be a man to whom I will warm: there was about him a Victorian earnestness which I find distasteful. Also, I find it intensely irritating that he thought his training as a scientist immediately fitted him to pronounce confidently on every other branch of scholarship, philosophy (my own subject!) particularly. But no person should be described as designed by nature to be a university administrator. More positively, as you will see, quite apart from the brave way in which he fought the most appalling depressions, there was a manliness about Huxley as he strove to realize his aims. He demands respect if not love (Desmond 1994; 1997).

Herbert Spencer is another matter. The more one learns about him and his private life, the more one fears one has been transferred into a strange fantasy land existing only in the imagination of Lewis Carroll. His neurotic concern for self truly knew no bounds. He never read a book with which he disagreed: the practice gave him headaches. At one level he had a wide circle of friends—he and Huxley were the closest of pals for many years— yet domestically he eschewed the company of normal people lest they disturb his thought processes, opting rather to live in a series of drab boarding houses where he could think in comfort. And what thinking! He preferred not to strain himself at the conscious level, rather letting the hidden depths of his thought processes do the work, with the ideas then pouring out to be dictated to a secretary between bouts of violent exercise on the tennis court. All of this was done and more, with the cocky self-confidence of one who has been thoroughly spoiled when young, which he was. At the same time, however, he had a massive influence on his fellow countrymen, on Americans ranging from Marxist socialists to the most powerful and rich of robber barons, and on people at just about every other point on the globe (Richards 1987; Pittenger 1993). In *fin de siècle* China, for instance, any forward-looking young intellectual could and would spout gems of Spencerian wisdom (Pusey 1983). Since the first edition of *The Darwinian Revolution,* there has been exponential growth in the dumbfounded amazement of historians at this man's peculiarities and wild speculations: approaching him from the late twentieth century and trying to understand him in our own terms gives a whole new dimension to theses about the incommensurability of rival thought processes. Yet our astonishment has been matched and exceeded by our appreciation of his influence.

Charles Darwin

I come to the central figure in the story. The most important primary-source additions since my book appeared are a magnificent new edition of all of the pertinent private notebooks, before and after the discovery of natural selection, and the first volumes of the collected edition of all of Darwin's correspondence, both letters to him and from him (Barrett et al 1987; Darwin 1985 –). This latter has now arrived at the post-*Origin* decade and, although it will take many more years and many more volumes to complete, it has probably already done most that will be done in the direction of revision: or rather, in confirmation of the portrait of Darwin sketched in *The Darwinian Revolution.* I saw Darwin primarily as a very professional scientist, trained by the leaders in the field to accept and use the proper norms of scientific method, and determined to push his ideas both (intellectually) through their intrinsic virtues and (sociologically) through the group of young supporters that he cultivated around himself. The correspondence particularly confirms this portrait.

We now realize indeed that Darwin's training as a scientist was more thorough than we had suspected, especially thanks to home tuition in chemistry with brother Erasmus and then in Edinburgh, where Darwin mixed regularly with students of the life sciences. We know also that Darwin's interaction with Robert Grant was extensive and surely meaningful: as a biologist and perhaps indeed as an evolutionist. The man who discovered natural selection and who then went on to write and publish the *Origin of Species* was a dedicated scientist almost from the first, one for whom evolution was probably always an issue (at least an idea to be considered), and Darwin's modesty and diffidence (particularly in his *Autobiography*) should not mislead.

Darwin's modesty should not mislead us in another sense either. He was a genuinely nice man, who cared for his family and his friends—demonstrating both a deep love for Hooker and a real concern for the rather fragile Huxley—but do not be deceived by his diffident nature. As remarked by the author of the most sensitive study which has yet appeared, there was a sliver of ice plunged through the heart of Charles Darwin (Browne 1995). He may have been self-deprecatory in conversation and letter, but he persisted without deviation in his plans and aims, and he used everyone to his own ends: gathering information, providing specimens, doing experiments, and much more. Moreover, one senses that Darwin used his illness, although genuinely difficult for him, to his advantage: he avoided conflicts and tedious tasks as needed, and justified the selective isolation that he needed for his own researches and writings. I use the word "selective" deliberately, for when it suited him Darwin was ready to go up to London or to have fellow scientists visit at his house in Downe village.

The illness continues to attract attention. I find myself relatively uninterested in the topic, partly because no one is going to solve it at this distance, and partly because its solution does not really matter—the Darwinian revolution was what it was regardless of whether Darwin was sick from physical or from psychological causes. If pressed, I would probably opt for physical causes, and not simply because that seems less demeaning to one of science's true heroes. My view of Darwin is of a man dedicated to his science, who became absolutely committed to evolution through selection, and who (to take the one issue that truly scared his contemporaries) never had any doubts whatsoever that we humans must be included in the picture. I do not see him as a man in any way torn on these issues, nor am I convinced, for all that he may sometimes have suggested otherwise, that he would let his family or his social situation or anything else trouble him on this matter. The neo-Marxist analysis of Darwin has him cowering in rural Kent, while England burned around him, wracked with guilt because he was betraying his class by contributing to the revolution (Desmond and Moore 1992). This strikes me as silly nonsense. Darwin was a rich man, with a comfortable position in society, known and liked by his fellow countrymen: he was after all "Darwin of the *Beagle*," author of one of the best travel books of the day in an age which loved travel books. This indicates an emotional security that showed in just about everything he thought and did, although Darwin was not an insensitive or unfeeling man. He could feel great psychological pain—intensely personal events like the death of his daughter Annie were extremely emotionally difficult for him. But evolution did not bring on stress of this kind.

In my portrait of Darwin, there are details that need correcting. It now

seems likely that it was the mockingbirds rather than the finches which so
excited Darwin when first he went to the Galapagos (Sulloway 1982). More
important is the question of Darwin's relationship to the concept of progress: the belief that society is capable of improvement through science
and the like, and the related belief that something similar can be seen in
evolution's history, as we get an upward path from the most simple (the
"monad") to the most complex and desirable (the "man"). In *The Darwinian Revolution,* I am cagy about linking Darwin and progress. I do see
that he was a biological progressionist of a kind, but I certainly do not
imply great enthusiasm on his part. I do not see him as obsessed with an
ever-higher passage up to our own species. Apart from certain progress-
questioning comments in the notebooks, my analysis was influenced
by my own conceptual understanding of natural selection and progress,
namely that any mechanism which stresses the relativism of change—
those that survive are those that at a particular time (which may be very
different from any other particular time) are able to survive—is going to
have difficulty explaining absolute change in a direction from worse to
better.

I now think, thanks particularly to brilliant analysis by the late Dov
Ospovat (1981) and to the ever-stimulating work of Robert J. Richards
(1992), that one must see Darwin's links to biological progressionism in a
more positive light. Interestingly (and, I shall suggest, significantly) Darwin's really enthusiastic attitude toward biological progressionism comes
through most forcefully in revisions that he made to the *Origin.* In particular, in the third edition (1861), he added passages that are favorable to progress, and tied it explicitly to *Homo sapiens:*

> If we look at the differentiation and specialisation of the several
> organs of each being when adult (and this will include the ad-
> vancement of the brain for intellectual purposes) as the best stan-
> dard of highness of organisation, natural selection clearly leads
> towards highness; for all physiologists admit that the specialisa-
> tion of organs, inasmuch as they perform in this state their func-
> tions better, is an advantage to each being; and hence the accu-
> mulation of variations tending towards specialisation is within
> the scope of natural selection. (Darwin 1959, 222)

How then do we account for the progess-questioning comments I cited
and why the possible reticence? The point is that Darwin was always worried about a popular and much-derided form of inevitable biological prog-

ress, one which was as it were preprogramed into early life. This is the kind of progress favored by the *Naturphilosophen* (and people like Chambers who drew on them), who saw evolution through the lens of embryology, and who saw development to the ultimate finished superior product as something which would follow necessarily once life had started. It is this belief that Darwin was criticizing in the passages cited. Darwin's evolution had no inner momentum, mimicking the development of individual organisms. In addition, Darwin realized that the presumption with selection is that change will be relative. It was therefore a matter of finding a way that such a mechanism can nevertheless produce lasting upward change. This came through an anticipation of what today's evolutionists call "arms races": lines of organisms compete, with the result that adaptations get refined and improved—the prey gets faster and so the predator gets faster—and that overall this leads to absolute change, particularly as some competitor develops weapons or defences using on-board computers (Dawkins 1986). For Darwin, there was no inevitability about the evolution of large brains and the consequent human success—but succeed we did, and thanks to natural selection.

Darwin's treatment of progress convinces me of two things (not that I really needed any convincing). First, he was very much a child of his time—Darwin's genius was not based on making things out of nothing or of total repudiation of his past and its influences. *The Darwinian Revolution* shows this fully in such areas as religion and philosophy, documenting as it does the extent to which Darwin was immersed in the ideas of his day and how he internalized and responded to them. It is the same with progress, and indeed with other ideas which had their origins in culture. Darwin took them all in and then created something new, a different picture, from what had existed before. Second, Darwin was simply ahead of virtually everyone when it came to thinking about evolution and its causes, although he was not right about everything. His troubles with heredity show this. But he grasped natural selection, and was unique in realizing its powers, its potentials, and the challenges that it posed for earlier ideas and assumptions. Darwin accepted biological progress, but he knew that one simply had to treat it in new and radical ways.

The Meaning of the Revolution

Moving to the post-Darwinian period, there is little I would add to the detailed discussion in my book. Darwin did grasp natural selection, and few others did. Most people looked at selection and decided that it was but a secondary part of the evolutionary scenario: as the chief mechanism, they preferred instead to endorse Lamarckism (inheritance of acquired characteristics), saltationism (evolution by jumps), orthogenesis (evolution through a kind of inner momentum), or some other force. This tendency is fully documented in *The Darwinian Revolution,* as is the fact that, the norm notwithstanding, there was not total rejection of or indifference to natural selection. Natural selection's codiscoverer Alfred Russel Wallace and his travelling mate Henry Walter Bates cared passionately about selection and used it most profitably in their studies of butterflies, and Darwin recognized and appreciated this concern.

The really interesting question is precisely why the scientific community (in particular) spurned Darwin's offer of a mechanism, opting instead for alternatives to selection. As I point out, part of the reason was (as one might expect from scientists) straightforwardly scientific: the age of the earth problem, for instance, seemed to demand a mechanism that would operate much more rapidly and efficiently than selection. It was not until this century and the discovery of the heating effects of radioactive decay that this obstacle to the effectiveness of natural selection was removed—and a similar tale can be told of other problems. But, as also hinted in the text, there was more to this rejection than the purely scientific. The sociological and personal was important: the ways in which people like Huxley controlled the scientific community (not to mention preached the gospel of evolution to the outside world) pointed to some significant factors that I will explore below. There were some unsuspected and very interesting (non-scientific) reasons why so many people, including Darwin's closest supporters, downgraded or rejected natural selection.

Before I take up these reasons, I want to stress the overall conclusions of *The Darwinian Revolution* in a slightly more forceful manner than I do in the short final chapter. This is not because I failed then to do my duty, but because the world outside my pages has changed somewhat. Since writing my book, the movement known as "social constructivism" has swept through the humanities, through the history of science in particular, like an Egyptian plague. In the eyes of many scholars, it is now no longer acceptable to talk in terms of truth or reality or progress, suggesting that our understanding gets ever closer to true knowledge of a real human-independent world. Everything, science particularly, must be see as a created (not discovered) epiphenomenon on the culture within which it is produced. Science is "constructed" from within society, and simply reflects our prejudices, aspirations, ideologies, and social status.

Let me say unequivocally that nothing in my book leads to such a conclusion. There are no grounds for concluding that organic evolution is but a human-created fiction, a massive edifice existing only in the minds of Darwin and his supporters, successful simply because the winners outsmarted their opponents. On the contrary: objectively speaking, the Darwinian revolution was a great and genuine movement forward in the history of science—my book is predicated on this assumption, and it confirms it. Darwin took us closer to knowledge of the real world. The Galapagos archipelago exists and so do its denizens. Darwin explained how the birds and reptiles come to have the differences that they do. Mammals exist and so do the homologies between their forelimbs. He also explained the existence of homologies between the forelimbs of mammals—and anyone who believes the isomorphisms are constructed should look again at figures 13, 14, and 15. Adaptations like the hand and the eye are truly objects of design, and natural selection explains why. Darwin showed us the basis for these similarities. Think of these phenomena and their explanations, think of analogous phenomena and their explanations, and then deny that Charles Darwin was telling us real truths about a real world.

Charles Darwin was not the first evolutionist, but it was he who made evolution a reasonable belief and it was also he who first found the mechanism which makes it all happen. Darwin was a twenty-four-carat gold genius, and thanks to him we know more about our world than we did before his labors. What we think about Darwin as an individual, the culture in which he was immersed, or the reception that his ideas had with his contemporaries and later generations—all of these factors fall away before the strength of his ideas. Darwin told us about origins, including our own, and he gave us a strong causal explanation for them. Charles Darwin is famous and rightly so, and his achievements justify the writing of such books as mine.

At the same time, none of this is to deny what I stress in *The Darwinian Revolution:* we must see the Darwinian revolution as part and parcel of general cultural movements in the nineteenth century, particularly in mid-century Victorian Britain (Young 1985). This was a time of change, when society and its members had to come to terms with the aftereffects of such developments as industrialism, urbanism, the collapse of traditional ways of thinking and acting (from the military to the ecclesiastical), the spread of universal education, and much more. In a way, the most significant facts in *The Darwinian Revolution* were those figures I presented, almost in passing, about the growth of cities. A society undergoing that kind of change simply could not stand still. It had to alter, to tear down, to build up, to adapt, to move onwards in some direction or another. There were people needing housing, occupations, education, medical care, entertainment, and everything else that makes a functioning and meaningful life. Rural feudal eighteenth-century Britain could not provide these things, and so the new men and women—Victorians like Thomas Henry Huxley—had to provide it.

The Darwinian revolution was part cause and part effect of all these social changes. On the one hand, Darwin was providing a secular story of origins that—if not necessarily replacing the Christian story of origins—was making everyone involved realize that old thought patterns and beliefs were no longer adequate, and, in many cases, were entirely false (Moore 1979). After Darwin you did not have to be an atheist or even an agnostic, but he was making it openly possible for people to be one or the other of these. He was making it possible for people who wanted to break from the religion of the day, and from religion's intrusion into virtually every aspect of private and public life. He was making it possible to strike out independently in a merit-based, tradition-breaking, secular fashion.

On the other hand, Darwin's revolution came because of the factors which were making such a change possible. Take the philosophers Herschel and Whewell—in a fashion they, the latter particularly, were preparing the way for their own demise: they were spelling out the criteria for good science, criteria that a bright young researcher might seize on and exploit, and at the same time they were laboring to make possible a career as a professional scientist. Through the British Association for the Advancement of Science they were articulating the norms for professional science, and at the same time they were raising funds for its support (Morrell and Thackray 1981). They could not themselves accept evolution (actually, Herschel accepted some form of God-directed change), but they did prepare the way for future acceptance. And the same is true of these men and their co-workers in other spheres, including religion itself. We have seen how Darwin and his circle drew on the religious heritage provided by the older men. And this was quite apart from the way that, thanks to such things as so-called "higher criticism," religion was itself weakening its defences against brash secular newcomers.

Britain—Europe and America also—were different at the end of the nineteenth century from the way that they were at the beginning of the nineteenth century. Darwin and his revolution were part of this change from beginning to end. The eighteenth century began the major demolition of the ancient world and its thought patterns and social practices. The twentieth century has completed this transformation, but it was the nine-teenth century that did the heavy-duty work. It was a remarkable change and Darwin's was a remarkable revolution.

The Bigger Picture

There is more to be said if we are to understand the full nature of the Darwinian revolution. We must try to set the revolution in an overall con-text against the full history of evolutionary thought, something which could not yet be done when *The Darwinian Revolution* was first published. The ideas in this and the following two sections are spelled out in full in two of my more recent books, *Monad to Man: The Concept of Progress in Evolutionary Biology* and *Mystery of Mysteries: Is Evolution a Social Con-struction?*

Begin with the basic facts of the history of evolutionism, a task which takes us back to the eighteenth century: we go to Erasmus Darwin in par-ticular as well as to one or two others who, since they were certainly edging around the idea, might best be described as "proto-evolutionists." The great French naturalist Georges Leclerc, Comte de Buffon, is just such a person (Roger 1997). From these beginnings, we then come down via people like Lamarck and Chambers to Charles Darwin and the *Origin of Species*. After Darwin, we have the acceptance of evolution and the hesi-tancy towards natural selection, attitudes which persisted right into the twentieth century. Then (mentioned but truly beyond the scope of my book), we get the development of an adequate theory of heredity: a theory based in part on the simple rules of transmission that even at the time of Darwin were being discovered by the Moravian monk Gregor Mendel, and also based in part on the new discoveries about the nature of the cell. Most significant of these latter discoveries were those about the chromosomes, string-like entities in the nucleus and identified as the carriers of the units of heredity, the genes.

After initial hesitancy, it was realized that this was the theory of heredity that Darwin needed but lacked. By the early 1930s, a number of mathe-matically gifted evolutionists—Ronald A. Fisher and J. B. S. Haldane in Britain, and Sewall Wright in the U. S.—had shown how to blend to-gether Mendelism and Darwinism into one synthesis (Provine 1971). Rap-idly thereafter, the empiricists and naturalists set to work. In England, these included E. B. Ford and his school of "ecological genetics." In America, one finds Theodosius Dobzhansky and his co-workers: the ornithologist and systematist Ernst Mayr, the paleontologist George Gaylord Simpson, and the botanist G. Ledyard Stebbins. Biological flesh was added to the mathematicians' formal skeletons. Thus around 1940 the new evolutionism was born—the so-called "Synthetic Theory of Evolution" or "Neo-Darwinism"(Cain 1993; Hull 1988).

This is essentially how things have remained down to this day, although there have been major refinements and additions. The turn to a Darwinian individual-based view of selection was very important, especially in the development (from about 1960) of a full and exciting perspective on the evolution of social behavior. This yielded a new (or newly refurbished) member of the evolutionary family: sociobiology. Although at first controversial because of the possible applications to humankind, it now takes its place alongside palaeontology, biogeography, embryology, and the other sub-disciplines (Ruse 1985; Cronin 1991). There have also been criticisms of Darwinism and suggestions of alternatives. Best known of the challengers is perhaps the neo-saltationary palaeontological theory of Stephen Jay Gould: he proposes a view of the fossil record which claims that most of the time there is little change ("stasis"), but that occasionally evolution is marked by events of rapid movement from one form to another (Eldredge and Gould 1972). I am not sure that today this theory of "punctuated equilibria" looks quite as scientifically significant as some thought it did once, but it has certainly garnered a massive amount of media publicity.

The Significance of Progress

Now, what do we make of this history? In one sense, it is just one fact after another, and that is the way that textbooks tend to treat it. But that is just chronology, not real history. In fact, there is a remarkably coherent and integrated tale to be told, an interpretation that makes much sense of the overall facts and throws light on the many problems *The Darwinian Revolution* left unexplored and unanswered. The key concept is the already-introduced notion of progress: social progress in the sense of improvement of knowledge, industry, society, and morals through human effort and intelligence; and biological progress in the sense of upward advance from the most primitive (the primordial blob) to the most sophisticated (humankind). In my book, this idea was certainly recognized as important and given major coverage, but it was not then realized just how significant it truly is. Today, thanks to much more new research, we know that it is no exaggeration to say that it is this notion of progress that has controlled evolutionary thought from the beginning to the end—or rather, progress in conjunction and (generally) in opposition with the social desire of evolutionists to be considered professional scientists of the highest standing: progress in an uneasy but fertile dance with professionalism.

There is no question that organic evolution is the child of the social doctrine of progress. Men like Erasmus Darwin and Jean Baptiste de Lamarck were ardent social progressionists and they read this doctrine into the world of animals and plants, and then usually read it right back out as confirmation of their social beliefs! Listen to Erasmus Darwin (who liked to express his thinking in verse):

> Organic Life beneath the shoreless waves
> Was born and nurs'd in Ocean's pearly caves;
> First forms minute, unseen by spheric glass,
> Move on the mud, or pierce the watery mass;
> These, as successive generations bloom,
> New powers acquire, and larger limbs assume;
> Whence countless groups of vegetation spring,
> And breathing realms of fin, and feet, and wing.
>
> Thus the tall Oak, the giant of the wood,
> Which bears Britannia's thunders on the flood;
> The Whale, unmeasured monster of the main,
> The lordly Lion, monarch of the plain,
> The Eagle soaring in the realms of air,
> Whose eye undazzled drinks the solar glare,
> Imperious man, who rules the bestial crowd,
> Of language, reason, and reflection proud,
> With brow erect who scorns this earthy sod,
> And styles himself the image of his God;
> Arose from rudiments of form and sense,
> An embryon point, or microscopic ens!
> (Darwin 1803, 1, 295–314.)

Early evolutionism was a vehicle for the ideology of progress: the secular ideology of progress, for the doctrine was taken (with good reason) to be antithetical to the Christian commitment to divine Providence. To the Providentialist, we are as worms and nothing is possible save we are blessed by the undeserved grace of God—to think otherwise, to think our puny efforts can make any difference, is to commit the Pelagian heresy of believing that we can buy off God with good works. To the progressionist, on the contrary, good works are all that count: virtually anything is possible if we ourselves make the effort. Hence, evolutionism was opposed by good church people, not so much because it went against a literal reading of Genesis—we know that people were moving well beyond that—but because of the inherent progressionism. This was coupled, I hasten to add, with such worries as the way that evolution apparently downgraded the status of humankind, not to mention all of those natural theological worries about the difficulty of getting design from processes driven by blind law.

There was, however, more to it than this, and here we must turn again to Cuvier. He was a Protestant in conservative Catholic France. He was not only concerned with the revolutionary implications of evolution (and there was good reason to think the philosophy of progress played its role in starting the French Revolution) or with the straight religious, philosophical, and scientific implications of evolution; he was also concerned about himself. It was vital to portray his area of expertise and power—scientific expertise and power—as neutral and non-threatening. This was the very area, in fact, where being a Protestant made and could make no difference whatsoever. (I am not saying that Cuvier was right in this belief, only that it was his strategy.) Hence, Cuvier was at pains to argue that genuine science is culture-value free and pays attention to those factors or norms that one prizes in such science: consistency with the rest of science, predictive fertility, unificatory power (Whewell's consilience of inductions), and more. Evolution as proposed by the evolutionists Lamarck and Geoffroy (particularly in Cuvier's sights) did none of this. It was, in short, a pseudo science. It was worthy only of contempt.

This was the status of evolution—that of a pseudo science—right down to the time of Darwin and the *Origin*. It was a cultural ideology, that of progress, translated into the animal and plant world and paying scant attention to anything that real scientists considered worthy in good science. Epistemologically, evolutionary thought was immature; socially, it was unprofessional. This is the status of the work produced by Robert Grant and Robert Chambers. Moreover, the critics like Cuvier and Sedgwick and others—whatever their personal motivations—had good reason to make such judgments. No one could seriously expect to make a genuine prediction on the basis of the *Philosophie Zoologique*, or to find simplicity and elegance, let alone consistency, in *Vestiges*. Evolution was junk science of a particularly radical secular disposition, and to pretend otherwise is to read what we know and believe today back into the past.

Given this background, let us think again about what Darwin accomplished. Evolution was no longer a mere pseudo science after his work. The consilience right at the heart of the *Origin* saw to that. His was not a system fuelled by and only by a cultural ideology: the idea of evolution *per se* was more than just an epiphenomenon on secular philosophies of progress. It was, as I have said already in this *Afterword,* a justified truth about the real world. And this was shown at once by the fact that every respectable person rapidly became an evolutionist after the *Origin:* even Christians could become evolutionists, for inasmuch as they were uncomfortable with progress (although one should note that by the end of the nineteenth century there were many Christians happy to commit the Pelagian heresy and downplay the significance of Providence), they could rightly say that evolution did not rest solely on progress for its support.

But, think now at another level. Turn from strict questions of truth and falsity to matters of social standing. Did Darwin raise evolution right up to the status of a fully mature and professional science, on a par, for example, with the physics of the day? Darwin did not expel progress from evolution, indeed, as we have now seen, he did anything but this. Does this tell us something, and are there other pertinent factors at play here? There is no doubt that Darwin—the young Darwin particularly—had hoped to make evolutionary studies into a mature professional science, the sort of thing that would be a university subject, attracting researchers and teachers and students. This was no idle or vain hope, for it was in the 1850s that English higher education first introduced real science degrees, with posts for professors and hopes of jobs for students. Darwin himself was eager to show how a selection-based evolutionism could function as a working and forward-looking science. Odd though it may seem, after the *Origin* the next book that Darwin penned was a monograph on orchids. Odd but not inexplicable, for both the content and the comments Darwin made (especially to his publisher) suggest strongly that the book was designed to show precisely what it would be like to work within the new "paradigm," to use a popular term of today (Kuhn 1962). And of course all of this ties in nicely with the encouragement Darwin showed towards Wallace and Bates in their selection-based researches on butterflies.

Yet full scientific success for evolution was not to be, at least not in the sense that Darwin hoped. Evolution became the popular science *par excellence,* but it struggled to become a fully mature professional science, with all that such status entailed. A second-rate morphological German-based evolutionism that was greatly influenced by Haeckel's biogenetic law (ontogeny recapitulates phylogeny) took root in the 1870s (Nyhart 1995). This evolutionism was obsessed with tracing paths, desperately searching for similarities (homologies) real and apparent between bodily parts (particularly embryological bodily parts) of organisms of widely different species, and basically indifferent towards natural selection, indeed regarding adaptation as a barrier to discerning true relationships (Bowler 1996). But it was second rate and, in reaction, the best young would-be evolutionary biologists realized this and moved off to other more profitable fields like cytology (the study of the cell) and (increasingly) genetics (Maienschein 1991). Evolution and evolutionists became firmly lodged in museums, palaces of public instruction and entertainment. Rather than subtle selection experiments, the epitome of evolutionary excellence became sweeping displays of those fabulous monsters that American fossil hunters of the 1870s and 1880s unearthed in the Far West. It was *Tyrannosaurus rex* not *Orchis mascula* that came to symbolize the Darwinian triumph (Rainger 1991).

Why did Evolution Remain Popular Science?

Darwin himself must take some of the blame, if that is the right word. Eager though he was to found a selection-based evolutionary science, he was prepared (or able) to go only so far. He did not push his way into the university structure, founding departments or research institutes of such science—places where the Wallaces and Bateses might do their work and attract attention and fellow evolutionists. At one level he was too sick to do this, and so "blame" is too strong a term. But, at another level, we have seen how Darwin used his sickness to isolate himself from life's stresses and obligations. Although this worked wonderfully well with respect to Darwin's own personal research programme—and no one ever worked harder than Charles Darwin—it did mean that he himself was not out there getting the really dirty jobs done. He was not out there building institutes, chairing committees, student teaching, setting exams, marking papers, and so forth. Significantly, when one man (a Glaswegian botanist) expressed an interest in selection studies, Darwin was ready to help him with his fare to India to take up a post, but recoiled in horror and fear at the thought of his coming down to Kent to work with the master himself. Professional scientist though Darwin was, there was always about him the air of the rich man who does not need to fight life's bloodiest battles. And so he did not. Perhaps there was that fatal English reserve about winning at any cost.

But more significant in this failure of evolution to professionalize fully and adequately was Darwin's lieutenant, his "bulldog," Thomas Henry

Huxley. There were two factors at play here: three if you include the ones discussed in my text, particularly that as a morphologist Huxley never had real need of selection and so as a scientist was never really excited by the prospect of a selection-based evolutionism, and yet as a scientist (especially as a paleontologist) would have felt drawn toward a German-style phylogeny-tracing evolutionism. Huxley and his co-workers really could not see the prospects for a viable professional selection-based evolutionism, referring here specifically to financial support, for researchers, teachers, or trained students. They could see the prospects for a professional biology, thinking now of physiology and morphology. The former they sold particularly to the medical profession which was desperate to upgrade socially and practically, no longer simply collecting fees but actually trying to heal people. Physiological training was just what the medics wanted and just want the biologists were happy to offer (Geison 1978). The latter, morphology, was sold to the teaching profession. It was to take the place of the classics, which Huxley and fellow scientists felt were no longer appropriate training in an industrial society. Evolution, unfortunately, did not cure pains and it was obviously too ideologically loaded to get onto school curricula, and so it got little support in this respect. If one looks at the lectures Huxley gave to his students in the years after the *Origin,* or at the textbooks he wrote, or at the examinations he and his assistants set, one finds that he simply avoided virtually all mention of evolution (Huxley and Martin 1875). It was not part of Huxley's programme for professional biology.

There was a second, more positive, factor. Huxley and friends were looking for a substitute for the Christianity that socially and intellectually represented the Britain (and Europe and America) that they were trying to overthrow (Desmond 1997). In my book, I joke that Huxley would have liked to have been Pope, and if not that, he would have settled for the see of Canterbury. In fact, the joke was on me a little, for it turns out that in the popular press he was called "Pope Huxley," referring precisely to his desire to possess the moral and spiritual authority that hitherto had belonged to church leaders. Socially and intellectually Huxley could never have been a Pope or an Archbishop, but he craved that kind of status and power, as a leader and exponent of what he hoped would become the dominant ideology—the secular religion—of the day. (My joke was that if Huxley could not have been Pope, he would have settled for the Archbishopric. It should have been the other way round. If Huxley could not have had Canterbury, he would have settled for Rome. Like all good Victorian gentlemen, he was deeply prejudiced against foreigners.)

Of course, we have long known that "Social Darwinism" played this ideological role, and I discuss it as such in my book. But historians have had a tendency—a tendency which I exemplify—to treat the socio-political system as something aside from the true evolutionary science, something a bit disreputable and down market. What we now realize is that the science itself and the ideology were never that far apart, even in the minds of the most respectable and influential of post-*Origin* evolutionists. Evolution as popular science was just what Huxley wanted. It was a doctrine to preach to working men's groups and to Americans and to anyone else who would listen. It was something on which one could hang moral and other edifying messages and proposals for correct action. It is no surprise that Spencer became (far more than Darwin) the public evolutionist *par excellence,* for it was Spencer more than anyone who was eager to read evolution as something with a moral message: the promotion of the ultra-progressive rise of the organic world as encouraged by cherishing the causal processes of such upward development. Interestingly, towards the end of his life, Huxley began to have doubts about this whole program. These doubts came less because he thought evolution-as-religion was wrong in principle, than because as a consummate and by then successful servant of the state, Huxley worried about how the evolutionary process could promote (and demand) the kind of group-preserving and promoting activities to which he had dedicated all of his efforts (Huxley 1989).

Inasmuch as evolution was a public and popular science, the support was forthcoming. Rich people were happy to support museums, especially since the evolutionary displays were bound up not just with the progress from monad to man, but from primitive man up to white European Protestant man. At the American Museum of Natural History on the side of Central Park in New York, generations of little immigrants from the Lower East Side were shipped up to marvel at the displays of Dimetrodon and Triceratops and Brontosaurus, to appreciate the efforts of the red-blooded (and white-skinned) fossil hunters, and to learn of the duty to show proper deference to one's Anglo-Saxon betters and elders. Interestingly and significantly the director of the museum for many years was Henry Fairfield Osborn, a sometime Huxley student—a heritage he shared with E. Ray Lankester, for some years director of the British Museum (Natural History).

I do not say these things about the failure to professionalize selection-based evolutionism to disparage Huxley. Indeed, part of my change of heart about him comes from my appreciation of the skill and determination with which he realized his aims. I do say these things to explain why evolution never became fully (or even partially) Darwinized, in the sense of forming a professional science with natural selection at its core. I say them to explain why evolution remained a popular science, outside the domain of top-flight research, and a vehicle on which people could hang their beliefs and aspirations and ideologies. And why Darwin felt able (perhaps even required) to salt the later editions of the *Origin* with progressivist discussions and endorsements. He saw the way the wind was starting to blow, and being himself a strong social progressivist, decided to go with the flow. This is perhaps also why the *Descent of Man,* published some twelve years later, is a far more popular-oriented book than the *Origin of Species.* Whatever else, the success of Huxley's strategy does explain—on the principle of if you cannot beat them then join them—why it was that the Anglican Church was willing to offer the hallowed ground of Westminster Abby as the final resting place of Charles Darwin. As science and its education became part of the fabric of Victorian Britain, it became more internally stable and less threatening and more appreciated for its real virtues by those outside the field. Honouring the memory of Darwin served the interests of both science and the Church (Moore 1982).

Moving On

Does all this mean that evolution today is no more than a popular science, something still lodged in the museums, failing to achieve the heights of physics and chemistry? Things have changed. With the coming of the synthetic theory in the 1930s and 1940s, there was a determined effort to upgrade the status of evolutionary studies. Evolutionists moved into the universities, they built research teams, they sought out students and grants, they founded professional journals and other outlets, and they did much more that we associate with functioning mature science (Cain 1994). At the same time as they experimented and studied and theorised, they worked hard to remove from their science the cultural aspects that had made evolution so attractive to the Victorians.

This move from popular culture was not always easy, for these people—Dobzhansky and his associates particularly—shared many of the Victorians' values about progress and so forth. They greatly appreciated the labors of Thomas Henry Huxley's grandson, Julian Huxley, who was determined to articulate a progressivist secular humanism, founded on evolutionary principles. Although Julian Huxley's *Religion without Revelation* was read and much admired, it was kept at a distance from the new professional evolutionism. The usual compromise was to write two sets of books, one professional and one popular, the latter with the mathematics removed— not much work here!—lots of culture added, and a firm label to the reader that this is popular. The palaeontologist Simpson was the master: *Tempo and Mode in Evolution* (1944), professional; *The Meaning of Evolution* (1949), popular, with a focus on progress and democracy and the virtues of the intellectual life; and then back to the stern *The Major Features of Evolution* (1953), professional.

Yet, for all the labors, there are times when I see the (Thomas Henry) Huxley legacy persisting. The child breaks from the parent, but occasionally in the half-light it is as if time stands still. A look or a gesture or a tone brings back many memories. It is certainly the case that evolution today is still very much more than a professional laboratory science. Witness the popularity of movies like *Jurassic Park,* which show that we in the late twentieth century are no less fascinated, amazed, and frightened by dinosaurs than our great-great-grandparents. Witness also the ways in which science and religion still clash, thanks to the anti-evolutionary efforts of the Biblical literalists, the fundamentalists or Creationists: efforts, for example, that led to the passing of laws in the U. S. demanding the teaching of Genesis alongside evolution shortly after the first publication of *The Darwinian Revolution* (Ruse 1988).

The fear remains that evolution is seen as rather less than a professional laboratory science in the minds of many people. Anyone who has had any input to college biology courses knows too well that there is always one more course in biochemistry, deemed essential for entry into medical school, pushing out the general course on evolution. Significantly, in the U. S. today, there are ten times as many departments of molecular biology as there are of evolution (which is usually linked with ecology). It is not just grants that are in short supply, but respect also—in the minds of many hard scientists, evolution is still altogether too "philosophical," using this term in the popular scientific sense of "second rate" and suitable only for the inadequate and aged.

We are now starting to get well beyond our story, so let me close by addressing one final worry. Leaving aside the present, in documenting the failure of evolution in Darwin's day to professionalize into a selection-based top-quality science, am I not undercutting the worth and the importance and the achievement of the Darwinian revolution? I do not think so; but, ultimately, you must be the judge. However, before you hand down your ruling, compare Charles Darwin to Nicholas Copernicus, the most famous and venerated natural philosopher (scientist) of all time. This sixteenth-century Polish cleric was not the first heliocentric (sun-centered) theorist of the universe: that honor goes to Aristarchus of Samos back in the third century before Christ. Copernicus did not get the details right: that was the task of his successors, Tycho Brahe (who mapped the heavens accurately) and Johannes Kepler (who saw the planetary orbits as elliptical) and Galileo Galilei (who used the telescope). Finally, this author of *De Revolutionibus Orbium Coelestium (On the Revolutions of the Heavenly Spheres)* did not provide the causal mechanism behind his theory. That was for Isaac Newton and his law of gravitational attraction. And the whole process took well over one hundred years (Kuhn 1957). Yet rightly we honor Copernicus, who was a truly great scientist whose work had implications far beyond his immediate studies. The same should be true also of Darwin, who, thanks to natural selection, was Newton to his own Copernicus. Darwin should be honored as a great scientist, whose work had implications far beyond his immediate studies. Not uncritically but with much respect: which is how I invite you to read what I have written.

Bibliography for the Afterword

Appel, T. A. 1987. *The Cuvier-Geoffroy Debate: French Biology in the Decades Before Darwin.* New York: Oxford University Press.

Barrett, P. H., Gautrey, P. J., Herbert, S., Kohn, D., and Smith, S., eds. 1987. *Charles Darwin's Notebooks, 1836–1844.* Ithaca: Cornell University Press.

Bowler, P. 1996. *Life's Splendid Drama.* Chicago: University of Chicago Press.

Browne, J. 1995. *Charles Darwin: Voyaging. Volume 1 of a Biography.* New York: Knopf.

Cain, J. A. 1993. Common problems and cooperative solutions: organizational activity in evolutionary studies 1936–1947. *Isis* 84:1–25.

———. 1994. Ernst Mayr as community architect: Launching the Society for the Study of Evolution and the journal *Evolution. Biology and Philosophy* 9 (3):387–428.

Cronin, H. 1991. *The Ant and the Peacock.* Cambridge: Cambridge University Press.

Darwin, C. 1959. *The Origin of Species by Charles Darwin: A Variorum Text,* ed. M. Peckham. Philadelphia: University of Pennsylvania Press.

———. 1985–. *The Correspondence of Charles Darwin,* ed. F. Burkhardt et al. Cambridge: Cambridge University Press.

Darwin, E. 1803. *The Temple of Nature.* London: J. Johnson.

Dawkins, R. 1986. *The Blind Watchmaker.* New York: Norton.

Desmond, A. 1989. *The Politics of Evolution: Morphology, Medicine and Reform in Radical London.* Chicago: University of Chicago Press.

———. 1994. *Huxley, the Devil's Disciple.* London: Michael Joseph.

———. 1997. *Huxley, Evolution's High Priest.* London: Michael Joseph.

Desmond, A., and Moore, J. 1992. *Darwin: The Life of a Tormented Evolutionist.* New York: Warner.

Eldredge, N., and Gould, S. J. 1972. Punctuated equilibria: an alternative to phyletic gradualism. *Models in Paleobiology,* ed. T. J. M. Schopf, 82–115. San Francisco: Freeman, Cooper.

Geison, G. L. 1978. *Michael Foster and the Cambridge School of Physiology: the Scientific Enterprise in Late Victorian Society.* Princeton: Princeton University Press.

Hegel, G. W. F. [1817] 1970. *Philosophy of Nature.* Oxford: Oxford University Press.

Hull, David. 1988. *Science as a Process.* Chicago: University of Chicago Press.

Huxley, J. S. 1927. *Religion Without Revelation.* London: Ernest Benn.

Huxley, T. H. 1989. *Evolution and Ethics with New Essays on its Victorian and Sociobiological Context,* eds. J. Paradis and G. C. Williams. Princeton: Princeton University Press.

Huxley, T. H., and Martin, H. N. 1875. *A Course of Practical Instruction in Elementary Biology.* London: Macmillan.

Kuhn, T. 1957. *The Copernican Revolution.* Cambridge: Harvard University Press.

———. 1962. *The Structure of Scientific Revolutions.* Chicago: University of Chicago Press.

Laurent, G. 1987. *Paléontologie et Evolution en France de 1800 à 1860. Une Histoire des Idées de Cuvier et Lamarck à Darwin.* Paris: Editions du C.T.H.S.

McNeill, M. 1987. *Under the Banner of Science: Erasmus Darwin and His Age.* Manchester: Manchester University Press.

Maienschein, J. 1991. *Transforming Traditions in American Biology. 1880–1915.* Baltimore: Johns Hopkins University Press.

Moore, J. 1979. *The Post-Darwinian Controversies: A Study of the Protestant Struggle to come to terms with Darwin in Great Britain and America, 1870–1900.* Cambridge: Cambridge University Press.

———. 1982. Charles Darwin lies in Westminster Abbey. *Biological Journal of the Linnean Society* 17:97–113.

Morrell, J., and Thackray, A. 1981. *Gentlemen of Science: Early Years of the British Association for the Advancement of Science.* Oxford: Oxford University Press.

Nyhart, L. K. 1995. *Biology Takes Form: Animal Morphology and the German Universities.* Chicago: University of Chicago Press.

Ospovat, D. 1981. *The Development of Darwin's Theory: Natural History, Natural*

Theology, and Natural Selection, 1838–1859. Cambridge: Cambridge University Press, reissue 1995.

Outram, D. 1984. *Georges Cuvier: Vocation, Science and Authority in Post Revolutionary France*. Manchester: Manchester University Press.

Owen, R. 1992. *The Hunterian Lectures in Comparative Anatomy, May and June 1837*, ed. P. R. Sloan. Chicago: Chicago University Press.

Pittenger, M. 1993. *American Socialists and Evolutionary Thought, 1870–1920*. Madison, Wisconsin: University of Wisconsin Press.

Provine, W. B. 1971. *The Origins of Theoretical Population Genetics*. Chicago: University of Chicago Press.

Pusey, J. R. 1983. *China and Charles Darwin*. Cambridge: Harvard University Press.

Rainger, R. 1991. *An Agenda for Antiquity: Henry Fairfield Osborn and Vertebrate Paleontology at the American Museum of Natural History, 1890–1935*. Tuscaloosa: University of Alabama Press.

Richards, R. J. 1987. *Darwin and the Emergence of Evolutionary Theories of Mind and Behavior*. Chicago: University of Chicago Press.

———. 1992. *The Meaning of Evolution: The Morphological Construction and Ideological Reconstruction of Darwin's Theory*. Chicago: University of Chicago Press.

Roger, J. 1997. *Buffon: A Life in Natural History*, trans. S. L. Bonnefoi. Ithaca, N.Y.: Cornell University Press.

Rupke, N. A. 1994. *Richard Owen: Victorian Naturalist*. New Haven: Yale University Press.

Ruse, M. 1985. *Sociobiology: Sense or Nonsense?* 2d ed. Dordrecht: Reidel.

———, ed. 1988. *But is it Science? The Philosophical Question in the Creation/Evolution Controversy*. Buffalo: Prometheus.

———. 1996a. The Darwin Industry: a guide. *Victorian Studies* 39 (2):217–35.

———. 1996b. *Monad to Man: The Concept of Progress in Evolutionary Biology*. Cambridge: Harvard University Press.

———. 1999. *Mystery of Mysteries: Is Evolution a Social Construction?* Cambridge: Harvard University Press.

Simpson, G. G. 1944. *Tempo and Mode in Evolution*. New York: Columbia University Press.

———. 1949. *The Meaning of Evolution* . New Haven: Yale University Press.

———. 1953. *The Major Features of Evolution*. New York: Columbia University Press.

Sulloway, F. J. 1982. Darwin and his finches: the evolution of a legend. *Journal of the History of Biology* 15:1–53.

Young, R. M. 1985. *Darwin's Metaphor: Nature's Place in Victorian Culture*. Cambridge: Cambridge University Press.

Index